THE OXFORD BOOK OF MODERN FAIRY TALES

THE OXFORD BOOK OF MODERN FAIRY TALES

Edited by ALISON LURIE

Oxford New York

OXFORD UNIVERSITY PRESS

1993

Oxford University Press, Walton Street, Oxford OX2 6DP
Oxford New York Toronto
Delhi Bombay Calcutta Madras Karachi
Kuala Lumpur Singapore Hong Kong Tokyo
Nairobi Dar es Salaam Cape Town
Melbourne Auckland Madrid
and associated companies in
Berlin Ibadan

Oxford is a trade mark of Oxford University Press

British Library Cataloguing in Publication Data
Data available

Library of Congress Cataloging in Publication Data
The Oxford book of modern fairy tales / edited by Alison Lurie.
p. cm.
1. Fairy tales—Great Britain. 2. Fairy tales—United States.
I. Lurie, Alison.
PR1309.F2609 1993 823'.0108—dc20 92–28007
ISBN 0–19–214218–6

1 3 5 7 9 10 8 6 4 2

Typeset by Best-set Typesetter Ltd., Hong Kong

Printed in Great Britain
on acid-free paper by
The Bath Press, Avon

EDITOR'S ACKNOWLEDGEMENTS

I HAVE been reading fairy tales for well over half a century, and often cannot remember where and when I first came across a particular story, or who first mentioned it to me. Nevertheless, I should like to thank everyone who remembers recommending one of these tales. Various experts also provided me with advice: among them are Doug Anderson of the Borealis Bookshop in Ithaca; Julia Briggs, the biographer of E. Nesbit; and Michael Colaccurcio of Cornell. I am especially indebted to Professor Jack Zipes of the University of Minnesota, probably the foremost American expert in this field, and to his fine anthologies, *Victorian Fairy Tales* and *Spells of Enchantment*.

My greatest debt, however, is to the researchers who searched library catalogues and stacks and did much of the preliminary screening and final copying. Without their patience, enthusiasm, and literary good taste this book would not be what it is. Susanna Gross in London (who negotiated the Victorian intricacies of the British Library), and in America Elizabeth Grove, Beth Lordan, and Micah Perks all made a vital, though invisible, contribution to the project.

Finally I should like to thank my editor at the Oxford University Press, Judith Luna, who first suggested the idea of an *Oxford Book of Modern Fairy Tales*, and has been its kind fairy godmother ever since.

A.L.

Key West, Florida
February 1992

CONTENTS

Contents

Contents

INTRODUCTION

THE stories of magic and transformation that we call 'fairy tales' (though they often contain no fairies) are one of the oldest known forms of literature, and also one of the most popular and enduring. Even today they are a central part of our imaginative world. We remember and refer to them all our lives; their themes and characters reappear in dreams, in songs, in films, in advertisements, and in casual speech: we say that someone is a giant-killer, or that theirs is a Cinderella story.

The fairy tale survives because it presents experience in vivid symbolic form. Sometimes we need to have the truth exaggerated and made more dramatic, even fantastic, in order to comprehend it. (The same sort of thing can occur in other ways, of course, as when at a costume party we suddenly recognize that one of our acquaintances is essentially a six-foot-tall white rabbit, or a dancing doll.)

'Hansel and Gretel', for instance, may dramatize the fact that some parents underfeed and abandon their children physically and/or emotionally, while others, like the witch, overfeed and try to possess and devour them. 'Beauty and the Beast' may suggest that a good man can seem at first like a dangerous wild animal, or that true love has power to soothe the savage heart. The message will probably be different for each reader; that is one of the great achievements of the fairy tale—traditional or modern.

For, though not everyone knows it, there are modern fairy tales. Though most people think of these stories as having come into existence almost magically long ago, they are in fact still being created, and not only in less urbanized parts of the world than our own. Over the last century and a half many famous authors have written tales of wonder and enchantment. In Britain and America they have included Nathaniel Hawthorne, Charles Dickens, Robert Louis Stevenson, Oscar Wilde, H. G. Wells, Carl Sandburg, James Thurber, Bernard Malamud, I. B. Singer, T. H. White, Angela Carter, and Louise Erdrich. Like other writers in other countries (notably Germany and France) they have used the characters and settings and events of the fairy tale to create new and marvellous stories—not only for children but for adults.

The traditional fairy tale was not read from a book, but passed on orally from one generation to the next, and its audience was not limited to children. Its heroes and heroines, most often, are not children but young people setting out to make their fortunes or find a mate, or both. Many of the stories in *The Oxford Book of Modern Fairy Tales* were also written for readers or listeners of all ages, while others were never intended for children. But even when they are principally directed to children and have child protagonists these tales often contain sophisticated comments and ironic asides directed to the adults who might be reading the story aloud.

Most of these modern stories, like the traditional folk tale, can be understood on many levels: they have one message for a 7 year old and another one, more complex and sometimes more melancholy, for a 17 year old or a 70 year old. George MacDonald's 'The Light Princess', for example, is on the face of it a traditional tale of enchantment and disenchantment, but it also has something to say about the dangers of loving too rashly or not at all.

Some of these stories, which today would be considered too lengthy and difficult even for an adolescent reader, were more accessible before television reduced literacy and shortened attention spans. When I and my friends were in primary school we read and loved many of the stories in this volume—including those of John Ruskin, Oscar Wilde, and Walter de la Mare, without any sense that they were beyond our capacity. But there are also tales here—for example, those by L. F. Baum, E. Nesbit, Richard Hughes, Joan Aiken, and Jay Williams—that any contemporary child can understand and enjoy, especially if they are read aloud.

The earliest attempts to compose modern fairy tales were tentative. At first, authors merely rewrote the traditional stories of Grimm and Perrault, sometimes in what now seems a ridiculous manner. In 1853 the Grimms' first English illustrator, George Cruickshank, began to publish revisions of the most popular tales from a teetotal point of view. The Giant in his 'Jack and the Beanstalk' is revealed to be an alcoholic, and Cinderella's wedding is celebrated by the destruction of all the drink in the Prince's castle.

Meanwhile other writers were beginning to go beyond revision to compose original tales, often in order to point an improving moral. The lesson, of course, varied with the concerns of the author. Catherine Sinclair's light-hearted 'Uncle David's Nonsensical Story about Giants and Fairies' (1839), suggested that idle and overfed children were apt to be eaten by giants, while Juliana Horatia Ewing's 'Good Luck Is Better Than Gold' (1882) and Howard Pyle's 'The Apple of Contentment' (1886) punished greed and rewarded kindness.

Many writers were concerned with more contemporary issues. Frances Browne's 'The Story of Fairyfoot' (1856), for instance, exposes the arbitrary nature of standards of Victorian beauty, imagining a kingdom where the larger your feet are, the more handsome you are thought to be. John Ruskin's famous ecological fable, *The King of the Golden River* (1850), promotes both his political and his aesthetic beliefs. The two wicked older brothers in this story are short-sighted capitalists who exploit both labour and natural resources, turning a once-fertile and dramatically beautiful valley into a barren wasteland. Their moods are so dark and their hearts so hard that it seems quite appropriate that they should eventually be transformed into two black stones, while little Gluck, who appreciates the sublime natural landscape and relieves the sufferings of the disadvantaged, restores the land to fertility.

In 'A Toy Princess' (1877) Mary De Morgan mounts a scathing attack on the ideal Victorian miss. The courtiers among whom her orphaned heroine grows up scold her for expressing her feelings, and much prefer the artificial doll-princess who never says anything but 'If you please', 'No thank you', 'Certainly', and 'Just so'. Nathaniel Hawthorne's 'Feathertop' (1846) is a disturbing male version of the Toy Princess: a fine gentleman made by a New England witch from a scarecrow. His vocabulary consists only of

'Really! Indeed! Pray tell me! Is it possible! Upon my word! By no means! O! Ah!', and 'Hem', but he is taken by the local population for a foreign nobleman and almost succeeds in winning the heart of a beautiful girl. Though both these stories end without any real damage having been done, they are full of the unease we feel in the presence of someone with fine clothes and impenetrably bland good manners.

More unsettling, and with a darker ending, is Lucy Lane Clifford's 'The New Mother' (1882), which tells of the awful fate of two innocent children who are repeatedly encouraged in naughty behaviour by a strange and charming young woman who may be an evil spirit. Eventually the children try their mother's patience so far that she threatens to leave them and 'send home a new mother, with glass eyes and a wooden tail'. Anyone who has ever seen a harassed parent appear to turn temporarily into a glassy-eyed monster will understand this story instinctively, and so will parents who have doubts about the moral qualities of their baby-sitters. The author was a good friend of Henry James, and it has been suggested that 'The New Mother' may be one of the sources of *The Turn of the Screw*.

After Perrault and Grimm the greatest influence on the literary fairy tale was Hans Christian Andersen, whose work was first translated into English in 1846. Andersen's romantic, spiritual narratives were echoed in the work of George MacDonald, Oscar Wilde, and Laurence Housman, among others. Often their tales seem remarkably modern. In MacDonald's 'The Light Princess' (1864) and Housman's 'The Rooted Lover' (1894) the hero is what my students at Cornell would call a post-feminist man. He does not fight giants and dragons but shows his courage and virtue through patient endurance for the sake of love. MacDonald's heroine is literally light: she is not subject to gravity. She is also, as my students put it, a complete airhead, incapable of serious thought or real affection.

In Wilde's 'The Selfish Giant' (1888) Christian morality replaces pagan, as it sometimes does in Andersen's tales. The traditional fairy-tale villain of the title is not slain but reformed by a child who turns out to be Christ. Other writers, also following Andersen's example, abandoned the usual happy ending of the fairy tale to create stories with a sad or disturbing conclusion, like Robert Louis Stevenson's 'The Song of the Morrow' (1894) in which a series of events is endlessly repeated—something that also occasionally happens in real life.

Not all late nineteenth-century fairy tales are this serious: many are mildly or broadly comic. There are good-natured burlesques like Charles Dickens's 'The Magic Fishbone' (1868) in which a scatty Micawber-like (or Dickens-like) family is saved by the patience and good sense of the oldest daughter; and there are gentle satires of social conformity and cowardice like Frank Stockton's 'The Griffin and the Minor Canon' (1887). Perhaps the best known of such stories is Kenneth Grahame's 'The Reluctant Dragon' (1898), possibly the first overtly pacifist fairy tale. It features a sentimental dragon who writes sonnets and only wishes to be admired by the villagers whom he has terrified; many readers will recognize the type.

The fashion for tales that were humorous and satirical as well as (or instead of) uplifting or improving continued into the early twentieth century. Many stories of this period take place in a partially modern setting and question accepted political and social beliefs. L. F. Baum's 'The Queen of Quok' (1901) contains a castle and royal personages; but Quok is essentially ruled by common sense and small-town American values. At one point the boy king has to borrow a dime from his chief counsellor to buy a ham sandwich. Love of money turns the would-be queen into a haggard old woman, while the insouciant young hero lives happily ever after.

E. Nesbit's 'The Book of Beasts' (1900) is a light-hearted fable about the magical power of art and literature. The volume of its title contains pictures of exotic creatures which come alive when the pages are opened. The boy who finds it releases first a butterfly, then a 'Blue Bird of Paradise', and finally a dragon who threatens to destroy the country. If any book is vivid enough, this story seems to say, its content will invade our world for good or evil.

For H. G. Wells, on the other hand, magic is allied with science. His rather spooky Magic Shop, in the story of the same name (1903), contains both traditional supernatural creatures, like a small angry red demon, and the actual inventions of the future, including a train that runs without steam.

Even further from the traditional mould are Carl Sandburg's Rootabaga stories, which reflect his love of American tall tales and deadpan humour. 'The Story of Blixie Bimber and the Power of the Gold Buckskin Whincher' (1922) is set in what is obviously the early twentieth-century Midwest, complete with hayrides, band concerts, and steeplejacks. But magic is still potent, and passion a kind of inexplicable spell. '"The first man you meet with an X in his name you must fall head over heels in love with him," said the silent power in the gold buckskin whincher', and Blixie Bimber does, the traditional three times over. James Thurber's famous comic fable, 'The Unicorn in the Garden' (1940), presents the triumph of a mild visionary over his would-be oppressors: the police, a psychiatrist, and a hostile, suspicious wife who thinks that anyone who sees unicorns must be mad.

In Britain, other twentieth-century writers composed more romantic tales. Some, like Walter de la Mare's 'The Lovely Myfanwy' (1925) and Sylvia Townsend Warner's witty 'Bluebeard's Daughter' (1940) have a fairy-story background of castles and princesses and rebuke old-fashioned faults—in the former case, possessive paternal love, in the latter, curiosity. Others are set in the contemporary world. John Collier's 'The Chaser' (1941) a very short story with a sting in its tail, takes place in modern London; Naomi Mitchison's 'In the Family' (1957) in a rural Scotland complete with buses and parish halls—and a fairy woman who warns the hero of a highway accident.

A few of these tales are interesting variations on earlier classics. Lord Dunsany's 'The Kith of the Elf-Folk' (1910) is a half-poetic, half-sardonic version of Andersen's 'The Little Mermaid', with a happier, though rather conservative, conclusion. In it a Wild Thing from the marshes ends by rejecting both her newly acquired human soul and a singing career in London. She returns to her former life and companions in the depths of the countryside—as other strange wild young women have sometimes done.

'The Courtship of Mr Lyon' (1979), a striking tale by Angela Carter, updates 'Beauty and the Beast'. Her Beast is the awkward, lonely, growling owner of a Palladian villa not far from London, equipped not only with invisible servants but with a telephone. Beauty temporarily abandons him to become a spoilt society girl who 'smiled at herself in mirrors a little too often', and she as well as he is transformed by the power of love.

Some modern authors of fairy tales, like these, revel in descriptions of exotic or luxurious settings. Others seem deliberately to choose the drabbest and most ordinary backgrounds, perhaps to remind us that strange and wonderful things can happen anywhere. In Philip K. Dick's 'The King of the Elves' (1953) the hero is an old man in charge of a run-down rural gas (petrol) station; and Joan Aiken's 'The Man Who Had Seen the Rope Trick' (1976) takes place in a dreary English seaside boarding house.

T. H. White, in 'The Troll' (1935), begins with a similarly pedestrian setting, 'a comfortable railway hotel', that rapidly becomes most uncomfortable. During his first night there the hero discovers that the professor in the next room is a troll who has eaten his wife. We accept this, and all that follows, not only because of White's literary skill but because we know that some men, even some professors, are really trolls; and that some husbands do psychologically devour their wives (and wives their husbands).

Some modern writers have taken the conventions of the folk tale or children's story and turned them upside down, as real life sometimes does. 'Gertrude's Child' (1940) by Richard Hughes is about a world in which toys own children; and in Richard Kennedy's 'The Porcelain Man' (1987) the heroine declines to rescue the enchanted hero. Strangest of all these reversals is the one that occurs in Ursula Le Guin's 'The Wife's Story' (1982), which can be read as a brief but terrifying fable about family love, madness, and social prejudice.

Some of the best recent literary fairy tales use the form to comment on twentieth-century events. In Bernard Malamud's 'The Jewbird' (1963) a talking crow flies into the Lower East Side apartment of a frozen-foods salesman and announces that he is fleeing from anti-Semites. To judge by what happens next, he may be one of those immigrant survivors of the Holocaust whom some American Jews, after the Second World War, found burdensome. Donald Barthelme's experimental 'The Glass Mountain' (1970) takes off from a story of the same name in Andrew Lang's *The Yellow Fairy Book*, and manages simultaneously to expose the callous ambition of New Yorkers, and the formulaic analysis of literary scholars.

In this century we have also begun to see tales of magic based upon previously untapped folk traditions. Many of I. B. Singer's stories, including 'Menaseh's Dream' (1968), draw on Jewish folk beliefs and make wise, if disguised, comments on Jewish life: in this case, on the power of memory and of family love. Louise Erdrich in 'Old Man Potchikoo' (1989) uses the Native American trickster tale as a starting-point for a celebration of Dionysian energy.

Recently several writers have produced fairy tales with a strong feminist slant. Among them are Tanith Lee's 'Prince Amilec' (1972), Jay Williams's 'Petronella' (1973), and Jeanne Desy's inventive 'The Princess Who Stood On Her Own Two Feet' (1982), in which a well-meaning young woman gives up more and more of her natural abilities in order to make her fiancé feel good—a procedure that unfortunately may still be observed in real life.

Another interesting example of the genre is Jane Yolen's 'The River Maid' (1982). The protagonists of Yolen's poetic fairy tales are usually pre-feminist: delicate, passive, and either victimized or self-sacrificing. But in 'The River Maid', though the eponymous heroine remains frail and helpless, the river of which she is the guardian spirit is strong. A greedy farmer dams and diverts the water to enrich his fields, and abducts and rapes the River Maid. The following spring the river rises, washes away the farm, and drowns the farmer. Afterwards it can be heard 'playing merrily over [his] bones', with a 'high, sweet, bubbling song. . . . full of freedom and a conquering joy'. Women may be imprisoned and abused, the story seems to say, but time and the forces of nature will avenge them.

Today the fairy tale is often dismissed as old fashioned, sentimental, and silly: a minor form of literature, appropriate only for children. To people who have been recently over-exposed to the bowdlerized and prettified cartoon versions of the classic stories, this view may seem justified.

But any reader who knows the authentic traditional tales, or these brilliant modern variations on their themes, will realize that fairy tales are not merely childish entertainments set in an unreal and irrelevant universe. It is true that they can and do entertain children—and adults; but we will also do well to listen seriously to what they tell us about the real world we live in.

The Oxford Book of Modern Fairy Tales

CATHERINE SINCLAIR

Uncle David's Nonsensical Story about Giants and Fairies

Pie-crust, and pastry-crust, that was the wall;
The windows were made of black-puddings and
white,
And slated with pancakes—you ne'er saw the like!

IN the days of yore, children were not all such clever, good, sensible people as they are now! Lessons were then considered rather a plague—sugar-plums were still in demand—holidays continued yet in fashion—and toys were not then made to teach mathematics, nor story-books to give instruction in chemistry and navigation. These were very strange times, and there existed at that period, a very idle, greedy, naughty boy, such as we never hear of in the present day. His papa and mamma were—no matter who, and he lived—no matter where. His name was Master No-book, and he seemed to think his eyes were made for nothing but to stare out of the windows, and his mouth for no other purpose but to eat. This young gentleman hated lessons like mustard, both of which brought tears into his eyes, and during school-hours he sat gazing at his books, pretending to be busy, while his mind wandered away to wish impatiently for dinner, and to consider where he could get the nicest pies, pastry, ices and jellies, while he smacked his lips at the very thoughts of them. I think he must have been first cousin to Peter Grey; but that is not perfectly certain.

Whenever Master No-book spoke, it was always to ask for something, and you might continually hear him say, in a whining tone of voice, 'Papa! may I take this piece of cake? Aunt Sarah! will you give me an apple? Mamma! do send me the whole of that plum-pudding!' Indeed, very frequently, when he did not get permission to gormandize, this naughty glutton helped himself without leave. Even his dreams were like his waking hours, for he had often a horrible nightmare about lessons, thinking he was smothered with Greek Lexicons, or pelted out of the school with a shower of English Grammars; while one night he fancied himself sitting down to devour an enormous plum-cake, and all on a sudden it became transformed into a Latin Dictionary!

One afternoon, Master No-book, having played truant all day from school, was lolling on his mamma's best sofa in the drawing-room, with his leather boots tucked up on the satin cushions, and nothing to do but to suck a few oranges, and nothing to think of but how much sugar to put upon them, when suddenly an event took place, which filled him with astonishment.

A sound of soft music stole into the room, becoming louder and louder the longer he listened, till at length, in a few moments afterwards, a large hole burst open in the wall of his room, and there stepped into his presence two magnificent fairies, just arrived from their castles in the air, to pay him a visit. They had travelled all the way on purpose to have some conversation with Master No-book, and immediately introduced themselves in a very ceremonious manner.

The fairy Do-nothing was gorgeously dressed with a wreath of flaming gas round her head, a robe of gold tissue, a necklace of rubies, and a bouquet in her hand of glittering diamonds. Her cheeks were rouged to the very eyes, her teeth were set in gold, and her hair was of a most brilliant purple; in short, so fine and fashionable-looking a fairy never was seen in a drawing-room before.

The fairy Teach-all, who followed next, was simply dressed in white muslin, with bunches of natural flowers in her light brown hair, and she carried in her hand a few neat small books, which Master No-book looked at with a shudder of aversion.

The two fairies now informed him, that they very often invited large parties of children to spend some time at their palaces, but as they lived in quite an opposite direction, it was necessary for their young guests to choose which it would be best to visit first; therefore now they had come to inquire of Master No-book whom he thought it would be most agreeable to accompany on the present occasion.

'In my house,' said the fairy Teach-all, speaking with a very sweet smile, and a soft, pleasing voice, 'you shall be taught to find pleasure in every sort of exertion; for I delight in activity and diligence. My young friends rise at seven every morning, and amuse themselves with working in a beautiful garden of flowers,—rearing whatever fruit they wish to eat,—visiting among the poor,—associating pleasantly together,—studying the arts and sciences,—and learning to know the world in which they live, and to fulfil the purposes for which they have been brought into it. In short, all our amusements tend to some useful object, either for our own improvement or the good of others, and you will grow wiser, better, and happier every day you remain in the palace of Knowledge.'

'But in Castle Needless, where I live,' interrupted the fairy Do-nothing, rudely pushing her companion aside, with an angry, contemptuous look, 'we never think of exerting ourselves for anything. You may put your head in your pocket, and your hands in your sides as long as you choose to stay. No one is ever even asked a question, that he may be spared the trouble of answering. We lead the most fashionable life imaginable, for nobody speaks to anybody! Each of my visitors is quite an exclusive, and sits with his back to as many of the company as possible, in the most comfortable armchair that can be contrived. There, if you are only so good as to take the trouble of wishing for anything, it is yours, without even turning an eye round to look where it comes from. Dresses are provided of the most magnificent kind, which go on themselves, without your having the smallest annoyance with either buttons or strings,—games which you can play without an effort of thought,—and dishes dressed by a French cook, smoking hot under your nose, from morning till night,—while any rain we have is either made of sherry, brandy, lemonade, or lavender water, and in winter it generally snows iced-punch for an hour during the forenoon.'

Nobody need be told which fairy Master No-book preferred; and quite charmed at his own good fortune in receiving so agreeable an invitation, he eagerly gave his hand to the splendid new acquaintance who promised him so much pleasure and ease, and gladly proceeded in a carriage lined with velvet, stuffed with downy pillows, and drawn by milk-white swans, to that magnificent residence, Castle Needless, which was lighted by a thousand windows during the day, and by a million of lamps every night.

Here Master No-book enjoyed a constant holiday and a constant feast, while a beautiful lady covered with jewels was ready to tell him stories from morning till night, and servants waited to pick up his playthings if they fell, or to draw out his purse or his pocket-handkerchief when he wished to use them.

Thus Master No-book lay dozing for hours and days on richly embroidered cushions, never stirring from his place, but admiring the view of trees covered with the richest burned-almonds, grottoes of sugar-candy, a *jet d'eau* of champagne, a wide sea which tasted of sugar instead of salt, and a bright clear pond, filled with goldfish, that let themselves be caught whenever he pleased. Nothing could be more complete; and yet, very strange to say, Master No-book did not seem particularly happy! This appears exceedingly unreasonable, when so much trouble was taken to please him, but the truth is, that every day he became more fretful and peevish. No sweetmeats were worth the trouble of eating, nothing was pleasant to play at, and in the end he wished it were possible to sleep all day, as well as all night.

Not a hundred miles from the fairy Do-nothing's palace, there lived a most cruel monster called the giant Snap-'em-up, who looked, when he stood up, like the tall steeple of a great church, raising his head so high that he could peep over the loftiest mountains, and was obliged to climb up a ladder to comb his own hair!

Every morning regularly, this prodigiously great giant walked round the world before breakfast for an appetite, after which he made tea in a large lake, used the sea as a slop-basin, and boiled his kettle on Mount Vesuvius. He lived in great style, and his dinners were most magnificent, consisting very often of an elephant roasted whole, ostrich patties, a tiger smothered in onions, stewed lions, and whale soup; but for a side dish his greatest favourite consisted of little boys, as fat as possible, fried in crumbs of bread, with plenty of pepper and salt.

No children were so well fed, or in such good condition for eating, as those in the fairy Do-nothing's garden, who was a very particular friend of the giant Snap-'em-up's, and who sometimes laughingly said she would give him a licence, and call her own garden his 'preserve,' because she allowed him to help himself, whenever he pleased, to as many of her visitors as he chose, without taking the trouble even to count them, and in return for such extreme civility, the giant very frequently invited her to dinner.

Snap-'em-up's favourite sport was to see how many brace of little boys he could bag in a morning; so in passing along the streets, he peeped into all the drawing-rooms without having occasion to get upon tiptoe, and picked up every young gentleman who was idly looking out of the windows, and even a few occasionally who were playing truant from school; but busy children seemed always somehow quite out of his reach.

One day, when Master No-book felt even more lazy, more idle, and more miserable than ever, he lay beside a perfect mountain of toys and cakes, wondering what to wish for next, and hating the very sight of everything and everybody. At last he gave so loud a yawn of weariness and disgust, that his jaw very nearly fell out of joint, and then he sighed so deeply, that the giant Snap-'em-up heard the sound as he passed along the road after breakfast, and instantly stepped into the garden, with his glass at his eye, to see what was the matter. Immediately on observing a large, fat, overgrown boy, as round as a dumpling, lying on a bed of roses, he gave a cry of delight, followed by a gigantic peal of laughter, which was heard three miles off, and picking up Master No-book between his finger and thumb, with a pinch that very nearly broke his ribs, he carried him rapidly towards his own castle, while the fairy Do-nothing laughingly shook her head as he passed, saying, 'That little man does me great credit! he has only been fed for a week, and is as fat already as a prize ox! What a dainty morsel he will be. When do you dine to-day, in case I should have time to look in upon you?'

On reaching home the giant immediately hung up Master No-book, by the hair of his head, on a prodigious hook in the larder, having first taken some large lumps of nasty suet, forcing them down his throat to make him become still fatter, and then stirring the fire, that he might be almost melted with heat, to make his liver grow larger. On a shelf quite near, Master No-book perceived the dead bodies of six other boys, whom he remembered to have seen fattening in the fairy Do-nothing's garden, while he recollected how some of them had rejoiced at the thoughts of leading a long, useless, idle life, with no one to please but themselves.

The enormous cook now seized hold of Master No-book,

brandishing her knife, with an aspect of horrible determination, intending to kill him, while he took the trouble of screaming and kicking in the most desperate manner, when the giant turned gravely round and said, that as pigs were considered a much greater dainty when whipped to death than killed in any other way, he meant to see whether children might not be improved by it also; therefore she might leave that great hog of a boy till he had time to try the experiment, especially as his own appetite would be improved by the exercise. This was a dreadful prospect for the unhappy prisoner; but meantime it prolonged his life a few hours, as he was immediately hung up again in the larder, and left to himself. There, in torture of mind and body,—like a fish upon a hook,—the wretched boy began at last to reflect seriously upon his former ways, and to consider what a happy home he might have had, if he could only have been satisfied with business and pleasure succeeding each other, like day and night, while lessons might have come in as a pleasant sauce to his play-hours, and his play-hours as a sauce to his lessons.

In the midst of many reflections, which were all very sensible, though rather too late, Master No-book's attention became attracted by the sound of many voices laughing, talking, and singing, which caused him to turn his eyes in a new direction, when, for the first time, he observed that the fairy Teach-all's garden lay upon a beautiful sloping bank not far off. There a crowd of merry, noisy, rosy-cheeked boys were busily employed, and seemed happier than the day was long; while poor Master No-book watched them during his own miserable hours, envying the enjoyment with which they raked the flower-borders, gathered the fruit, carried baskets of vegetables to the poor, worked with carpenter's tools, drew pictures, shot with bows and arrows, played at cricket, and then sat in the sunny arbours learning their tasks, or talking agreeably together, till at length, a dinner-bell having been rung, the whole party sat merrily down with hearty appetites, and cheerful good humour, to an entertainment of plain roast meat and pudding, where the fairy Teach-all presided herself, and helped her guests moderately, to as much as was good for each.

Large tears rolled down the cheeks of Master No-book while watching this scene; and remembering that if he had known what was best for him, he might have been as happy as the happiest of these excellent boys, instead of suffering *ennui* and weariness, as he had done at the fairy Do-nothing's, ending in a miserable death; but his attention was soon after most alarmingly roused by hearing the giant Snap-'em-up again in conversation with his cook; who said, that if he wished for a good large dish of scalloped children at dinner, it would be necessary to catch a few more, as those he had already provided would scarcely be a mouthful.

As the giant kept very fashionable hours, and always waited dinner for himself till nine o'clock, there was still plenty of time; so, with a loud grumble about the trouble, he seized a large basket in his hand, and set off at a rapid pace towards the fairy Teach-all's garden. It was very seldom that Snap-'em-up ventured to think of foraging in this direction, as he never once succeeded in carrying off a single captive from the enclosure, it was so well fortified and so bravely defended; but on this occasion, being desperately hungry, he felt as bold as a lion, and walked with outstretched hands, straight towards the fairy Teach-all's dinner-table, taking such prodigious strides, that he seemed almost as if he would trample on himself.

A cry of consternation arose the instant this tremendous giant appeared; and as usual on such occasions, when he had made the same attempt before, a dreadful battle took place. Fifty active little boys bravely flew upon the enemy, armed with their dinner knives, and looked like a nest of hornets, stinging him in every direction, till he roared with pain, and would have run away, but the fairy Teach-all, seeing his intention, rushed forward with the carving-knife, and brandishing it high over her head, she most courageously stabbed him to the heart!

If a great mountain had fallen to the earth, it would have seemed like nothing in comparison with the giant Snap-'em-up, who crushed two or three houses to powder beneath him, and upset several fine monuments that were to have made people remembered for ever; but all this would have seemed scarcely worth mentioning, had it not been for a still greater event which occurred on the occasion, no less than the death of the fairy Do-nothing, who had been indolently looking on at this great battle without taking the trouble to interfere, or even to care who was victorious; but being also lazy about running away, when the giant fell, his sword came with so violent a stroke on her head, that she instantly expired.

Thus, luckily for the whole world, the fairy Teach-all got possession of immense property, which she proceeded without delay to make the best use of in her power.

In the first place, however, she lost no time in liberating Master No-book from his hook in the larder, and gave him a lecture on activity, moderation, and good conduct, which he never afterwards forgot; and it was astonishing to see the change that took place immediately in his whole thought and actions. From this very hour, Master No-book became the most diligent, active, happy boy in the fairy Teach-all's garden; and on returning home a month afterwards, he astonished all the masters at school by his extraordinary reformation. The most difficult lessons were a pleasure to him,—he scarcely ever stirred without a book in his hand,—never lay on a sofa again,—would scarcely even sit on a chair with a back to it, but preferred a three-legged stool,—detested holidays,—never thought any exertion a trouble,—preferred climbing over the top of a hill to creeping round the bottom,—always ate the plainest food in very small quantities, joined a Temperance Society!—and never tasted a morsel till he had worked very hard and got an appetite.

Not long after this, an old uncle, who had formerly been ashamed of Master No-book's indolence and gluttony, became so pleased at the wonderful change, that, on his death, he left him a magnificent estate, desiring that he should take his name; therefore, instead of being any longer one of the No-book family, he is now called Sir Timothy Bluestocking,—a pattern to the whole country round, for the good he does to every one, and especially for his extraordinary activity, appearing as if he could do twenty things at once. Though generally very good-natured and agreeable, Sir Timothy is occasionally observed in a violent passion, laying about him with his walking-stick in the most terrific manner, and beating little boys within an inch of their lives; but on inquiry, it invariably appears that he has found them out to be lazy, idle, or greedy, for all the industrious boys in the parish are sent to get employment from him, while he assures them that they are far happier breaking stones on the road, than if they were sitting idly in a drawing-room with nothing to do. Sir Timothy cares very little for poetry in general; but the following are his favourite verses, which he has placed over the chimney-piece at a school that he built for the poor, and every scholar is obliged, the very day he begins his education, to learn them:—

> Some people complain they have nothing to do,
> And time passes slowly away;
> They saunter about, with no object in view,
> And long for the end of the day.
>
> In vain are the trifles and toys they desire,
> For nothing they truly enjoy;
> Of trifles, and toys, and amusements they tire,
> For want of some useful employ.
>
> Although for transgression the ground was accursed,
> Yet gratefully man must allow,
> 'Twas really a blessing which doomed him, at first,
> To live by the sweat of his brow.

1839

NATHANIEL HAWTHORNE

Feathertop A Moralized Legend

'DICKON,' cried Mother Rigby, 'a coal for my pipe!'

The pipe was in the old dame's mouth when she said these words. She had thrust it there after filling it with tobacco, but without stooping to light it at the hearth, where indeed there was no appearance of a fire having been kindled that morning. Forthwith, however, as soon as the order was given, there was an intense red glow out of the bowl of the pipe, and a whiff of smoke from Mother Rigby's lips. Whence the coal came, and how brought thither by an invisible hand, I have never been able to discover.

'Good!' quoth Mother Rigby, with a nod of her head. 'Thank ye, Dickon! And now for making this scarecrow. Be within call, Dickon, in case I need you again.'

The good woman had risen thus early (for as yet it was scarcely sunrise), in order to set about making a scarecrow, which she intended to put in the middle of her cornpatch. It was now the latter week of May, and the crows and blackbirds had already discovered the little, green, rolled-up leaf of the Indian corn just peeping out of the soil. She was determined, therefore, to contrive as lifelike a scarecrow as ever was seen, and to finish it immediately, from top to toe, so that it should begin its sentinel's duty that very morning. Now Mother Rigby (as every body must have heard), was one of the most cunning and potent witches in New England, and might, with very little trouble, have made a scarecrow ugly enough to frighten the minister himself. But on this occasion, as she had awakened in an uncommonly pleasant humor, and was further dulcified by her pipe of tobacco, she resolved to produce something fine, beautiful, and splendid, rather than hideous and horrible.

'I don't want to set up a hobgoblin in my own corn patch, and almost at my own doorstep,' said Mother Rigby to herself, puffing out a whiff of smoke; 'I could do it if I pleased, but I'm tired of doing marvellous things, and so I'll keep within the bounds of everyday business, just for variety's sake. Besides, there is no use in scaring the little children for a mile roundabout, though 'tis true I'm a witch.'

It was settled, therefore, in her own mind, that the scarecrow should represent a fine gentleman of the period, so far as the materials at hand would allow. Perhaps it may be as well to enumerate the chief of the articles that went to the composition of this figure.

The most important item of all, probably, although it made so little show, was a certain broomstick, on which Mother Rigby had taken many an airy gallop at midnight, and which now served the scarecrow by way of a spinal column, or, as the unlearned phrase it, a backbone. One of its arms was a disabled flail which used to be wielded by Goodman Rigby, before his spouse worried him out of this troublesome world; the other, if I mistake not, was composed of the pudding stick and a broken rung of a chair, tied loosely together at the elbow. As for its legs, the right was a hoe handle, and the left an undistinguished and miscellaneous stick from the woodpile. Its lungs, stomach, and other affairs of that kind were nothing better than a meal bag stuffed with straw. Thus we have made out the skeleton and entire corporcity of the scarecrow, with the exception of its head; and this was admirably supplied by a somewhat withered and shrivelled pumpkin, in which Mother Rigby cut two holes for the eyes, and a slit for the mouth, leaving a bluish-colored knob in the middle to pass for a nose. It was really quite a respectable face.

'I've seen worse ones on human shoulders, at any rate,' said Mother Rigby. 'And many a fine gentleman has a pumpkin head, as well as my scarecrow.'

But the clothes, in this case, were to be the making of the man. So the good old woman took down from a peg an ancient plum-colored coat of London make, and with relics of embroidery on its seams, cuffs, pocket flaps, and button holes, but lamentably worn and faded, patched at the elbows, tattered at the skirts, and threadbare all over. On the left breast was a round hole, whence either a star of nobility had been rent away, or else the hot heart of some former wearer had scorched it through and through. The neighbors said that this rich garment belonged to the Black Man's wardrobe, and that he kept it at Mother Rigby's cottage for the convenience of slipping it on whenever he wished to make a grand appearance at the governor's table. To match the coat there was a velvet waistcoat of very ample size and formerly embroidered with foliage that had been as brightly golden as the maple leaves in October, but which had now quite vanished out of the substance of the velvet. Next came a pair of scarlet breeches, once worn by the French governor of Louisbourg, and the knees of which had touched the lower step of the throne of Louis le Grand. The Frenchman had given these smallclothes to an Indian powwow, who parted with them to the old witch for a gill of strong waters, at one of their dances in the forest. Furthermore, Mother Rigby produced a pair of silk stockings and put them on the figure's legs, where they showed as unsubstantial as a dream, with the wooden reality of the two sticks making itself miserably apparent through the holes. Lastly, she put her dead husband's wig on the bare scalp of the pumpkin, and surmounted the whole with a dusty three-cornered hat, in which was stuck the longest tail feather of a rooster.

Then the old dame stood the figure up in a corner of her cottage and chuckled to behold its yellow semblance of a visage, with its nobby little nose thrust into the air. It had a strangely self-satisfied aspect, and seemed to say, 'Come look at me!'

'And you are well worth looking at, that's a fact!' quoth Mother Rigby, in admiration at her own handiwork. 'I've made many a puppet since I've been a witch; but methinks this is the finest of them all. 'Tis almost too good for a scarecrow. And, by the by, I'll just fill a fresh pipe of tobacco, and then take him out to the corn patch.'

While filling her pipe, the old woman continued to gaze with almost motherly affection at the figure in the corner. To say the truth, whether it were chance, or skill, or downright witchcraft, there was something wonderfully human in this ridiculous shape, bedizened with its tattered finery; and as for the countenance, it appeared to shrivel its yellow surface into a grin—a funny kind of expression betwixt scorn and merriment, as if it understood itself to be a jest at mankind. The more Mother Rigby looked the better she was pleased.

'Dickon,' cried she sharply, 'another coal for my pipe!'

Hardly had she spoken, than, just as before, there was a red-glowing coal on the top of the tobacco. She drew in a long whiff and puffed it forth again into the bar of morning sunshine which struggled through the one dusty pane of her cottage window. Mother Rigby always liked to flavor her pipe with a coal of fire from the particular chimney corner whence this had been brought. But where that chimney corner might be, or who brought the coal from it—further than that the invisible messenger seemed to respond to the name of Dickon—I cannot tell.

'That puppet yonder,' thought Mother Rigby, still with her eyes fixed on the scarecrow, 'is too good a piece of work to stand all summer in a corn patch, frightening away the crows and black-birds. He's capable of better things. Why, I've danced with a worse one, when partners happened to be scarce, at our witch meetings in the forest! What if I should let him take his chance among the other men of straw and empty fellows who go bustling about the world?'

The old witch took three or four more whiffs of her pipe and smiled.

'He'll meet plenty of his brethren at every street corner!' continued she. 'Well; I didn't mean to dabble in witchcraft to-day, further than the lighting of my pipe; but a witch I am, and a witch I'm likely to be, and there's no use trying to shirk it. I'll make a man of my scarecrow, were it only for the joke's sake!'

While muttering these words Mother Rigby took the pipe from her own mouth and thrust it into the crevice which represented the same feature in the pumpkin visage of the scarecrow.

'Puff, darling, puff!' said she. 'Puff away, my fine fellow! your life depends on it!'

This was a strange exhortation, undoubtedly, to be addressed to a mere thing of sticks, straw, and old clothes, with nothing better than a shrivelled pumpkin for a head; as we know to have been the scarecrow's case. Nevertheless, as we must carefully hold in remembrance, Mother Rigby was a witch of singular power and dexterity; and, keeping this fact duly before our minds, we shall see nothing beyond credibility in the remarkable incidents of our story. Indeed, the great difficulty will be at once got over, if we can only bring ourselves to believe that, as soon as the old dame bade him puff, there came a whiff of smoke from the scarecrow's mouth. It was the very feeblest of whiffs, to be sure; but it was followed by another and another, each more decided than the preceding one.

'Puff away, my pet! puff away, my pretty one!' Mother Rigby kept repeating, with her pleasantest smile. 'It is the breath of life to ye; and that you may take my word for.'

Beyond all question the pipe was bewitched. There must have been a spell either in the tobacco or in the fiercely-glowing coal that so mysteriously burned on top of it, or in the pungently-aromatic smoke which exhaled from the kindled weed. The figure, after a few doubtful attempts, at length blew forth a volley of smoke extending all the way from the obscure corner into the bar of sunshine. There it eddied and melted away among the motes of dust. It seemed a convulsive effort; for the two or three next whiffs were fainter, although the coal still glowed and threw a gleam over the scarecrow's visage. The old witch clapped her skinny hands together, and smiled encouragingly upon her handiwork. She saw that the charm worked well. The shrivelled, yellow face, which heretofore had been no face at all, had already a thin, fantastic haze, as it were, of human likeness, shifting to and fro across it; sometimes vanishing entirely, but growing more perceptible than ever with the next whiff from the pipe. The whole figure, in like manner, assumed a show of life, such as we impart to ill-defined shapes among the clouds, and half deceive ourselves with the pastime of our own fancy.

If we must needs pry closely into the matter, it may be doubted whether there was any real change, after all, in the sordid, worn-out, worthless, and ill-jointed substance of the scarecrow; but merely a spectral illusion, and a cunning effect of light and shade so colored and contrived as to delude the eyes of most men. The miracles of witchcraft seem always to have had a very shallow subtlety; and, at least, if the above explanation do not hit the truth of the process, I can suggest no better.

'Well puffed, my pretty lad!' still cried old Mother Rigby. 'Come, another good stout whiff, and let it be with might and main. Puff for thy life, I tell thee! Puff out of the very bottom of thy heart; if any heart thou hast, or any bottom to it! Well done, again! Thou didst suck in that mouthful as if for the pure love of it.'

And then the witch beckoned to the scarecrow, throwing so much magnetic potency into her gesture that it seemed as if it must inevitably be obeyed, like the mystic call of the loadstone when it summons the iron.

'Why lurkest thou in the corner, lazy one?' said she. 'Step forth! Thou hast the world before thee!'

Upon my word, if the legend were not one which I heard on my grandmother's knee, and which had established its place among things credible before my childish judgment could analyze its probability, I question whether I should have the face to tell it now.

In obedience to Mother Rigby's word, and extending its arm as if to reach her outstretched hand, the figure made a step forward—a kind of hitch and jerk, however, rather than a step—then tottered and almost lost its balance. What could the witch expect? It was nothing, after all, but a scarecrow stuck upon two sticks. But the strong-willed old beldam scowled, and beckoned, and flung the energy of her purpose so forcibly at this poor combination of rotten wood, and musty straw, and ragged garments, that it was compelled to show itself a man, in spite of the reality of things. So it stepped into the bar of sunshine. There it stood—poor devil of a contrivance that it was!—with only the thinnest vesture of human similitude about it, through which was evident the stiff, ricketty, incongruous, faded, tattered, good-for-nothing patchwork of its substance, ready to sink in a heap upon the floor, as conscious of its own unworthiness to be erect. Shall I confess the truth? At its present point of vivification, the scarecrow reminds me of some of the lukewarm and abortive characters, composed of heterogeneous materials, used for the thousandth time, and never worth using, with which romance writers (and myself, no doubt, among the rest), have so overpeopled the world of fiction.

But the fierce old hag began to get angry and show a glimpse of her diabolic nature (like a snake's head, peeping with a hiss out of her bosom), at this pusillanimous behavior of the thing which she had taken the trouble to put together.

'Puff away, wretch!' cried she, wrathfully. 'Puff, puff, puff, thou thing of straw and emptiness! thou rag or two! thou meal bag! thou pumpkin head! thou nothing! Where shall I find a name vile enough to call thee by? Puff, I say, and suck in thy fantastic life along with the smoke; else I snatch the pipe from thy mouth and hurl thee where that red coal came from.'

Thus threatened, the unhappy scarecrow had nothing for it but to puff away for dear life. As need was, therefore, it applied itself lustily to the pipe and sent forth such abundant volleys of tobacco smoke that the small cottage kitchen became all vaporous. The one sunbeam struggled mistily through, and could but imperfectly define the image of the cracked and dusty window pane on the opposite wall. Mother Rigby, meanwhile, with one brown arm akimbo and the other stretched towards the figure, loomed grimly amid the obscurity with such port and expression as when she was wont to heave a ponderous nightmare on her victims and stand at the bedside to enjoy their agony. In fear and trembling did this poor scarecrow puff. But its efforts, it must be acknowledged, served an excellent purpose; for, with each successive whiff, the figure lost more and more of its dizzy and perplexing tenuity and seemed to take denser substance. Its very garments, moreover, partook of the magical change, and shone with the gloss of novelty and glistened with the skilfully embroidered gold that had long ago been rent away. And, half revealed among the smoke, a yellow visage bent its lustreless eyes on Mother Rigby.

At last the old witch clinched her fist and shook it at the figure. Not that she was positively angry, but merely acting on the principle—perhaps untrue, or not the only truth, though as high a one as Mother Rigby could be expected to attain—that feeble and torpid natures, being incapable of better inspiration, must be stirred up by fear. But here was the crisis. Should she fail in what she now sought to effect, it was her ruthless purpose to scatter the miserable simulacre into its original elements.

'Thou hast a man's aspect,' said she, sternly. 'Have also the echo and mockery of a voice! I bid thee speak!'

The scarecrow gasped, struggled, and at length emitted a murmur, which was so incorporated with its smoky breath that you could scarcely tell whether it were indeed a voice or only a whiff of tobacco. Some narrators of this legend hold the opinion that Mother Rigby's conjurations and the fierceness of her will had compelled a familiar spirit into the figure, and that the voice was his.

'Mother,' mumbled the poor stifled voice, 'be not so awful with me! I would fain speak; but being without wits, what can I say?'

'Thou canst speak, darling, canst thou?' cried Mother Rigby, relaxing her grim countenance into a smile. 'And what shalt thou say, quotha! Say, indeed! Art thou of the brotherhood of the empty skull, and demandest of me what thou shalt say? Thou shalt say a thousand things, and saying them a thousand times over, thou shalt still have said nothing! Be not afraid, I tell thee! When thou comest into the world (whither I purpose sending thee forthwith), thou shalt not lack the wherewithal to talk. Talk! Why, thou shalt babble like a mill stream, if thou wilt. Thou hast brains enough for that, I trow!'

'At your service, mother,' responded the figure.

'And that was well said, my pretty one,' answered Mother Rigby. 'Then thou spakest like thyself, and meant nothing. Thou shalt have a hundred such set phrases, and five hundred to the boot of them. And now, darling, I have taken so much pains with thee, and thou art so beautiful, that, by my troth, I love thee better than any witch's puppet in the world; and I've made them of all sorts—clay, wax, straw, sticks, night fog, morning mist, sea foam, and chimney smoke. But thou art the very best. So give heed to what I say.'

'Yes, kind mother,' said the figure, 'with all my heart!'

'With all thy heart!' cried the old witch, setting her hands to her sides and laughing loudly. 'Thou hast such a pretty way of speaking. With all thy heart! And thou didst put thy hand to the left side of thy waistcoat as if thou really hadst one!'

So now, in high good humor with this fantastic contrivance of hers, Mother Rigby told the scarecrow that it must go and play its part in the great world, where not one man in a hundred, she affirmed, was gifted with more real substance than itself. And, that he might hold up his head with the best of them, she endowed him, on the spot, with an unreckonable amount of wealth. It consisted partly of a gold mine in Eldorado, and of ten thousand shares in a broken bubble, and of half a million acres of vineyard at the North Pole, and of a castle in the air, and a chateau in Spain, together with all the rents and income therefrom accruing. She further made over to him the cargo of a certain ship, laden with salt of Cadiz, which she herself by her necromantic arts, had caused to founder, ten years before, in the deepest part of mid ocean. If the salt were not dissolved, and could be brought to market, it would fetch a pretty penny among the fishermen. That he might not lack ready money, she gave him a copper farthing of Birmingham manufacture, being all the coin she had about her, and likewise a great deal of brass, which she applied to his forehead, thus making it yellower than ever.

'With that brass alone,' quoth Mother Rigby, 'thou canst pay thy way all over the earth. Kiss me, pretty darling! I have done my best for thee.'

Furthermore, that the adventurer might lack no possible advantage towards a fair start in life, this excellent old dame gave him a token by which he was to introduce himself to a certain magistrate, member of the council, merchant, and elder of the church (the four capacities constituting but one man), who stood at the head of society in the neighboring metropolis. The token was neither more nor less than a single word, which Mother Rigby whispered to the scarecrow, and which the scarecrow was to whisper to the merchant.

'Gouty as the old fellow is, he'll run thy errands for thee, when once thou hast given him that word in his ear,' said the old witch. 'Mother Rigby knows the worshipful Justice Gookin, and the worshipful Justice knows Mother Rigby!'

Here the witch thrust her wrinkled face close to the puppet's, chuckling irrepressibly, and fidgeting all through her system, with delight at the idea which she meant to communicate.

'The worshipful Master Gookin,' whispered she, 'hath a comely maiden to his daughter. And hark ye, my pet! Thou hast a fair outside, and a pretty wit enough of thine own. Yea, a pretty wit enough! Thou wilt think better of it when thou hast seen more of other people's wits. Now, with thy outside and thy inside, thou art the very man to win a young girl's heart. Never doubt it! I tell thee it shall be so. Put but a bold face on the matter, sigh, smile, flourish thy hat, thrust forth thy leg like a dancing master, put thy right hand to the left side of thy waistcoat, and pretty Polly Gookin is thine own!'

All this while the new creature had been sucking in and exhaling the vapory fragrance of his pipe, and seemed now to continue this occupation as much for the enjoyment it afforded as because it was an essential condition of his existence. It was wonderful to see how exceedingly like a human being it behaved. Its eyes (for it appeared to possess a pair), were bent on Mother Rigby, and at suitable junctures it nodded or shook its head. Neither did it lack words proper for the occasion: 'Really! Indeed! Pray tell me! Is it possible! Upon my word! By no means! O! Ah! Hem!' and other such weighty utterances as imply attention, inquiry, acquiescence, or dissent on the part of the auditor. Even had you stood by and seen the scarecrow made you could scarcely have resisted the conviction that it perfectly understood the cunning counsels which the old witch poured into its counterfeit of an ear. The more earnestly it applied its lips to the pipe the more distinctly was its human likeness stamped among visible realities, the more sagacious grew its expression, the more lifelike its gestures and movements, and the more intelligibly audible its voice. Its garments, too, glistened so much the brighter with an illusory magnificence. The very pipe, in which burned the spell of all this wonderwork, ceased to appear as a smoke-blackened earthen stump, and became a meerschaum, with painted bowl and amber mouthpiece.

It might be apprehended, however, that as the life of the illusion seemed identical with the vapor of the pipe, it would terminate simultaneously with the reduction of the tobacco to ashes. But the beldam foresaw the difficulty.

'Hold thou the pipe, my precious one,' said she, 'while I fill it for thee again.'

It was sorrowful to behold how the fine gentleman began to fade back into a scarecrow while Mother Rigby shook the ashes out of the pipe and proceeded to replenish it from her tobacco box.

'Dickon,' cried she, in her high, sharp tone, 'another coal for this pipe!'

No sooner said than the intensely red speck of fire was glowing within the pipe bowl; and the scarecrow, without waiting for the witch's bidding, applied the tube to his lips and drew in a few short, convulsive whiffs, which soon, however, became regular and equable.

'Now, mine own heart's darling,' quoth Mother Rigby, 'whatever may happen to thee, thou must stick to thy pipe. Thy life is in it; and that, at least, thou knowest well, if thou knowest nought besides. Stick to thy pipe, I say! Smoke, puff, blow thy cloud; and tell the people, if any question be made, that it is for thy health, and that so the physician orders thee to do. And, sweet one, when thou shalt find thy pipe getting low, go apart into some corner, and (first filling thyself with smoke), cry sharply, "Dickon, a fresh pipe of tobacco!" and, "Dickon, another coal for my pipe!" and have it into thy pretty mouth as speedily as may be. Else, instead of a gallant gentleman in a gold-laced coat, thou wilt be but a jumble of sticks and tattered clothes, and a bag of straw, and a withered pumpkin! Now depart, my treasure, and good luck go with thee!'

'Never fear, mother!' said the figure, in a stout voice, and sending forth a courageous whiff of smoke. 'I will thrive, if an honest man and a gentleman may!'

'O, thou wilt be the death of me!' cried the old witch, convulsed with laughter. 'That was well said. If an honest man and a gentleman may! Thou playest thy part to perfection. Get along with thee for a smart fellow; and I will wager on thy head, as a man of pith and substance, with a brain, and what they call a heart, and all else that a man should have, against any other thing on two legs. I hold myself a better witch than yesterday, for thy sake. Did not I make thee? And I defy any witch in New England to make such another! Here; take my staff along with thee!'

The staff, though it was but a plain oaken stick, immediately took the aspect of a gold-headed cane.

'That gold head has as much sense in it as thine own,' said Mother Rigby, 'and it will guide thee straight to worshipful Master Gookin's door. Get thee gone, my pretty pet, my darling, my precious one, my treasure; and if any ask thy name, it is Feathertop. For thou hast a feather in thy hat, and I have thrust a handful of feathers into the hollow of thy head, and thy wig too is of the fashion they call Feathertop,—so be Feathertop thy name!'

And, issuing from the cottage, Feathertop strode manfully towards town. Mother Rigby stood at the threshold, well pleased to see how the sunbeams glistened on him, as if all his magnificence were real, and how diligently and lovingly he smoked his pipe, and how handsomely he walked, in spite of a little stiffness of his legs. She watched him until out of sight, and threw a witch benediction after her darling, when a turn of the road snatched him from her view.

Betimes in the forenoon, when the principal street of the neighboring town was just at its acme of life and bustle, a stranger of very distinguished figure was seen on the sidewalk. His port as well as his garments betokened nothing short of nobility. He wore a richly-embroidered plum-colored coat, a waistcoat of costly velvet magnificently adorned with golden foliage, a pair of splendid scarlet breeches, and the finest and glossiest of white silk stockings. His head was covered with a peruke, so daintily powdered and adjusted that it would have been sacrilege to disorder it with a hat; which, therefore (and it was a gold-laced hat, set off with a snowy feather), he carried beneath his arm. On the breast of his coat glistened a star. He managed his gold-headed cane with an airy grace peculiar to the fine gentlemen of the period; and, to give the highest possible finish to his equipment, he had lace ruffles at his wrist, of a most ethereal delicacy, sufficiently avouching how idle and aristocratic must be the hands which they half concealed.

It was a remarkable point in the accoutrement of this brilliant personage, that he held in his left hand a fantastic kind of a pipe, with an exquisitely painted bowl and an amber mouthpiece. This he applied to his lips as often as every five or six paces, and inhaled a deep whiff of smoke, which, after being retained a moment in his lungs, might be seen to eddy gracefully from his mouth and nostrils.

As may well be supposed, the street was all astir to find out the stranger's name.

'It is some great nobleman, beyond question,' said one of the townspeople. 'Do you see the star at his breast?'

'Nay; it is too bright to be seen,' said another. 'Yes; he must needs be a nobleman, as you say. But by what conveyance, think you, can his lordship have voyaged or travelled hither? There has been no vessel from the old country for a month past; and if he have arrived overland from the southward, pray where are his attendants and equipage?'

'He needs no equipage to set off his rank,' remarked a third. 'If he came among us in rags, nobility would shine through a hole in his elbow. I never saw such dignity of aspect. He has the old Norman blood in his veins, I warrant him.'

'I rather take him to be a Dutchman, or one of your high Germans,' said another citizen. 'The men of those countries have always the pipe at their mouths.'

'And so has a Turk,' answered his companion. 'But, in my judgment, this stranger hath been bred at the French court, and hath there learned politeness and grace of manner, which none understand so well as the nobility of France. That gait, now! A vulgar spectator might deem it stiff—he might call it a hitch and jerk—but, to my eye, it hath an unspeakable majesty, and must have been acquired by constant observation of the deportment of the Grand Monarque. The stranger's character and office are evident enough. He is a French ambassador, come to treat with our rulers about the cession of Canada.'

'More probably a Spaniard,' said another, 'and hence his yellow complexion; or, most likely, he is from the Havana, or from some port on the Spanish main, and comes to make investigation about the piracies which our governor is thought to connive at. Those settlers in Peru and Mexico have skins as yellow as the gold which they dig out of their mines.'

'Yellow or not,' cried a lady, 'he is a beautiful man!—so tall, so slender! such a fine, noble face, with so well-shaped a nose, and all that delicacy of expression about the mouth! And, bless me, how bright his star is! It positively shoots out flames!'

'So do your eyes, fair lady,' said the stranger, with a bow and a flourish of his pipe; for he was just passing at the instant. 'Upon my honor, they have quite dazzled me.'

'Was ever so original and exquisite a compliment?' murmured the lady, in an ecstasy of delight.

Amid the general admiration excited by the stranger's appearance, there were only two dissenting voices. One was that of an impertinent cur, which, after snuffing at the heels of the glistening figure, put its tail between its legs and skulked into its master's back yard, vociferating an execrable howl. The other dissentient was a young child, who squalled at the fullest stretch of his lungs, and babbled some unintelligible nonsense about a pumpkin.

Feathertop meanwhile pursued his way along the street. Except for the few complimentary words to the lady, and now and then a slight inclination of the head in requital of the profound reverences of the bystanders, he seemed wholly absorbed in his pipe. There needed no other proof of his rank and consequence than the perfect equanimity with which he comported himself, while the curiosity and admiration of the town swelled almost into clamor around him. With a crowd gathering behind his footsteps, he finally reached the mansion house of the worshipful Justice Gookin, entered the gate, ascended the steps of the front door, and knocked. In the interim, before his summons was answered, the stranger was observed to shake the ashes out of his pipe.

'What did he say in that sharp voice?' inquired one of the spectators.

'Nay, I know not,' answered his friend. 'But the sun dazzles my eyes strangely. How dim and faded his lordship looks all of a sudden! Bless my wits, what is the matter with me?'

'The wonder is,' said the other, 'that his pipe, which was out only an instant ago, should be all alight again, and with the reddest coal I ever saw. There is something mysterious about this stranger. What a whiff of smoke was that! Dim and faded did you call him? Why, as he turns about the star on his breast is all ablaze.'

'It is, indeed,' said his companion; 'and it will go near to dazzle pretty Polly Gookin, whom I see peeping at it out of the chamber window.'

The door being now opened, Feathertop turned to the crowd, made a stately bend of his body like a great man acknowledging the reverence of the meaner sort, and vanished into the house. There was a mysterious kind of a smile, if it might not better be called a grin or grimace, upon his visage; but, of all the throng that beheld him, not an individual appears to have possessed insight enough to detect the illusive character of the stranger except a little child and a cur dog.

Our legend here loses somewhat of its continuity, and, passing over the preliminary explanation between Feathertop and the merchant, goes in quest of the pretty Polly Gookin. She was a damsel of a soft, round figure, with light hair and blue eyes, and a fair, rosy face, which seemed neither very shrewd nor very simple. This young lady had caught a glimpse of the glistening stranger while standing at the threshold, and had forthwith put on a laced cap, a string of beads, her finest kerchief, and her stiffest damask petticoat, in preparation for the interview. Hurrying from her chamber to the parlor, she had ever since been viewing herself in the large looking glass and practising pretty airs—now a smile, now a ceremonious dignity of aspect, and now a softer smile than the former, kissing her hand likewise, tossing her head, and managing her fan; while within the mirror an unsubstantial little maid repeated every gesture and did all the foolish things that Polly did, but without making her ashamed of them. In short, it was the fault of pretty Polly's ability rather than her will if she failed to be as complete an artifice as the illustrious Feathertop himself; and, when she thus tampered with her own simplicity, the witch's phantom might well hope to win her.

No sooner did Polly hear her father's gouty footsteps approaching the parlor door, accompanied with the stiff clatter of Feathertop's high-heeled shoes, than she seated herself bolt upright and innocently began warbling a song.

'Polly! daughter Polly!' cried the old merchant. 'Come hither, child.'

Master Gookin's aspect, as he opened the door, was doubtful and troubled.

'This gentleman,' continued he, presenting the stranger, 'is the Chevalier Feathertop,—nay, I beg his pardon, my Lord Feathertop,—who hath brought me a token of remembrance from an ancient friend of mine. Pay your duty to his lordship, child, and honor him as his quality deserves.'

After these few words of introduction the worshipful magistrate immediately quitted the room. But, even in that brief moment, had the fair Polly glanced aside at her father instead of devoting herself wholly to the brilliant guest, she might have taken warning of some mischief nigh at hand. The old man was nervous, fidgety, and very pale. Purposing a smile of courtesy, he had deformed his face with a sort of galvanic grin, which, when Feathertop's back was turned, he exchanged for a scowl, at the same time shaking his fist and stamping his gouty foot—an incivility which brought its retribution along with it. The truth appears to have been, that Mother Rigby's word of introduction, whatever it might be, had operated far more on the rich merchant's fears than on his good will. Moreover, being a man of wonderfully acute observation, he had noticed that the painted figures on the bowl of Feathertop's pipe were in motion. Looking more closely, he became convinced that these figures were a party of little demons, each duly provided with horns and a tail, and dancing hand in hand, with gestures of diabolical merriment, round the circumference of the pipe bowl. As if to confirm his suspicions, while Master Gookin ushered his guest along a dusky passage from his private room to the parlor, the star on Feathertop's breast had scintillated actual flames, and threw a flickering gleam upon the wall, the ceiling, and the floor.

With such sinister prognostics manifesting themselves on all hands, it is not to be marvelled at that the merchant should have felt that he was committing his daughter to a very questionable acquaintance. He cursed, in his secret soul, the insinuating elegance of Feathertop's manners, as this brilliant personage bowed, smiled, put his hand on his heart, inhaled a long whiff from his pipe, and enriched the atmosphere with the smoky vapor of a fragrant and visible sigh. Gladly would poor Master Gookin have thrust his dangerous guest into the street; but there was a constraint and terror within him. This respectable old gentleman, we fear, at an earlier period of life, had given some pledge or other to the evil principle, and perhaps was now to redeem it by the sacrifice of his daughter.

It so happened that the parlor door was partly of glass, shaded by a silken curtain, the folds of which hung a little awry. So strong was the merchant's interest in witnessing what was to ensue between the fair Polly and the gallant Feathertop that after quitting the room he could by no means refrain from peeping through the crevice of the curtain.

But there was nothing very miraculous to be seen; nothing—except the trifles previously noticed—to confirm the idea of a supernatural peril environing the pretty Polly. The stranger, it is true, was evidently a thorough and practised man of the world, systematic and self-possessed, and therefore the sort of a person to whom a parent ought not to confide a simple, young girl without due watchfulness for the result. The worthy magistrate, who had been conversant with all degrees and qualities of mankind, could not but perceive every motion and gesture of the distinguished Feathertop came in its proper place; nothing had been left rude or native in him; a well-digested conventionalism had incorporated itself thoroughly with his substance and transformed him into a work of art. Perhaps it was this peculiarity that invested him with a species of ghastliness and awe. It is the effect of any thing completely and consummately artificial, in human shape, that the person impresses us as an unreality and as having hardly pith enough to cast a shadow upon the floor. As regarded Feathertop, all this resulted in a wild, extravagant, and fantastical impression, as if his life and being were akin to the smoke that curled upward from his pipe.

But pretty Polly Gookin felt not thus. The pair were now promenading the room; Feathertop with his dainty stride and no less dainty grimace; the girl with a native maidenly grace, just touched, not spoiled, by a slightly affected manner, which seemed caught from the perfect artifice of her companion. The longer the interview continued, the more charmed was pretty Polly, until, within the first quarter of an hour (as the old magistrate noted by his watch), she was evidently beginning to be in love. Nor need it have been witchcraft that subdued her in such a hurry; the poor child's heart, it may be, was so very fervent that it melted her with its own warmth as reflected from the hollow semblance of a lover. No matter what Feathertop said, his words found depth and reverberation in her ear; no matter what he did, his action was heroic to her eye. And by this time it is to be supposed there was a blush on Polly's cheek, a tender smile about her mouth, and a liquid softness in her glance; while the star kept coruscating on Feathertop's breast, and the little demons careered with more frantic merriment than ever about the circumference of his pipe bowl. O pretty Polly Gookin, why should these imps rejoice so madly that a silly maiden's heart was about to be given to a shadow! Is it so unusual a misfortune, so rare a triumph?

By and by Feathertop paused, and, throwing himself into an imposing attitude, seemed to summon the fair girl to survey his figure and resist him longer if she could. His star, his embroidery, his buckles glowed at that instant with unutterable splendor; the picturesque hues of his attire took a richer depth of coloring; there was a gleam and polish over his whole presence betokening the perfect witchery of well-ordered manners. The maiden raised her eyes and suffered them to linger upon her companion with a bashful and admiring gaze. Then, as if desirous of judging what value her own simple comeliness might have side by side with so much brilliancy, she cast a glance towards the full-length looking glass in front of which they happened to be standing. It was one of the truest plates in the world, and incapable of flattery. No sooner did the images therein reflected meet Polly's eye than she shrieked, shrank from the stranger's side, gazed at him for a moment in the wildest dismay, and sank insensible upon the floor. Feathertop likewise had looked towards the mirror, and there beheld, not the glittering mockery of his outside show, but a picture of the sordid patchwork of his real composition, stripped of all witchcraft.

The wretched simulacrum! We almost pity him. He threw up his arms with an expression of despair that went further than any of his previous manifestations towards vindicating his claims to be reckoned human; for, perchance the only time since this so often empty and deceptive life of mortals began its course, an illusion had seen and fully recognized itself.

Mother Rigby was seated by her kitchen hearth in the twilight of this eventful day, and had just shaken the ashes out of a new pipe, when she heard a hurried tramp along the road. Yet it did not seem so much the tramp of human footsteps as the clatter of sticks or the rattling of dry bones.

'Ha!' thought the old witch, 'what step is that? Whose skeleton is out of its grave now, I wonder?'

A figure burst headlong into the cottage door. It was Feathertop! His pipe was still alight; the star still flamed upon his breast; the embroidery still glowed upon his garments; nor had he lost, in any degree or manner that could be estimated, the aspect that assimilated him with our mortal brotherhood. But yet, in some indescribable way (as is the case with all that has deluded us when once found out), the poor reality was felt beneath the cunning artifice.

'What has gone wrong?' demanded the witch. 'Did yonder sniffling hypocrite thrust my darling from his door? The villain! I'll set twenty fiends to torment him till he offer thee his daughter on his bended knees!'

'No, mother,' said Feathertop despondingly; 'it was not that.'

'Did the girl scorn my precious one?' asked Mother Rigby, her fierce eyes glowing like two coals of Tophet. 'I'll cover her face with pimples! Her nose shall be as red as the coal in thy pipe! Her front teeth shall drop out! In a week hence she shall not be worth thy having!'

'Let her alone, mother,' answered poor Feathertop; 'the girl was half won; and methinks a kiss from her sweet lips might have made me altogether human. But,' he added, after a brief pause and then a howl of self-contempt, 'I've seen myself, mother! I've seen myself for the wretched, ragged, empty thing I am! I'll exist no longer!'

Snatching the pipe from his mouth, he flung it with all his might against the chimney, and at the same instant sank upon the floor, a medley of straw and tattered garments, with some sticks protruding from the heap, and a shrivelled pumpkin in the midst. The eyeholes were now lustreless; but the rudely-carved gap, that just before had been a mouth, still seemed to twist itself into a despairing grin, and was so far human.

'Poor fellow!' quoth Mother Rigby, with a rueful glance at the relics of her ill-fated contrivance. 'My poor, dear, pretty

Feathertop! There are thousands upon thousands of coxcombs and charlatans in the world, made up of just such a jumble of wornout, forgotten, and good-for-nothing trash as he was! Yet they live in fair repute, and never see themselves for what they are. And why should my poor puppet be the only one to know himself and perish for it?'

While thus muttering, the witch had filled a fresh pipe of tobacco, and held the stem between her fingers, as doubtful whether to thrust it into her own mouth or Feathertop's.

'Poor Feathertop!' she continued. 'I could easily give him another chance and send him forth again tomorrow. But no; his feelings are too tender, his sensibilities too deep. He seems to have too much heart to bustle for his own advantage in such an empty and heartless world. Well! well! I'll make a scarecrow of him after all. 'Tis an innocent and a useful vocation, and will suit my darling well; and, if each of his human brethren had as fit a one, 'twould be the better for mankind; and as for this pipe of tobacco, I need it more than he.'

So saying, Mother Rigby put the stem between her lips. 'Dickon!' cried she, in her high, sharp tone, 'another coal for my pipe!'

1846

JOHN RUSKIN

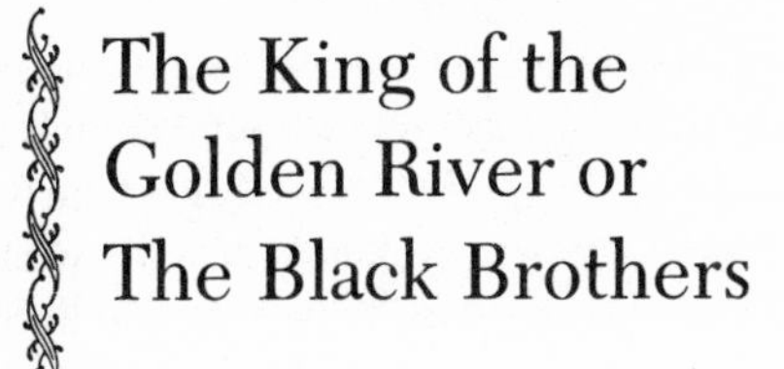

The King of the Golden River or The Black Brothers

CHAPTER I

How the Agricultural System of the Black Brothers was interfered with by South West Wind, Esquire

IN a secluded and mountainous part of Stiria, there was, in old time, a valley of the most surprising and luxuriant fertility. It was surrounded, on all sides, by steep and rocky mountains, rising into peaks, which were always covered with snow, and from which a number of torrents descended in constant cataracts. One of these fell westward, over the face of a crag so high, that, when the sun had set to everything else, and all below was darkness, his beams still shone full upon this waterfall, so that it looked like a shower of gold. It was, therefore, called by the people of the neighbourhood, the Golden River. It was strange that none of these streams fell into the valley itself. They all descended on the other side of the mountains, and wound away through broad plains and by populous cities. But the clouds were drawn so constantly to the snowy hills, and rested so softly in the circular hollow, that in time of drought and heat, when all the country round was burnt up, there was still rain in the little valley; and its crops were so heavy, and its hay so high, and its apples so red, and its grapes so blue, and its wine so rich, and its honey so sweet, that it was a marvel to every one who beheld it, and was commonly called the Treasure Valley. The whole of this little valley belonged to three brothers, called Schwartz, Hans, and Gluck. Schwartz and Hans, the two elder brothers, were very ugly men, with over-hanging eyebrows and small dull eyes, which were always half shut, so that you couldn't see into *them*, and always fancied they saw very far into *you*. They lived by farming the Treasure Valley, and very good farmers they were. They killed everything that did not pay for its eating. They shot the blackbirds, because they pecked the fruit; and killed the hedgehogs, lest they should suck the cows; they poisoned the crickets for eating the crumbs in the kitchen, and smothered the cicadas, which used to sing all summer in the lime trees. They worked their servants without any wages, till they would not work any more, and then quarrelled with them, and turned them out of doors without paying them. It would have been very odd, if with such a farm, and such a system of farming, they hadn't got very rich; and very rich they *did* get. They generally contrived to keep their corn by them till it was very dear, and then sell it for twice its value; they had heaps of gold lying about on their floors, yet it was never known that they had given so much as a penny or a crust in charity; they never went to mass; grumbled perpetually at paying tithes; and were, in a word, of so cruel and grinding a temper, as to receive from all those with whom they had any dealings, the nick-name of the 'Black Brothers.' The youngest brother, Gluck, was as completely opposed, in both appearance and character, to his seniors as could possibly be imagined or desired. He was not above twelve years old, fair, blue eyed, and kind in temper to every living thing. He did not, of course, agree particularly well with his brothers, or rather, they did not agree with *him*. He was usually appointed to the honourable office of turnspit, when there was anything to roast, which was not often; for, to do the brothers justice, they were hardly less sparing upon themselves than upon other people. At other times he used to clean the shoes, floors, and sometimes the plates, occasionally getting what was left on them, by way of encouragement, and a wholesome quantity of dry blows, by way of education.

Things went on in this manner for a long time. At last came a very wet summer, and everything went wrong in the country round. The hay had hardly been got in, when the haystacks were floated bodily down to the sea by an inundation; the vines were cut to pieces with the hail; the corn was all killed by a black blight; only in the Treasure Valley, as usual, all was safe. As it had rain when there was rain no where else, so it had sun when there was sun no where else. Every body came to buy corn at the farm, and went away pouring maledictions on the Black Brothers. They asked what they liked, and got it, except from the poor people, who could only beg, and several of whom were starved at their very door, without the slightest regard or notice.

It was drawing towards winter, and very cold weather, when one day the two elder brothers had gone out, with their usual warning to little Gluck, who was left to mind the roast, that he was to let nobody in, and give nothing out. Gluck sat down quite close to the fire, for it was raining very hard, and the kitchen walls were by no means dry or comfortable looking. He turned and turned, and the roast got nice and brown. 'What a pity,' thought Gluck, 'my brothers never ask any body to dinner. I'm sure, when they've got such a nice piece of mutton as this, and nobody else has got so much as a piece of dry bread, it would do their hearts good to have somebody to eat it with them.'

Just as he spoke, there came a double knock at the house door, yet heavy and dull, as though the knocker had been tied up—more like a puff than a knock. 'It must be the wind,' said Gluck; 'nobody else would venture to knock double knocks at our door.' No it wasn't the wind: there it came again very hard, and what was particularly astounding, the knocker seemed to be in a hurry, and not to be in the least afraid of the consequences. Gluck went to the window, opened it, and put his head out to see who it was. It was the most extraordinary looking little gentleman he had ever seen in his life. He had a very long nose, slightly brass-coloured, and expanding towards its termination into a development not unlike the lower extremity of a key bugle. His cheeks were very round, and very red, and might have warranted a supposition that he had been blowing a refractory fire for the last eight-and-forty hours. His eyes twinkled merrily through long silky eyelashes, his moustaches curled twice round like a cork-screw on each side of his mouth, and his hair, of a curious mixed pepper and salt colour, descended far over his shoulders. He was about four feet six in height, and wore a conical pointed cap of nearly the same altitude, decorated with a black feather some three feet long. His doublet was prolonged behind into something resembling a violent exaggeration of what is now termed a 'swallow tail,' but was much obscured by the swelling folds of an enormous black, glossy looking cloak, which must have been very much too long in calm weather, as the wind, whistling round the old house, carried it clear out from the wearer's shoulders to about four times his own length.

Gluck was so perfectly paralyzed by the singular appearance of his visitor, that he remained fixed without uttering a word, until the old gentleman, having performed another, and a more energetic concerto on the knocker, turned round to look after his fly-away cloak. In so doing he caught sight of Gluck's little yellow head jammed in the window, with its mouth and eyes very wide open indeed.

'Hollo!' said the little gentleman, 'that's not the way to answer the door: I'm wet, let me in.'

To do the little gentleman justice, he *was* wet. His feather hung down between his legs like a beaten puppy's tail, dripping like an umbrella; and from the ends of his moustaches the water was running into his waistcoat pockets, and out again like a mill stream.

'I beg pardon, sir,' said Gluck, 'I'm very sorry, but I really can't.'

'Can't what?' said the old gentleman.

'I can't let you in, sir,—I can't indeed; my brothers would beat me to death, sir, if I thought of such a thing. What do you want, sir?'

'Want?' said the old gentleman petulantly. 'I want fire, and shelter; and there's your great fire there blazing, crackling, and dancing on the walls, with nobody to feel it. Let me in I say; I only want to warm myself.'

Gluck had had his head, by this time, so long out of the window, that he began to feel it was really unpleasantly cold, and when he turned, and saw the beautiful fire rustling and roaring, and throwing long bright tongues up the chimney, as if it were licking its chops at the savoury smell of the leg of mutton, his heart melted within him that it should be burning away for nothing. 'He does look *very* wet,' said little Gluck; 'I'll just let him in for a quarter of an hour.' Round he went to the door, and opened it; and as the little gentleman walked in, there came a gust of wind through the house, that made the old chimneys totter.

'That's a good boy,' said the little gentleman. 'Never mind your brothers. I'll talk to them.'

'Pray, sir, don't do any such thing;' said Gluck. 'I can't let you stay till they come; they'd be the death of me.'

'Dear me,' said the old gentleman, 'I'm very sorry to hear that. How long may I stay?'

'Only till the mutton's done, sir,' replied Gluck, 'and it's very brown.'

Then the old gentleman walked into the kitchen, and sat himself down on the hob, with the top of his cap accommodated up the chimney, for it was a great deal too high for the roof.

'You'll soon dry there, sir,' said Gluck, and sat down again to turn the mutton. But the old gentleman did *not* dry there, but went on drip, drip, dripping among the cinders, and the fire fizzed, and sputtered, and began to look very black, and uncomfortable; never was such a cloak; every fold in it ran like a gutter.

'I beg pardon, sir,' said Gluck at length, after watching the water spreading in long, quick-silver like streams over the floor for a quarter of an hour; 'mayn't I take your cloak?'

'No, thank you,' said the old gentleman.

'Your cap, sir?'

'I am all right, thank you,' said the old gentleman rather gruffly.

'But,—sir,—I'm very sorry,' said Gluck, hesitatingly; 'but—really, sir,—you're—putting the fire out.'

'It'll take longer to do the mutton then,' replied his visitor drily.

Gluck was very much puzzled by the behaviour of his guest; it was such a strange mixture of coolness and humility. He turned away at the string meditatively for another five minutes.

'That mutton looks very nice,' said the old gentleman at length. 'Can't you give me a little bit?'

'Impossible, sir,' said Gluck.

'I'm very hungry,' continued the old gentleman. 'I've had nothing to eat yesterday, nor to-day. They surely couldn't miss a bit from the knuckle!'

He spoke in so very melancholy a tone, that it quite melted Gluck's heart. 'They promised me one slice to-day, sir,' said he. 'I can give you that, but not a bit more.'

'That's a good boy,' said the old gentleman again.

Then Gluck warmed a plate, and sharpened a knife. 'I don't care if I do get beaten for it,' thought he. Just as he had cut a large slice out of the mutton, there came a tremendous rap at the door. The old gentleman jumped off the hob, as if it had suddenly become inconveniently warm. Gluck fitted the slice into the mutton again, with desperate efforts at exactitude, and ran to open the door.

'What did you keep us waiting in the rain for?' said Schwartz, as he walked in, throwing his umbrella in Gluck's face. 'Ay! what for, indeed, you little vagabond?' said Hans, administering an educational box on the ear, as he followed his brother into the kitchen.

'Bless my soul!' said Schwartz when he opened the door.

'Amen,' said the little gentleman, who had taken his cap off, and was standing in the middle of the kitchen, bowing with the utmost possible velocity.

'Who's that?' said Schwartz, catching up a rolling-pin, and turning to Gluck with a fierce frown.

'I don't know, indeed, brother,' said Gluck in great terror.

'How did he get in?' roared Schwartz.

'My dear brother,' said Gluck, deprecatingly, 'he was so *very* wet!'

The rolling-pin was descending on Gluck's head; but, at the instant, the old gentleman interposed his conical cap, on which it crashed with a shock that shook the water out of it all over the room. What was very odd, the rolling-pin no sooner touched the cap, than it flew out of Schwartz's hand, spinning like a straw in a high wind, and fell into the corner at the further end of the room.

'Who are you, sir?' demanded Schwartz, turning upon him.

'What's your business?' snarled Hans.

'I'm a poor old man, sir,' the little gentleman began very modestly, 'and I saw your fire through the window, and begged shelter for a quarter of an hour.'

'Have the goodness to walk out again, then,' said Schwartz. 'We've quite enough water in our kitchen, without making it a drying house.'

'It is a cold day to turn an old man out in, sir; look at my grey hairs.' They hung down to his shoulders, as I told you before.

'Ay!' said Hans, 'there are enough of them to keep you warm. Walk!'

'I'm very, very hungry, sir; couldn't you spare me a bit of bread before I go?'

'Bread, indeed!' said Schwartz; 'do you suppose we've nothing to do with our bread, but to give it to such red-nosed fellows as you?'

'Why don't you sell your feather?' said Hans, sneeringly. 'Out with you.'

'A little bit,' said the old gentleman.

'Be off!' said Schwartz.

'Pray, gentlemen.'

'Off, and be hanged!' cried Hans, seizing him by the collar. But he had no sooner touched the old gentleman's collar, than away he went after the rolling-pin, spinning round and round, till he fell into the corner on the top of it. Then Schwartz was very angry, and ran at the old gentleman to turn him out; but he also had hardly touched him, when away he went after Hans and the rolling-pin, and hit his head against the wall as he tumbled into the corner. And so there they lay, all three.

Then the old gentleman spun himself round with velocity in the opposite direction; continued to spin until his long cloak was all wound neatly about him; clapped his cap on his head, very much on one side (for it could not stand upright without going through the ceiling), gave an additional twist to his cork-screw moustaches, and replied with perfect coolness: 'Gentlemen, I wish you a very good morning. At twelve o'clock to-night, I'll call again; after such a refusal of hospitality as I have just experienced, you will not be surprised if that visit is the last I ever pay you.'

'If ever I catch you here again,' muttered Schwartz, coming half frightened, out of the cornner—but, before he could finish his sentence, the old gentleman had shut the house door behind him with a great bang: and there drove past the window, at the same instant, a wreath of ragged cloud, that whirled and rolled away down the valley in all manner of shapes; turning over and over in the air; and melting away at last in a gush of rain.

'A very pretty business, indeed, Mr Gluck!' said Schwartz. 'Dish the mutton, sir. If ever I catch you at such a trick again—bless me, why the mutton's been cut!'

'You promised me one slice, brother, you know,' said Gluck.

'Oh! and you were cutting it hot, I suppose, and going to catch all the gravy. It'll be long before I promise you such a thing again. Leave the room, sir; and have the kindness to wait in the coal-cellar till I call you.'

Gluck left the room, melancholy enough. The brothers ate as much mutton as they could, locked the rest in the cupboard, and proceeded to get very drunk after dinner.

Such a night as it was! Howling wind, and rushing rain, without intermission. The brothers had just sense enough left to put up all the shutters, and double bar the door, before they went to bed. They usually slept in the same room. As the clock struck twelve, they were both awakened by a tremendous crash. Their door burst open with a violence that shook the house from top to bottom.

'What's that?' cried Schwartz, starting up in his bed.

'Only I,' said the little gentleman.

The two brothers sat up on their bolster, and stared into the darkness. The room was full of water, and by a misty moon-beam, which found its way through a hole in the shutter, they could see in the midst of it, an enormous foam globe, spinning round, and bobbing up and down like a cork, on which, as on a most luxurious cushion, reclined the little old gentleman, cap and all. There was plenty of room for it now, for the roof was off.

'Sorry to incommode you,' said their visitor, ironically. 'I'm afraid your beds are dampish; perhaps you had better go to your brother's room: I've left the ceiling on, there.'

They required no second admonition, but rushed into Gluck's room, wet through, and in an agony of terror.

'You'll find my card on the kitchen table,' the old gentleman called after them. 'Remember, the *last* visit.'

'Pray Heaven it may!' said Schwartz, shuddering. And the foam globe disappeared.

Dawn came at last, and the two brothers looked out of Gluck's little window in the morning. The Treasure Valley was one mass of ruin, and desolation. The inundation had swept away trees, crops, and cattle, and left in their stead, a waste of red sand, and grey mud. The two brothers crept shivering and horror-struck into the kitchen. The water had gutted the whole first floor; corn, money, almost every moveable thing had been swept away, and there was left only a small white card on the kitchen table. On it, in large, breezy, long-legged letters, were engraved the words:—

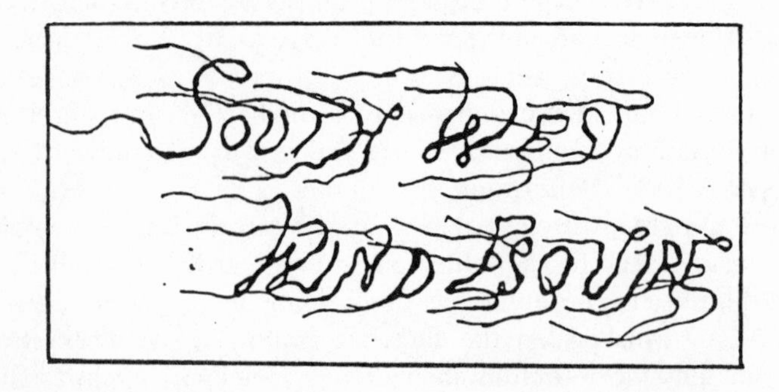

CHAPTER II

Of the Proceedings of the Three Brothers after the Visit of South West Wind, Esquire; and how little Gluck had an Interview with the King of the Golden River

South West Wind, Esquire, was as good as his word. After the momentous visit above related, he entered the Treasure Valley no more; and, what was worse, he had so much influence with his relations, the West Winds in general, and used it so effectually, that they all adopted a similar line of conduct. So no rain fell in the valley from one year's end to another. Though everything remained green and flourishing in the plains below, the inheritance of the Three Brothers was a desert. What had once been the richest soil in the kingdom, became a shifting heap of red sand; and the brothers, unable longer to contend with the adverse skies, abandoned their valueless patrimony in despair, to seek some means of gaining a livelihood among the cities and people of the plains. All their money was gone, and they had nothing left but some curious old-fashioned pieces of gold plate, the last remnants of their ill-gotten wealth.

'Suppose we turn goldsmiths?' said Schwartz to Hans, as they entered the large city. 'It is a good knave's trade; we can put a great deal of copper into the gold, without any one's finding it out.'

The thought was agreed to be a very good one; they hired a furnace, and turned goldsmiths. But two slight circumstances affected their trade: the first, that people did not approve of the coppered gold; the second, that the two elder brothers, whenever they had sold anything, used to leave little Gluck to mind the furnace, and go and drink out the money in the ale-house next door. So they melted all their gold, without making money enough to buy more, and were at last reduced to one large drinking-mug, which an uncle of his had given to little Gluck, and which he was very fond of, and would not have parted with for the world; though he never drank anything out of it but milk and water. The mug was a very odd mug to look at. The handle was formed of two wreaths of flowing golden hair, so finely spun that it looked more like silk than metal, and these wreaths descended into, and mixed with a beard and whiskers, of the same exquisite workmanship, which surrounded and decorated a very fierce little face, of the reddest gold imaginable, right in the front of the mug, with a pair of eyes in it which seemed to command its whole circumference. It was impossible to drink out of the mug without being subjected to an intense gaze out of the side of these eyes; and Schwartz positively averred, that once, after emptying it, full of Rhenish seventeen times, he had seen them wink! When it came to the mug's turn to be made into spoons, it half broke poor little Gluck's heart; but the brothers only laughed at him, tossed the mug into the melting-pot, and staggered out to the ale-house; leaving him, as usual, to pour the gold into bars, when it was all ready.

When they were gone, Gluck took a farewell look at his old friend in the melting-pot. The flowing hair was all gone; nothing remained but the red nose, and the sparkling eyes, which looked more malicious than ever. 'And no wonder,' thought Gluck, 'after being treated in that way.' He sauntered disconsolately to the window, and sat himself down to catch the fresh evening air, and escape the hot breath of the furnace. Now this window commanded a direct view of the range of mountains, which, as I told you before, overhung the Treasure Valley, and more especially of the peak from which fell the Golden River. It was just at the close of the day, and, when Gluck sat down at the window, he saw the rocks of the mountain tops, all crimson, and purple with the sunset; and there were bright tongues of fiery cloud burning and quivering about them; and the river, brighter than all, fell, in a waving column of pure gold, from precipice to precipice, with the double arch of a broad purple rainbow stretched across it, flushing and fading alternately in the wreaths of spray.

'Ah!' said Gluck aloud, after he had looked at it for a little while, 'if that river were really all gold, what a nice thing it would be.'

'No it wouldn't, Gluck,' said a clear metallic voice, close at his ear.

'Bless me, what's that?' exclaimed Gluck, jumping up. There was nobody there. He looked round the room, and under the table, and a great many times behind him, but there was certainly nobody there, and he sat down again at the window. This time he didn't speak, but he couldn't help thinking again that it would be very convenient if the river was really all gold.

'Not at all, my boy,' said the same voice, louder than before.

'Bless me!' said Gluck again, 'what *is* that?' He looked again into all the corners, and cupboards, and then began turning round, and round, as fast as he could, in the middle of the room, thinking there was somebody behind him, when the same voice struck again on his ear. It was singing now very merrily 'Lala-lira-la'; no words, only a soft running effervescent melody, something like that of a kettle on the boil. Gluck looked out of the window. No, it was certainly in the house. Up stairs, and down stairs. No, it was certainly in that very room, coming in quicker time, and clearer notes, every moment. 'Lala-lira-la.' All at once it struck Gluck, that it sounded louder near the furnace. He ran to the opening, and looked in: yes, he saw right, it seemed to be coming, not only out of the furnace, but out of the pot. He uncovered it, and ran back in a great fright, for the pot was certainly singing! He stood in the farthest corner of the room, with his hands up, and his mouth open, for a minute or two, when the singing stopped, and the voice became clear, and pronunciative.

'Hollo!' said the voice.

Gluck made no answer.

'Hollo! Gluck, my boy,' said the pot again.

Gluck summoned all his energies, walked straight up to the crucible, drew it out of the furnace, and looked in. The gold was all melted, and its surface as smooth and polished as a river; but instead of reflecting little Gluck's head, as he looked in, he saw meeting his glance, from beneath the gold, the red nose, and sharp eyes of his old friend of the mug, a thousand times redder, and sharper than ever he had seen them in his life.

'Come Gluck, my boy,' said the voice out of the pot again, 'I'm all right; pour me out.'

But Gluck was too much astonished to do anything of the kind.

'Pour me out, I say,' said the voice rather gruffly.

Still Gluck couldn't move.

'*Will* you pour me out?' said the voice passionately, 'I'm too hot.'

By a violent effort, Gluck recovered the use of his limbs, took hold of the crucible, and sloped it, so as to pour out the gold. But instead of a liquid stream, there came out, first, a pair of pretty little yellow legs, then some coat tails, then a pair of arms stuck a kimbo, and, finally, the well known head of his friend the mug; all which articles, uniting as they rolled out, stood up energetically on the floor, in the shape of a little golden dwarf, about a foot and a half high.

'That's right!' said the dwarf, stretching out first his legs, and then his arms, and then shaking his head up and down, and as far round as it would go, for five minutes, without stopping; apparently with the view of ascertaining if he were quite correctly put together, while Gluck stood contemplating him in speechless amazement. He was dressed in a slashed doublet of spun gold, so fine in its texture, that the prismatic colours gleamed over it, as if on a surface of mother of pearl; and, over this brilliant doublet, his hair and beard fell full half way to the ground, in waving curls, so exquisitely delicate, that Gluck could hardly tell where they ended; they seemed to melt into air. The features of the face, however, were by no means finished with the same delicacy; they were rather coarse, slightly inclining to coppery in complexion, and indicative, in expression, of a very pertinacious and intractable disposition in their small proprietor. When the dwarf had finished his self examination, he turned his small sharp eyes full on Gluck, and stared at him deliberately for a minute or two. 'No it wouldn't, Gluck, my boy,' said the little man.

This was certainly rather an abrupt, and unconnected mode of commencing conversation. It might indeed be supposed to refer to the course of Gluck's thoughts, which had first produced the dwarf's observations out of the pot; but whatever it referred to, Gluck had no inclination to dispute the dictum.

'Wouldn't it, sir?' said Gluck, very mildly, and submissively indeed.

'No,' said the dwarf, conclusively. 'No it wouldn't.' And with that, the dwarf pulled his cap hard over his brows, and took two turns, of three feet long, up and down the room, lifting his legs up very high, and setting them down very hard. This pause gave time for Gluck to collect his thoughts a little, and, seeing no great reason to view his diminutive visitor with dread, and feeling his curiosity overcome his amazement, he ventured on a question of peculiar delicacy.

'Pray, sir,' said Gluck, rather hesitatingly, 'were you my mug?'

On which the little man turned sharp round, walked straight up to Gluck, and drew himself up to his full height. 'I,' said the little man, 'am the King of the Golden River.' Whereupon he turned about again, and took two more turns, some six feet long, in order to allow time for the consternation which this announcement produced in his auditor to evaporate. After which, he again walked up to Gluck and stood still, as if expecting some comment on his communication.

Gluck determined to say something at all events. 'I hope your majesty is very well,' said Gluck.

'Listen!' said the little man, deigning no reply to this polite inquiry. 'I am the King of what you mortals call the Golden River. The shape you saw me in, was owing to the malice of a stronger king, from whose enchantments you have this instant freed me. What I have seen of you, and your conduct to your wicked brothers, renders me willing to serve you; therefore attend to what I tell you. Whoever shall climb to the top of that mountain from which you see the Golden River issue, and shall cast into the stream at its source, three drops of holy water, for him, and for him only, the river shall turn to gold. But no one failing in his first, can succeed in a second attempt; and if any one shall cast unholy water into the river, it will overwhelm him, and he will become a black stone.' So saying, the King of the Golden River turned away, and deliberately walked into the centre of the hottest flame of the furnace. His figure became red, white, transparent, dazzling—a blaze of intense light—rose, trembled, and disappeared. The King of the Golden River had evaporated.

'Oh!' cried poor Gluck, running to look up the chimney after him; 'Oh, dear, dear, dear me! My mug! my mug! my mug!'

CHAPTER III

How Mr Hans set off on an Expedition to the Golden River, and how he prospered therein

The King of the Golden River had hardly made the extraordinary exit related in the last chapter, before Hans and Schwartz came roaring into the house, very savagely drunk. The discovery of the total loss of their last piece of plate had the effect of sobering them just enough to enable them to stand over Gluck, beating him very steadily for a quarter of an hour; at the expiration of which period they dropped into a couple of chairs, and requested to know what he had got to say for himself. Gluck told them his story, of which of course they did not believe a word. They beat him again, till their arms were tired, and staggered to bed. In the morning, however, the steadiness with which he adhered to his story obtained him some degree of credence; the immediate consequence of which was, that the two brothers, after wrangling a long time on the knotty question, which of them should try his fortune first, drew their swords, and began fighting. The noise of the fray alarmed the neighbours, who, finding they could not pacify the combatants, sent for the constable.

Hans, on hearing this, contrived to escape, and hid himself; but Schwartz was taken before the magistrate, fined for breaking the peace, and, having drunk out his last penny the evening before, was thrown into prison till he should pay.

When Hans heard this, he was much delighted, and determined to set out immediately for the Golden River. How to get the holy water, was the question. He went to the priest, but the priest could not give any holy water to so abandoned a character. So Hans went to vespers in the evening for the first time in his life, and, under pretence of crossing himself, stole a cupful, and returned home in triumph.

Next morning he got up before the sun rose, put the holy water into a strong flask, and two bottles of wine and some meat in a basket, slung them over his back, took his alpine staff in his hand, and set off for the mountains.

On his way out of the town he had to pass the prison, and as he looked in at the windows, whom should he see but Schwartz himself peeping out of the bars, and looking very disconsolate.

'Good morning, brother,' said Hans; 'have you any message for the King of the Golden River?'

Schwartz gnashed his teeth with rage, and shook the bars with all his strength; but Hans only laughed at him, and, advising him to make himself comfortable till he came back again, shouldered his basket, shook the bottle of holy water in Schwartz's face till it frothed again, and marched off in the highest spirits in the world.

It was, indeed, a morning that might have made any one happy, even with no Golden River to seek for. Level lines of dewy mist lay stretched along the valley, out of which rose the massy mountains—their lower cliffs in pale grey shadow, hardly distinguishable from the floating vapour, but gradually ascending till they caught the sunlight, which ran in sharp touches of ruddy colour, along the angular crags, and pierced, in long level rays, through their fringes of spear-like pine. Far above, shot up red splintered masses of castellated rock, jagged and shivered into myriads of fantastic forms, with here and there a streak of sunlit snow, traced down their chasms like a line of forked lightning; and, far beyond, and far above all these, fainter than the morning cloud, but purer and changeless, slept, in the blue sky, the utmost peaks of the eternal snow.

The Golden River, which sprang from one of the lower and snowless elevations, was now nearly in shadow; all but the uppermost jets of spray, which rose like slow smoke above the undulating line of the cataract, and floated away in feeble wreaths upon the morning wind.

On this object, and on this alone, Hans' eyes and thoughts were fixed; forgetting the distance he had to traverse, he set off at an imprudent rate of walking, which greatly exhausted him before he had scaled the first range of the green and low hills. He was, moreover, surprised, on surmounting them, to find that a large glacier, of whose existence, notwithstanding his previous knowledge of the mountains, he had been absolutely ignorant, lay between him and the source of the Golden River. He entered on it with the boldness of a practised mountaineer; yet he thought he had never traversed so strange, or so dangerous a glacier in his life. The ice was excessively slippery, and out of all its chasms came wild sounds of gushing water; not monotonous or low, but changeful and loud, rising occasionally into drifting passages of wild melody, then breaking off into short melancholy tones, or sudden shrieks, resembling those of human voices in distress or pain. The ice was broken into thousands of confused shapes, but none, Hans thought, like the ordinary forms of splintered ice. There seemed a curious *expression* about all their outlines—a perpetual resemblance to living features, distorted and scornful. Myriads of deceitful shadows, and lurid lights, played and floated about and through the pale blue pinnacles, dazzling and confusing the sight of the traveller; while his ears grew dull and his head giddy with the constant gush and roar of the concealed waters. These painful circumstances increased upon him as he advanced; the ice crashed and yawned into fresh chasms at his feet, tottering spires nodded around him, and fell thundering across his path; and though he had repeatedly faced these dangers on the most terrific glaciers, and in the wildest weather, it was with a new and oppressive feeling of panic terror that he leaped the last chasm, and flung himself, exhausted and shuddering, on the firm turf of the mountain.

He had been compelled to abandon his basket of food, which

became a perilous incumbrance on the glacier, and had now no means of refreshing himself but by breaking off and eating some of the pieces of ice. This, however, relieved his thirst; an hour's repose recruited his hardy frame, and, with the indomitable spirit of avarice, he resumed his laborious journey.

His way now lay straight up a ridge of bare red rocks, without a blade of grass to ease the foot, or a projecting angle to afford an inch of shade from the south sun. It was past noon, and the rays beat intensely upon the steep path, while the whole atmosphere was motionless, and penetrated with heat. Intense thirst was soon added to the bodily fatigue with which Hans was now afflicted; glance after glance he cast on the flask of water which hung at his belt. 'Three drops are enough,' at last thought he; 'I may, at least, cool my lips with it.'

He opened the flask, and was raising it to his lips, when his eye fell on an object lying on the rock beside him; he thought it moved. It was a small dog, apparently in the last agony of death from thirst. Its tongue was out, its jaws dry, its limbs extended lifelessly, and a swarm of black ants were crawling about its lips and throat. Its eye moved to the bottle which Hans held in his hand. He raised it, drank, spurned the animal with his foot, and passed on. And he did not know how it was, but he thought that a strange shadow had suddenly come across the blue sky.

The path became steeper and more rugged every moment; and the high hill air, instead of refreshing him, seemed to throw his blood into a fever. The noise of the hill cataracts sounded like mockery in his ears: they were all distant, and his thirst increased every moment. Another hour passed, and he again looked down to the flask at his side; it was half empty, but there was much more than three drops in it. He stopped to open it, and again, as he did so, something moved in the path above him. It was a fair child, stretched nearly lifeless on the rock, its breast heaving with thirst, its eyes closed, and its lips parched and burning. Hans eyed it deliberately, drank, and passed on. And a dark grey cloud came over the sun, and long, snake-like shadows crept up along the mountain sides. Hans struggled on. The sun was sinking, but its descent seemed to bring no coolness; the leaden weight of the dead air pressed upon his brow and heart, but the goal was near. He saw the cataract of the Golden River springing from the hill-side, scarcely five hundred feet above him. He paused for a moment to breathe, and sprang on to complete his task.

At this instant a faint cry fell on his ear. He turned, and saw a grey haired old man extended on the rocks. His eyes were sunk, his features deadly pale, and gathered into an expression of despair. 'Water!' he stretched his arms to Hans, and cried feebly, 'Water! I am dying.'

'I have none,' replied Hans; 'thou hast had thy share of life.' He strode over the prostrate body, and darted on. And a flash of blue lightning rose out of the East, shaped like a sword; it shook thrice over the whole heaven, and left it dark with one heavy, impenetrable shade. The sun was setting; it plunged towards the horizon like a red-hot ball.

The roar of the Golden River rose on Hans' ear. He stood at the brink of the chasm through which it ran. Its waves were filled with the red glory of the sunset: they shook their crests like tongues of fire, and flashes of bloody light gleamed along their foam. Their sound came mightier and mightier on his senses; his brain grew giddy with the prolonged thunder. Shuddering, he drew the flask from his girdle, and hurled it into the centre of the torrent. As he did so, an icy chill shot through his limbs; he staggered, shrieked, and fell. The waters closed over his cry. And the moaning of the river rose wildly into the night, as it gushed over

THE BLACK STONE.

CHAPTER IV

How Mr Schwartz set off on an Expedition to the Golden River, and how he prospered therein

Poor little Gluck waited very anxiously alone in the house, for Hans' return. Finding he did not come back, he was terribly frightened, and went and told Schwartz in the prison, all that had happened. Then Schwartz was very much pleased, and said that Hans must certainly have been turned into a black stone, and he should have all the gold to himself. But Gluck was very sorry, and cried all night. When he got up in the morning, there was no bread in the house, nor any money; so Gluck went, and hired himself to another goldsmith, and he worked so hard, and so neatly, and so long every day, that he soon got money enough together, to pay his brother's fine, and he went, and gave it all to

Schwartz, and Schwartz got out of prison. Then Schwartz was quite pleased, and said he should have some of the gold of the river. But Gluck only begged he would go and see what had become of Hans.

Now when Schwartz had heard that Hans had stolen the holy water, he thought to himself that such a proceeding might not be considered altogether correct by the King of the Golden River, and determined to manage matters better. So he took some more of Gluck's money, and went to a bad priest, who gave him some holy water very readily for it. Then Schwartz was sure it was all quite right. So Schwartz got up early in the morning before the sun rose, and took some bread and wine, in a basket, and put his holy water in a flask, and set off for the mountains. Like his brother he was much surprised at the sight of the glacier, and had great difficulty in crossing it, even after leaving his basket behind him. The day was cloudless, but not bright: there was a heavy purple haze hanging over the sky, and the hills looked lowering and gloomy. And as Schwartz climbed the steep rock path, the thirst came upon him, as it had upon his brother, until he lifted his flask to his lips to drink. Then he saw the fair child lying near him on the rocks, and it cried to him, and moaned for water.

'Water indeed,' said Schwartz; 'I haven't half enough for myself,' and passed on. And as he went he thought the sunbeams grew more dim, and he saw a low bank of black cloud rising out of the West; and, when he had climbed for another hour, the thirst overcame him again, and he would have drunk. Then he saw the old man lying before him on the path, and heard him cry out for water. 'Water, indeed,' said Schwartz, 'I haven't half enough for myself,' and on he went.

Then again the light seemed to fade from before his eyes, and he looked up, and, behold, a mist, of the colour of blood, had come over the sun; and the bank of black cloud had risen very high, and its edges were tossing and tumbling like the waves of the angry sea. And they cast long shadows, which flickered over Schwartz's path.

Then Schwartz climbed for another hour, and again his thirst returned; and as he lifted his flask to his lips, he thought he saw his brother Hans lying exhausted on the path before him, and, as he gazed, the figure stretched its arms to him, and cried for water. 'Ha, ha,' laughed Schwartz, 'are you there? remember the prison bars, my boy. Water, indeed! do you suppose I carried it all the way up here for *you*?' And he strode over the figure; yet, as he passed, he thought he saw a strange expression of mockery about its lips. And, when he had gone a few yards farther, he looked back; but the figure was not there.

And a sudden horror came over Schwartz, he knew not why; but the thirst for gold prevailed over his fear, and he rushed on. And the bank of black cloud rose to the zenith, and out of it came bursts of spiry lightning, and waves of darkness seemed to heave and float, between their flashes, over the whole heavens. And the sky where the sun was setting was all level, and like a lake of blood; and a strong wind came out of that sky, tearing its crimson clouds into fragments, and scattering them far into the darkness. And when Schwartz stood by the brink of the Golden River, its waves were black, like thunder clouds, but their foam was like fire; and the roar of the waters below, and the thunder above met, as he cast the flask into the stream. And, as he did so, the lightning glared in his eyes, and the earth gave way beneath him, and the waters closed over his cry. And the moaning of the river rose wildly into the night, as it gushed over the

TWO BLACK STONES.

CHAPTER V

How little Gluck set off on an Expedition to the Golden River, and how he prospered therein; with other matters of interest

When Gluck found that Schwartz did not come back, he was very sorry, and did not know what to do. He had no money, and was obliged to go and hire himself again to the goldsmith, who worked him very hard, and gave him very little money. So, after a month, or two, Gluck grew tired, and made up his mind to go and try his fortune with the Golden River. 'The little king looked very kind,' thought he. 'I don't think he will turn me into a black stone.' So he went to the priest, and the priest gave him some holy water as soon as he asked for it. Then Gluck took some bread in his basket, and the bottle of water, and set off very early for the mountains.

If the glacier had occasioned a great deal of fatigue to his brothers, it was twenty times worse for him, who was neither so strong nor so practised on the mountains. He had several very bad falls, lost his basket, and bread, and was very much frightened at the strange noises under the ice. He lay a long time to rest on the grass, after he had got over, and began to climb the hill just in the hottest part of the day. When he had climbed for an hour, he got dreadfully thirsty, and was going to drink like his brothers, when he saw an old man coming down the path above him, looking very feeble, and leaning on a staff. 'My son,' said the old man, 'I am faint with thirst, give me some of that water.' Then Gluck looked at him, and when he saw that he was pale and weary, he gave him the water: 'Only pray don't drink it all,' said Gluck. But the old man drank a great deal, and gave him back the bottle two-thirds empty. Then he bade him good speed, and Gluck went on again merrily. And the path became easier to his feet, and two or three blades of grass appeared upon it, and some grasshoppers began singing on the bank beside it; and Gluck thought he had never heard such merry singing.

Then he went on for another hour, and the thirst increased on him so that he thought he should be forced to drink. But, as he raised the flask, he saw a little child lying panting by the road-side, and it cried out piteously for water. Then Gluck struggled with himself, and determined to bear the thirst a little longer; and he put the bottle to the child's lips, and it drank it all but a few drops. Then it smiled on him, and got up, and ran down the hill; and Gluck looked after it, till it became as small as a little star, and then turned, and began climbing again. And then there were all kinds of sweet flowers growing on the rocks, bright green moss, with pale pink starry flowers, and soft belled gentians, more blue than the sky at its deepest, and pure white transparent lilies. And crimson and purple butterflies darted hither and thither, and the sky sent down such pure light, that Gluck had never felt so happy in his life.

Yet, when he had climbed for another hour, his thirst became intolerable again; and, when he looked at his bottle, he saw that there were only five or six drops left in it, and he could not venture to drink. And, as he was hanging the flask to his belt again, he saw a little dog lying on the rocks, gasping for breath—just as Hans had seen it on the day of his ascent. And Gluck stopped and looked at it, and then at the Golden River, not five hundred yards above him; and he thought of the dwarf's words, 'that no one could succeed, except in their first attempt'; and he tried to pass the dog, but it whined piteously, and Gluck stopped again, 'Poor beastie,' said Gluck, 'it'll be dead when I come down again, if I don't help it.' Then he looked closer and closer at it, and its eye turned on him so mournfully, that he could not stand it. 'Confound the King and his gold too,' said Gluck; and he opened the flask, and poured all the water into the dog's mouth.

The dog sprang up and stood on its hind legs. Its tail disappeared, its ears became long, longer, silky, golden; its nose became very red, its eyes became very twinkling; in three seconds the dog was gone, and before Gluck stood his old acquaintance, the King of the Golden River.

'Thank you,' said the monarch, 'but don't be frightened, it's all right'; for Gluck showed manifest symptoms of consternation at this unlooked-for reply to his last observation. 'Why didn't you come before,' continued the dwarf, 'instead of sending me those rascally brothers of yours, for me to have the trouble of turning into stones? Very hard stones they make too.'

'Oh dear me!' said Gluck, 'have you really been so cruel?'

'Cruel!' said the dwarf, 'they poured unholy water into my stream: do you suppose I'm going to allow that?'

'Why,' said Gluck, 'I am sure, sir—your majesty, I mean—they got the water out of the church font.'

'Very probably,' replied the dwarf; 'but,' and his countenance grew stern as he spoke, 'the water which has been refused to the cry of the weary and dying, is unholy, though it had been blessed by every saint in heaven; and the water which is found in the vessel of mercy is holy, though it had been defiled with corpses.'

So saying, the dwarf stooped and plucked a lily that grew at his feet. On its white leaves there hung three drops of clear dew. And the dwarf shook them into the flask which Gluck held in his hand. 'Cast these into the river,' he said, 'and descend on the other side of the mountains into the Treasure Valley. And so good speed.'

As he spoke, the figure of the dwarf became indistinct. The playing colours of his robe formed themselves into a prismatic mist of dewy light: he stood for an instant veiled with them as with the belt of a broad rainbow. The colours grew faint, the mist rose into the air; the monarch had evaporated.

And Gluck climbed to the brink of the Golden River, and its waves were as clear as crystal, and as brilliant as the sun. And, when he cast the three drops of dew into the stream, there opened where they fell, a small circular whirlpool, into which the waters descended with a musical noise. Gluck stook watching it for some time, very much disappointed, because not only the river was not turned into gold, but its waters seemed much diminished in quantity. Yet he obeyed his friend the dwarf, and descended the other side of the mountains, towards the Treasure Valley; and, as he went, he thought he heard the noise of water working its way under the ground. And, when he came in sight of the Treasure Valley, behold, a river, like the Golden River, was springing from a new cleft of the rocks above it, and was flowing in innumerable streams among the dry heaps of red sand. And as Gluck gazed, fresh grass sprang beside the new streams, and creeping plants grew, and climbed among the moistening soil. Young flowers opened suddenly along the river sides, as stars leap out when twilight is deepening, and thickets of myrtle, and tendrils of vine, cast lengthening shadows over the valley as they grew. And thus the Treasure Valley became a garden again, and the inheritance, which had been lost by cruelty, was regained by love. And Gluck went, and dwelt in the valley, and the poor were never driven from his door; so that his barns became full of corn, and his house of treasure. And, for him, the river had, according to the dwarf's promise, become a River of Gold. And, to this day, the inhabitants of the valley point out the place, where the three drops of holy dew were cast into the stream, and trace the course of the Golden River under the ground, until it emerges in the Treasure Valley. And, at the top of the cataract of the Golden River, are still to be seen two BLACK STONES, round which the waters howl mournfully every day at sunset; and these stones are still called by the people of the valley,

THE BLACK BROTHERS.

1850

FRANCES BROWNE

The Story of Fairyfoot

ONCE upon a time there stood far away in the west country a town called Stumpinghame. It contained seven windmills, a royal palace, a market place, and a prison, with every other convenience befitting the capital of a kingdom. A capital city was Stumpinghame, and its inhabitants thought it the only one in the world. It stood in the midst of a great plain, which for three leagues round its walls was covered with corn, flax, and orchards. Beyond that lay a great circle of pasture land, seven leagues in breadth, and it was bounded on all sides by a forest so thick and old that no man in Stumpinghame knew its extent; and the opinion of the learned was, that it reached to the end of the world.

There were strong reasons for this opinion. First, that forest was known to be inhabited time out of mind by the fairies, and no hunter cared to go beyond its borders—so all the west country believed it to be solidly full of old trees to the heart. Secondly, the people of Stumpinghame were no travellers—man, woman, and child had feet so large and heavy that it was by no means convenient to carry them far. Whether it was the nature of the place or the people, I cannot tell, but great feet had been the fashion there time immemorial, and the higher the family the larger were they. It was, therefore, the aim of everybody above the degree of shepherds, and such-like rustics, to swell out and enlarge their feet by way of gentility; and so successful were they in these undertakings that, on a pinch, respectable people's slippers would have served for panniers.

Stumpinghame had a king of its own, and his name was Stiffstep; his family was very ancient and large-footed. His subjects called him Lord of the World, and he made a speech to them every year concerning the grandeur of his mighty empire. His queen,

Hammerheel, was the greatest beauty in Stumpinghame. Her majesty's shoe was not much less than a fishing-boat; their six children promised to be quite as handsome, and all went well with them till the birth of their seventh son.

For a long time nobody about the palace could understand what was the matter—the ladies-in-waiting looked so astonished, and the king so vexed; but at last it was whispered through the city that the queen's seventh child had been born with such miserably small feet that they resembled nothing ever seen or heard of in Stumpinghame, except the feet of the fairies.

The chronicles furnished no example of such an affliction ever before happening in the royal family. The common people thought it portended some great calamity to the city; the learned men began to write books about it; and all the relations of the king and queen assembled at the palace to mourn with them over their singular misfortune. The whole court and most of the citizens helped in this mourning, but when it had lasted seven days they all found out it was of no use. So the relations went to their homes, and the people took to their work. If the learned men's books were written, nobody ever read them; and to cheer up the queen's spirits, the young prince was sent privately out to the pasture lands, to be nursed among the shepherds.

The chief man there was called Fleecefold, and his wife's name was Rough Ruddy. They lived in a snug cottage with their son Blackthorn and their daughter Brownberry, and were thought great people, because they kept the king's sheep. Moreover, Fleecefold's family were known to be ancient; and Rough Ruddy boasted that she had the largest feet in all the pastures. The shepherds held them in high respect, and it grew still higher when the news spread that the king's seventh son had been sent to their cottage. People came from all quarters to see the young prince, and great were the lamentations over his misfortune in having such small feet.

The king and queen had given him fourteen names, beginning with Augustus—such being the fashion in that royal family; but the honest country people could not remember so many; besides, his feet were the most remarkable thing about the child, so with one accord they called him Fairyfoot. At first it was feared this might be high-treason, but when no notice was taken by the king or his ministers, the shepherds concluded it was no harm, and the boy never had another name throughout the pastures. At court it was not thought polite to speak of him at all. They did not keep his birthday, and he was never sent for at Christmas, because the queen and her ladies could not bear the sight. Once a year the undermost scullion was sent to see how he did, with a bundle of his next brother's cast-off clothes; and, as the king grew old and cross, it was said he had thoughts of disowning him.

So Fairyfoot grew in Fleecefold's cottage. Perhaps the country air made him fair and rosy—for all agreed that he would have been a handsome boy but for his small feet, with which nevertheless he learned to walk, and in time to run and to jump, thereby amazing everybody, for such doings were not known among the children of Stumpinghame. The news of court, however, travelled to the shepherds, and Fairyfoot was despised among them. The old people thought him unlucky; the children refused to play with him. Fleecefold was ashamed to have him in his cottage, but he durst not disobey the king's orders. Moreover, Blackthorn wore most of the clothes brought by the scullion. At last, Rough Ruddy found out that the sight of such horrid jumping would make her children vulgar; and, as soon as he was old enough, she sent Fairyfoot every day to watch some sickly sheep that grazed on a wild, weedy pasture, hard by the forest.

Poor Fairyfoot was often lonely and sorrowful; many a time he wished his feet would grow larger, or that people wouldn't notice them so much; and all the comfort he had was running and jumping by himself in the wild pasture, and thinking that none of the shepherds' children could do the like, for all their pride of their great feet.

Tired of this sport, he was lying in the shadow of a mossy rock one warm summer's noon, with the sheep feeding around, when a robin, pursued by a great hawk, flew into the old velvet cap which lay on the ground beside him. Fairyfoot covered it up, and the hawk, frightened by his shout, flew away.

'Now you may go, poor robin!' he said, opening the cap: but instead of the bird, out sprang a little man dressed in russet-brown, and looking as if he were an hundred years old. Fairyfoot could not speak for astonishment, but the little man said—

'Thank you for your shelter, and be sure I will do as much for you. Call on me if you are ever in trouble, my name is Robin Goodfellow'; and darting off, he was out of sight in an instant. For days the boy wondered who that little man could be, but he told nobody, for the little man's feet were as small as his own, and it was clear he would be no favourite in Stumpinghame. Fairyfoot kept the story to himself, and at last midsummer came. That evening was a feast among the shepherds. There were bonfires on the hills, and fun in the villages. But Fairyfoot sat alone beside his sheepfold, for the children of his village had refused to let him dance with them about the bonfire, and he had gone there to bewail the size of his feet, which came between him and so many good things. Fairyfoot and never felt so lonely in all his life, and remembering the little man, he plucked up spirit, and cried—

'Ho! Robin Goodfellow!'

'Here I am,' said a shrill voice at his elbow; and there stood the little man himself.

'I am very lonely, and no one will play with me, because my feet are not large enough,' said Fairyfoot.

'Come then and play with us,' said the little man. 'We lead the merriest lives in the world, and care for nobody's feet; but all companies have their own manners, and there are two things you must mind among us: first, do as you see the rest doing; and secondly, never speak of anything you may hear or see, for we and the people of this country have had no friendship ever since large feet came in fashion.'

'I will do that, and anything more you like,' said Fairyfoot; and the little man taking his hand, led him over the pasture into the forest, and along a mossy path among old trees wreathed with ivy (he never knew how far), till they heard the sound of music, and came upon a meadow where the moon shone as bright as day, and all the flowers of the year—snowdrops, violets, primroses, and cowslips—bloomed together in the thick grass. There were a crowd of little men and women, some clad in russet colour, but far more in green, dancing round a little well as clear as crystal. And under great rose-trees which grew here and there in the meadow, companies were sitting round low tables covered with cups of milk, dishes of honey, and carved wooden flagons filled with clear red wine. The little man led Fairyfoot up to the nearest table, handed him one of the flagons, and said—

'Drink to the good company!'

Wine was not very common among the shepherds of Stumpinghame, and the boy had never tasted such drink as that before; for scarcely had it gone down, when he forgot all his troubles—how Blackthorn and Brownberry wore his clothes, how Rough Ruddy sent him to keep the sickly sheep, and the children would not dance with him: in short, he forgot the whole misfortune of his feet, and it seemed to his mind that he was a king's son, and all was well with him. All the little people about the well cried—

'Welcome! welcome!' and every one said—'Come and dance with me!' So Fairyfoot was as happy as a prince, and drank milk and ate honey till the moon was low in the sky, and then the little man took him by the hand, and never stopped nor stayed till he was at his own bed of straw in the cottage corner.

Next morning Fairyfoot was not tired for all his dancing. Nobody in the cottage had missed him, and he went out with the sheep as usual; but every night all that summer, when the shepherds were safe in bed, the little man came and took him away to dance in the forest. Now he did not care to play with the shepherds' children, nor grieve that his father and mother had forgotten him, but watched the sheep all day singing to himself or plaiting rushes; and when the sun went down, Fairyfoot's heart rejoiced at the thought of meeting that merry company.

The wonder was that he was never tired nor sleepy, as people are apt to be who dance all night; but before the summer was ended Fairyfoot found out the reason. One night, when the moon was full, and the last of the ripe corn rustling in the fields, Robin Goodfellow came for him as usual, and away they went to the flowery green. The fun there was high, and Robin was in haste. So he only pointed to the carved cup from which Fairyfoot every night drank the clear red wine.

'I am not thirsty, and there is no use losing time,' thought the boy to himself, and he joined the dance; but never in all his life did Fairyfoot find such hard work as to keep pace with the company. Their feet seemed to move like lightning; the swallows did not fly so fast or turn so quickly. Fairyfoot did his best, for he never gave in easily, but at length, his breath and strength being spent, the boy was glad to steal away, and sit down behind a mossy oak, where his eyes closed for very weariness. When he awoke the dance was nearly over, but two little ladies clad in green talked close beside him.

'What a beautiful boy!' said one of them. 'He is worthy to be a king's son. Only see what handsome feet he has!'

'Yes,' said the other, with a laugh that sounded spiteful; 'they are just like the feet Princess Maybloom had before she washed them in the Growing Well. Her father has sent far and wide throughout the whole country searching for a doctor to make them small again, but nothing in this world can do it except the water of the Fair Fountain, and none but I and the nightingales know where it is.'

'One would not care to let the like be known,' said the first little lady: 'there would come such crowds of these great coarse creatures of mankind, nobody would have peace for leagues round. But you will surely send word to the sweet princess!—she was so kind to our birds and butterflies, and danced so like one of ourselves!'

'Not I, indeed!' said the spiteful fairy. 'Her old skinflint of a father cut down the cedar which I loved best in the whole forest, and made a chest of it to hold his money in; besides, I never liked the princess—everybody praised her so. But come, we shall be too late for the last dance.'

When they were gone, Fairyfoot could sleep no more with astonishment. He did not wonder at the fairies admiring his feet, because their own were much the same; but it amazed him that Princess Maybloom's father should be troubled at hers growing large. Moreover, he wished to see that same princess and her country, since there were really other places in the world than Stumpinghame.

When Robin Goodfellow came to take him home as usual he durst not let him know that he had overheard anything; but never was the boy so unwilling to get up as on that morning, and all day he was so weary that in the afternoon Fairyfoot fell asleep, with his head on a clump of rushes. It was seldom that any one thought of looking after him and the sickly sheep; but it so happened that towards evening the old shepherd, Fleecefold, thought he would see how things went on in the pastures. The shepherd had a bad temper and a thick staff, and no sooner did he catch sight of Fairyfoot sleeping, and his flock straying away, than shouting all the ill names he could remember, in a voice which woke up the boy, he ran after him as fast as his great feet would allow; while Fairyfoot, seeing no other shelter from his fury, fled into the forest, and never stopped nor stayed till he reached the banks of a little stream.

Thinking it might lead him to the fairies' dancing-ground, he followed that stream for many an hour, but it wound away into the heart of the forest, flowing through dells, falling over mossy rocks and at last leading Fairyfoot, when he was tired and the night had fallen, to a grove of great rose-trees, with the moon shining on it as bright as day, and thousands of nightingales singing in the branches. In the midst of that grove was a clear spring, bordered with banks of lilies, and Fairyfoot sat down by it to rest himself and listen. The singing was so sweet he could have listened for ever, but as he sat the nightingales left off their songs, and began to talk together in the silence of the night—

'What boy is that,' said one on a branch above him, 'who sits so lonely by the Fair Fountain? He cannot have come from Stumpinghame with such small and handsome feet.'

'No, I'll warrant you,' said another, 'he has come from the west country. How in the world did he find the way?'

'How simple you are!' said a third nightingale. 'What had he to do but follow the ground-ivy which grows over height and hollow, bank and bush, from the lowest gate of the king's kitchen garden to the root of this rose-tree? He looks a wise boy, and I hope he will keep the secret, or we shall have all the west country here, dabbling in our fountain, and leaving us no rest to either talk or sing.'

Fairyfoot sat in great astonishment at this discourse, but by and by, when the talk ceased and the songs began, he thought it might be as well for him to follow the ground-ivy, and see the Princess Maybloom, not to speak of getting rid of Rough Ruddy, the sickly sheep, and the crusty old shepherd. It was a long journey; but he went on, eating wild berries by day, sleeping in the hollows of old trees by night, and never losing sight of the ground-ivy, which led him over height and hollow, bank and bush, out of the forest, and along a noble high road, with fields and villages on every side, to a great city, and a low old-fashioned gate of the king's kitchen-garden, which was thought too mean for the scullions, and had not been opened for seven years.

There was no use knocking—the gate was overgrown with tall weeds and moss; so, being an active boy, he climbed over, and walked through the garden, till a white fawn came frisking by, and he heard a soft voice saying sorrowfully—

'Come back, come back, my fawn! I cannot run and play with you now, my feet have grown so heavy'; and looking round he saw the loveliest young princess in the world, dressed in snow-white, and wearing a wreath of roses on her golden hair; but walking slowly, as the great people did in Stumpinghame, for her feet were as large as the best of them.

After her came six young ladies, dressed in white and walking slowly, for they could not go before the princess; but Fairyfoot was amazed to see that their feet were as small as his own. At once he guessed that this must be the Princess Maybloom, and made her an humble bow, saying—

'Royal princess, I have heard of your trouble because your feet have grown large: in my country that's all the fashion. For seven years past I have been wondering what would make mine grow, to no purpose; but I know of a certain fountain that will make yours smaller and finer than ever they were, if the king, your father, gives you leave to come with me, accompanied by two of your maids that are the least given to talking, and the most prudent officer in all his household; for it would grievously offend the fairies and the nightingales to make that fountain known.'

When the princess heard that, she danced for joy in spite of her large feet, and she and her six maids brought Fairyfoot before the king and queen, where they sat in their palace hall, with all the courtiers paying their morning compliments. The lords were very much astonished to see a ragged, bare-footed boy brought in among them, and the ladies thought Princess Maybloom must have gone mad; but Fairyfoot, making an humble reverence, told his message to the king and queen, and offered to set out with the princess that very day. At first the king would not believe that there could be any use in his offer, because so many great physicians had failed to give any relief. The courtiers laughed Fairyfoot to scorn, the pages wanted to turn him out for an impudent impostor, and the prime-minister said he ought to be put to death for high-treason.

Fairyfoot wished himself safe in the forest again, or even keeping the sickly sheep; but the queen, being a prudent woman, said—

'I pray your majesty to notice what fine feet this boy has. There may be some truth in his story. For the sake of our only daughter, I will choose two maids who talk the least of all our train, and my chamberlain, who is the most discreet officer in our household. Let them go with the princess: who knows but our sorrow may be lessened?'

After some persuasion the king consented, though all his councillors advised the contrary. So the two silent maids, the discreet chamberlain, and her fawn, which would not stay behind, were sent with Princess Maybloom, and they all set out after dinner. Fairyfoot had hard work guiding them along the track of the ground-ivy. The maids and the chamberlain did not like the brambles and rough roots of the forest—they thought it hard to eat berries and sleep in hollow trees; but the princess went on with good courage, and at last they reached the grove of rose-trees, and the spring bordered with lilies.

The chamberlain washed—and though his hair had been grey, and his face wrinkled, the young courtiers envied his beauty for years after. The maids washed—and from that day they were esteemed the fairest in all the palace. Lastly, the princess washed also—it could make her no fairer, but the moment her feet touched the water they grew less, and when she had washed and dried them three times, they were as small and finely-shaped as Fairyfoot's own. There was great joy among them, but the boy said sorrowfully—

'Oh! if there had been a well in the world to make my feet large, my father and mother would not have cast me off, nor sent me to live among the shepherds.'

'Cheer up your heart,' said the Princess Maybloom; 'if you want large feet, there is a well in this forest that will do it. Last summer time, I came with my father and his foresters to see a great cedar cut down, of which he meant to make a money chest. While they were busy with the cedar, I saw a bramble branch covered with berries. Some were ripe and some were green, but it was the longest bramble that ever grew; for the sake of the berries, I went on and on to its root, which grew hard by a muddy-looking well, with banks of dark green moss, in the deepest part of the forest. The day was warm and dry, and my feet were sore with the rough ground, so I took off my scarlet shoes, and washed my feet in the well; but as I washed they grew larger every minute, and nothing could ever make them less again. I have seen the bramble this day; it is not far off, and as you have shown me the Fair Fountain, I will show you the Growing Well.'

Up rose Fairyfoot and Princess Maybloom, and went together till they found the bramble, and came to where its root grew, hard by the muddy-looking well, with banks of dark green moss in the deepest dell of the forest. Fairyfoot sat down to wash, but at that minute he heard a sound of music, and knew it was the fairies going to their dancing ground.

'If my feet grow large,' said the boy to himself, 'how shall I dance with them?' So, rising quickly, he took the Princess Maybloom by the hand. The fawn followed them; the maids and the chamberlain followed it, and all followed the music through the forest. At last they came to the flowery green. Robin Goodfellow welcomed the company for Fairyfoot's sake, and gave every one a drink of the fairies' wine. So they danced there from sunset till the grey morning, and nobody was tired; but before the lark sang, Robin Goodfellow took them all safe home, as he used to take Fairyfoot.

There was great joy that day in the palace because Princess Maybloom's feet were made small again. The king gave Fairyfoot all manner of fine clothes and rich jewels; and when they heard his wonderful story, he and the queen asked him to live with them and be their son. In process of time Fairyfoot and Princess Maybloom were married, and still live happily. When they go to visit at Stumpinghame, they always wash their feet in the Growing Well, lest the royal family might think them a disgrace, but when they come back, they make haste to the Fair Fountain; and the fairies and the nightingales are great friends to them, as well as the maids and the chamberlain, because they have told nobody about it, and there is peace and quiet yet in the grove of rose-trees.

1856

GEORGE MACDONALD

The Light Princess

1. *What! No Children?*

ONCE upon a time, so long ago that I have quite forgotten the date, there lived a king and queen who had no children.

And the king said to himself, 'All the queens of my acquaintance have children, some three, some seven, and some as many as twelve; and my queen has not one. I feel ill-used.' So he made up his mind to be cross with his wife about it. But she bore it all like a good patient queen as she was. Then the king grew very cross indeed. But the queen pretended to take it all as a joke, and a very good one too.

'Why don't you have any daughters, at least?' said he. 'I don't say *sons*; that might be too much to expect.'

'I am sure, dear king, I am very sorry,' said the queen.

'So you ought to be,' retorted the king; 'you are not going to make a virtue of *that*, surely.'

But he was not an ill-tempered king, and in any matter of less moment would have let the queen have her own way with all his heart. This, however, was an affair of state.

The queen smiled.

'You must have patience with a lady, you know, dear king,' said she.

She was, indeed, a very nice queen, and heartily sorry that she could not oblige the king immediately.

2. *Won't I, Just?*

The king tried to have patience, but he succeeded very badly. It was more than he deserved therefore, when, at last, the queen gave him a daughter—as lovely a little princess as ever cried.

The day drew near when the infant must be christened. The

king wrote all the invitations with his own hand. Of course somebody was forgotten.

Now it does not generally matter if somebody *is* forgotten, only you must mind who. Unfortunately, the king forgot without intending to forget; and so the chance fell upon the Princess Makemnoit, which was awkward. For the princess was the king's own sister; and he ought not to have forgotten her. But she had made herself so disagreeable to the old king, their father, that he had forgotten her in making his will; and so it was no wonder that her brother forgot her in writing his invitations. But poor relations don't do anything to keep you in mind of them. Why don't they? The king could not see into the garret she lived in, could he?

She was a sour, spiteful creature. The wrinkles of contempt crossed the wrinkles of peevishness, and made her face as full of wrinkles as a pat of butter. If ever a king could be justified in forgetting anybody, this king was justified in forgetting his sister, even at a christening. She looked very odd, too. Her forehead was as large as all the rest of her face, and projected over it like a precipice. When she was angry, her little eyes flashed blue. When she hated anybody, they shone yellow and green. What they looked like when she loved anybody, I do not know; for I never heard of her loving anybody but herself, and I do not think she could have managed that if she had not somehow got used to herself. But what made it highly imprudent in the king to forget her was—that she was awfully clever. In fact, she was a witch; and when she bewitched anybody, he very soon had enough of it; for she beat all the wicked fairies in wickedness, and all the clever ones in cleverness. She despised all the modes we read of in history, in which offended fairies and witches have taken their revenges; and therefore, after waiting and waiting in vain for an invitation, she made up her mind at last to go without one, and make the whole family miserable, like a princess as she was.

So she put on her best gown, went to the palace, was kindly received by the happy monarch, who forgot that he had forgotten her, and took her place in the procession to the royal chapel. When they were all gathered about the font, she contrived to get next to it, and throw something into the water; after which she maintained a very respectful demeanour till the water was applied to the child's face. But at that moment she turned round in her place three times, and muttered the following words, loud enough for those beside her to hear:—

> 'Light of spirit, by my charms,
> Light of body, every part,
> Never weary human arms—
> Only crush thy parents' heart!'

They all thought she had lost her wits, and was repeating some foolish nursery rhyme; but a shudder went through the whole of them notwithstanding. The baby, on the contrary, began to laugh and crow; while the nurse gave a start and a smothered cry, for she thought she was struck with paralysis: she could not feel the baby in her arms. But she clasped it tight and said nothing.

The mischief was done.

3. *She Can't Be Ours*

Her atrocious aunt had deprived the child of all her gravity. If you ask me how this was effected, I answer, 'In the easiest way in the world. She had only to destroy gravitation.' For the princess was a philosopher, and knew all the *ins* and *outs* of the laws of gravitation as well as the *ins* and *outs* of her boot-lace. And being a witch as well, she could abrogate those laws in a moment; or at least so clog their wheels and rust their bearings, that they would not work at all. But we have more to do with what followed than with how it was done.

The first awkwardness that resulted from this unhappy privation was, that the moment the nurse began to float the baby up and down, she flew from her arms towards the ceiling. Happily, the resistance of the air brought her ascending career to a close within a foot of it. There she remained, horizontal as when she left her nurse's arms, kicking and laughing amazingly. The nurse in terror flew to the bell, and begged the footman, who answered it, to bring up the house-steps directly. Trembling in every limb, she climbed upon the steps, and had to stand upon the very top, and reach up, before she could catch the floating tail of the baby's long clothes.

When the strange fact came to be known, there was a terrible commotion in the palace. The occasion of its discovery by the king was naturally a repetition of the nurse's experience. Astonished that he felt no weight when the child was laid in his arms, he began to wave her up and—not down, for she slowly ascended to the ceiling as before, and there remained floating in perfect comfort and satisfaction, as was testified by her peals of tiny laughter. The king stood staring up in speechless amazement, and trembled so that his beard shook like grass in the wind. At last, turning to the queen, who was just as horror-struck as himself, he said, gasping, staring, and stammering,—

'She *can't* be ours, queen!'

Now the queen was much cleverer than the king, and had begun already to suspect that 'this effect defective came by cause.'

'I am sure she is ours,' answered she. 'But we ought to have taken better care of her at the christening. People who were never invited ought not to have been present.'

'Oh, ho!' said the king, tapping his forehead with his forefinger, 'I have it all. I've found her out. Don't you see it, queen? Princess Makemnoit has bewitched her.'

'That's just what I say,' answered the queen.

'I beg your pardon, my love; I did not hear you.—John! bring the steps I get on my throne with.'

For he was a little king with a great throne, like many other kings.

The throne-steps were brought, and set upon the dining-table, and John got upon the top of them. But he could not reach the little princess, who lay like a baby-laughter-cloud in the air, exploding continuously.

'Take the tongs, John,' said his Majesty; and getting up on the table, he handed them to him.

John could reach the baby now, and the little princess was handed down by the tongs.

4. *Where Is She?*

One fine summer day, a month after these her first adventures, during which time she had been very carefully watched, the princess was lying on the bed in the queen's own chamber, fast asleep. One of the windows was open, for it was noon, and the day was so sultry that the little girl was wrapped in nothing less ethereal than slumber itself. The queen came into the room, and not observing that the baby was on the bed, opened another window. A frolicsome fairy wind, which had been watching for a chance of mischief, rushed in at the one window, and taking its way over the bed where the child was lying, caught her up, and rolling and floating her along like a piece of flue, or a dandelion seed, carried her with it through the opposite window, and away. The queen went down-stairs, quite ignorant of the loss she had herself occasioned.

When the nurse returned, she supposed that her Majesty had carried her off, and, dreading a scolding, delayed making inquiry about her. But hearing nothing, she grew uneasy, and went at length to the queen's boudoir, where she found her Majesty.

'Please, your Majesty, shall I take the baby?' said she.

'Where is she?' asked the queen.

'Please forgive me. I know it was wrong.'

'What do you mean?' said the queen, looking grave.

'Oh! don't frighten me, your Majesty!' exclaimed the nurse, clasping her hands.

The queen saw that something was amiss, and fell down in a faint. The nurse rushed about the palace, screaming, 'My baby! my baby!'

Every one ran to the queen's room. But the queen could give no orders. They soon found out, however, that the princess was missing, and in a moment the palace was like a beehive in a garden; and in one minute more the queen was brought to herself by a great shout and a clapping of hands. They had found the princess fast asleep under a rose-bush, to which the elvish little wind-puff had carried her, finishing its mischief by shaking a shower of red rose-leaves all over the little white sleeper. Startled by the noise the servants made, she woke, and, furious with glee, scattered the rose-leaves in all directions, like a shower of spray in the sunset.

She was watched more carefully after this, no doubt; yet it would be endless to relate all the odd incidents resulting from this peculiarity of the young princess. But there never was a baby in a house, not to say a palace, that kept the household in such constant good humour, at least below-stairs. If it was not easy for her nurses to hold her, at least she made neither their arms nor their hearts ache. And she was so nice to play at ball with! There was positively no danger of letting her fall. They might throw her down, or knock her down, or push her down, but couldn't *let* her down. It is true, they might let her fly into the fire or the coal-hole, or through the window; but none of these accidents had happened as yet. If you heard peals of laughter resounding from some unknown region, you might be sure enough of the cause. Going down into the kitchen, or *the room*, you would find Jane and Thomas, and Robert and Susan, all and sum, playing at ball with the little princess. She was the ball herself, and did not enjoy it the less for that. Away she went, flying from one to another, screeching with laughter. And the servants loved the ball itself better even than the game. But they had to take some care how they threw her, for if she received an upward direction, she would never come down again without being fetched.

5. *What Is to Be Done?*

But above-stairs it was different. One day, for instance, after breakfast, the king went into his counting-house, and counted out his money.

The operation gave him no pleasure.

'To think,' said he to himself, 'that every one of these gold sovereigns weighs a quarter of an ounce, and my real, live, flesh-and-blood princess weighs nothing at all!'

And he hated his gold sovereigns, as they lay with a broad smile of self-satisfaction all over their yellow faces.

The queen was in the parlour, eating bread and honey. But at the second mouthful she burst out crying, and could not swallow it.

The king heard her sobbing. Glad of anybody, but especially of his queen, to quarrel with, he clashed his gold sovereigns into his money-box, clapped his crown on his head, and rushed into the parlour.

'What is all this about?' exclaimed he. 'What are you crying for, queen?'

'I can't eat it,' said the queen, looking ruefully at the honey-pot.

'No wonder!' retorted the king. 'You've just eaten your breakfast—two turkey eggs, and three anchovies.'

'Oh, that's not it!' sobbed her Majesty. 'It's my child, my child!'

'Well, what's the matter with your child? She's neither up the chimney nor down the draw-well. Just hear her laughing.'

Yet the king could not help a sigh, which he tried to turn into a cough, saying—

'It is a good thing to be light-hearted, I am sure, whether she be ours or not.'

'It is a bad thing to be light-headed,' answered the queen, looking with prophetic soul far into the future.

''Tis a good thing to be light-handed,' said the king.

''Tis a bad thing to be light-fingered,' answered the queen.

''Tis a good thing to be light-footed,' said the king.

''Tis a bad thing—' began the queen; but the king interrupted her.

'In fact,' said he, with the tone of one who concludes an argument in which he has had only imaginary opponents, and in which, therefore, he has come off triumphant—'in fact, it is a good thing altogether to be light-bodied.'

'But it is a bad thing altogether to be light-minded,' retorted the queen, who was beginning to lose her temper.

This last answer quite discomfited his Majesty, who turned on his heel, and betook himself to his counting-house again. But he was not half-way towards it, when the voice of his queen overtook him.

'And it's a bad thing to be light-haired,' screamed she, determined to have more last words, now that her spirit was roused.

The queen's hair was black as night; and the king's had been, and his daughter's was, golden as morning. But it was not this reflection on his hair that arrested him; it was the double use of the word *light*. For the king hated all witticisms, and punning especially. And besides, he could not tell whether the queen meant light-*haired* or light-*heired*; for why might she not aspirate her vowels when she was ex-asperated herself?

He turned upon his other heel, and rejoined her. She looked angry still, because she knew that she was guilty, or, what was much the same, knew that he thought so.

'My dear queen,' said he, 'duplicity of any sort is exceedingly objectionable between married people of any rank, not to say kings and queens; and the most objectionable form duplicity can assume is that of punning.'

'There!' said the queen, 'I never made a jest, but I broke it in the making. I am the most unfortunate woman in the world!'

She looked so rueful, that the king took her in his arms; and they sat down to consult.

'Can you bear this?' said the king.

'No, I can't,' said the queen.

'Well, what's to be done?' said the king.

'I'm sure I don't know,' said the queen. 'But might you not try an apology?'

'To my old sister, I suppose you mean?' said the king.

'Yes,' said the queen.

'Well, I don't mind,' said the king.

So he went the next morning to the house of the princess, and making a very humble apology, begged her to undo the spell. But the princess declared, with a grave face, that she knew nothing at all about it. Her eyes, however, shone pink, which was a sign that she was happy. She advised the king and queen to have patience, and to mend their ways. The king returned disconsolate. The queen tried to comfort him.

'We will wait till she is older. She may then be able to suggest something herself. She will know at least how she feels, and explain things to us.'

'But what if she should marry?' exclaimed the king, in sudden consternation at the idea.

'Well, what of that?' rejoined the queen.

'Just think! If she were to have children! In the course of a hundred years the air might be as full of floating children as of gossamers in autumn.'

'That is no business of ours,' replied the queen. 'Besides, by that time they will have learned to take care of themselves.'

A sigh was the king's only answer.

He would have consulted the court physicians; but he was afraid they would try experiments upon her.

6. *She Laughs Too Much*

Meantime, notwithstanding awkward occurrences, and griefs that she brought upon her parents, the little princess laughed and grew—not fat, but plump and tall. She reached the age of seventeen, without having fallen into any worse scrape than a chimney; by rescuing her from which, a little bird-nesting urchin got fame and a black face. Nor, thoughtless as she was, had she committed anything worse than laughter at everybody and everything that came in her way. When she was told, for the sake of experiment, that General Clanrunfort was cut to pieces with all his troops, she laughed, when she heard that the enemy was on his way to besiege her papa's capital, she laughed hugely; but when she was told that the city would certainly be abandoned to the mercy of the enemy's soldiery—why, then she laughed immoderately. She never could be brought to see the serious side of anything. When her mother cried, she said,—

'What queer faces mamma makes! And she squeezes water out of her cheeks? Funny mamma!'

And when her papa stormed at her, she laughed, and danced round and round him, clapping her hands, and crying—

'Do it again, papa. Do it again! It's such fun! Dear, funny papa!'

And if he tried to catch her, she glided from him in an instant, not in the least afraid of him, but thinking it part of the game not to be caught. With one push of her foot, she would be floating in the air above his head; or she would go dancing backwards and forwards and sideways, like a great butterfly. It happened several times, when her father and mother were holding a consultation about her in private, that they were interrupted by vainly repressed outbursts of laughter over their heads; and looking up with indignation, saw her floating at full length in the air above them, whence she regarded them with the most comical appreciation of the position.

One day an awkward accident happened. The princess had come out upon the lawn with one of her attendants, who held her by the hand. Spying her father at the other side of the lawn, she snatched her hand from the maid's, and sped across to him. Now when she wanted to run alone, her custom was to catch up a stone in each hand, so that she might come down again after a bound. Whatever she wore as part of her attire had no effect in this way: even gold, when it thus became as it were a part of herself, lost all its weight for the time. But whatever she only held in her hands retained its downward tendency. On this occasion she could see nothing to catch up but a huge toad, that was walking across the lawn as if he had a hundred years to do it in. Not knowing what disgust meant, for this was one of her peculiarities, she snatched up the toad and bounded away. She had almost reached her father, and he was holding out his arms to receive her, and take from her lips the kiss which hovered on them like a butterfly on a rosebud, when a puff of wind blew her aside into the arms of a young page, who had just been receiving a message from his Majesty. Now it was no great peculiarity in the princess that, once she was set agoing, it always cost her time and trouble to check herself. On this occasion there was no time. She *must* kiss—and she kissed the page. She did not mind it much; for she had no shyness in her composition; and she knew, besides, that she could not help it. So she only laughed, like a musical box. The poor page fared the worst. For the princess, trying to correct the unfortunate tendency of the kiss, put out her hands to keep her off the page; so that, along with the kiss, he received, on the other cheek, a slap with the huge black toad, which she poked right into his eye. He tried to laugh, too, but the attempt resulted in such an odd contortion of countenance, as showed that there was no danger of his pluming himself on the kiss. As for the king, his dignity was greatly hurt, and he did not speak to the page for a whole month.

I may here remark that it was very amusing to see her run, if her mode of progression could properly be called running. For first she would make a bound; then, having alighted, she would run a few steps, and make another bound. Sometimes she would fancy she had reached the ground before she actually had, and her feet would go backwards and forwards, running upon nothing at all, like those of a chicken on its back. Then she would laugh like the very spirit of fun; only in her laugh there was something missing. What it was, I find myself unable to describe. I think it was a certain tone, depending upon the possibility of sorrow—*morbidezza*, perhaps. She never smiled.

7. *Try Metaphysics*

After a long avoidance of the painful subject, the king and queen resolved to hold a council of three upon it; and so they sent for the princess. In she came, sliding and flitting and gliding from one piece of furniture to another, and put herself at last in an armchair, in a sitting posture. Whether she could be said *to sit*, seeing she received no support from the seat of the chair, I do not pretend to determine.

'My dear child,' said the king, 'you must be aware by this time that you are not exactly like other people.'

'Oh, you dear funny papa! I have got a nose, and two eyes, and all the rest. So have you. So has mamma.'

'Now be serious, my dear, for once,' said the queen.

'No, thank you, mamma; I had rather not.'

'Would you not like to be able to walk like other people?' said the king.

'No indeed, I should think not. You only crawl. You are such slow coaches!'

'How do you feel, my child?' he resumed, after a pause of discomfiture.

'Quite well, thank you.'

'I mean, what do you feel like?'

'Like nothing at all, that I know of.'

'You must feel like something.'

'I feel like a princess with such a funny papa, and such a dear pet of a queen-mamma!'

'Now really!' began the queen; but the princess interrupted her.

'Oh yes,' she added, 'I remember. I have a curious feeling sometimes, as if I were the only person that had any sense in the whole world.'

She had been trying to behave herself with dignity; but now she burst into a violent fit of laughter, threw herself backwards over the chair, and went rolling about the floor in an ecstasy of enjoyment. The king picked her up easier than one does a down quilt, and replaced her in her former relation to the chair. The exact preposition expressing this relation I do not happen to know.

'Is there nothing you wish for?' resumed the king, who had learned by this time that it was useless to be angry with her.

'Oh, you dear papa!—yes,' answered she.

'What is it, my darling?'

'I have been longing for it—oh, such a time!—ever since last night.'

'Tell me what it is.'

'Will you promise to let me have it?'

The king was on the point of saying *Yes*, but the wiser queen checked him with a single motion of her head.

'Tell me what it is first,' said he.

'No no. Promise first.'

'I dare not. What is it?'

'Mind, I hold you to your promise.—It is—to be tied to the end of a string—a very long string indeed, and be flown like a kite. Oh, such fun! I would rain rose-water, and hail sugar-plums, and snow whipped-cream, and—and—and—'

A fit of laughing checked her; and she would have been off again over the floor, had not the king started up and caught her just in time. Seeing nothing but talk could be got out of her, he rang the bell, and sent her away with two of her ladies-in-waiting.

'Now, queen,' he said, turning to her Majesty, 'what *is* to be done?'

'There is but one thing left,' answered she. 'Let us consult the college of Metaphysicians.'

'Bravo!' cried the king; 'we will.'

Now at the head of this college were two very wise Chinese philosophers—by name Hum-Drum, and Kopy-Keck. For them the king sent; and straightway they came. In a long speech he communicated to them what they knew very well already—as who did not?—namely, the peculiar condition of his daughter in relation to the globe on which she dwelt; and requested them to consult together as to what might be the cause and probable cure of her *infirmity*. The king laid stress upon the word, but failed to discover his own pun. The queen laughed; but Hum-Drum and Kopy-Keck heard with humility and retired in silence.

The consultation consisted chiefly in propounding and supporting, for the thousandth time, each his favourite theories. For the condition of the princess afforded delightful scope for the discussion of every question arising from the division of thought—in fact, of all the Metaphysics of the Chinese Empire. But it is only justice to say that they did not altogether neglect the discussion of the practical question, *what was to be done*.

Hum-Drum was a Materialist, and Kopy-Keck was a Spiritualist. The former was slow and sententious; the latter was quick and flighty: the latter had generally the first word; the former the last.

'I reassert my former assertion,' began Kopy-Keck, with a plunge. 'There is not a fault in the princess, body or soul; only they are wrong put together. Listen to me now, Hum-Drum, and I will tell you in brief what I think. Don't speak. Don't answer me. I *won't* hear you till I have done.—At that decisive moment, when souls seek their appointed habitations, two eager souls met, struck, rebounded, lost their way, and arrived each at the wrong place. The soul of the princess was one of those, and she went far astray. She does not belong by rights to this world at all, but to some other planet, probably Mercury. Her proclivity to her true sphere destroys all the natural influence which this orb would otherwise possess over her corporeal frame. She cares for nothing here. There is no relation between her and this world.

'She must therefore be taught, by the sternest compulsion, to take an interest in the earth as the earth. She must study every department of its history—its animal history; its vegetable history; its mineral history; its social history; its moral history; its political history; its scientific history; its literary history; its musical history; its artistical history; above all, its metaphysical history. She must begin with the Chinese dynasty and end with Japan. But first of all she must study geology, and especially the history of the extinct races of animals—their natures, their habits, their loves, their hates, their revenges. She must—'

'Hold, h-o-o-old!' roared Hum-Drum. 'It is certainly my turn now. My rooted and insubvertible conviction is, that the causes of the anomalies evident in the princess's condition are strictly and solely physical. But that is only tantamount to acknowledging that they exist. Hear my opinion.—From some cause or other, of no importance to our inquiry, the motion of her heart has been reversed. That remarkable combination of the suction and the force-pump works the wrong way—I mean in the case of the unfortunate princess: it draws in where it should force out, and forces out where it should draw in. The offices of the auricles and the ventricles are subverted. The blood is sent forth by the veins, and returns by the arteries. Consequently it is running the wrong way through all her corporeal organism—lungs and all. Is it then at all mysterious, seeing that such is the case, that on the other particular of gravitation as well, she should differ from normal humanity? My proposal for the cure is this:—

'Phlebotomize until she is reduced to the last point of safety. Let it be effected, if necessary, in a warm bath. When she is reduced to a state of perfect asphyxy, apply a ligature to the left ankle, drawing it as tight as the bone will bear. Apply, at the same moment, another of equal tension around the right wrist. By means of plates constructed for the purpose, place the other foot and hand under the receivers of two air-pumps. Exhaust the receivers. Exhibit a pint of French brandy, and await the result.'

'Which would presently arrive in the form of grim Death,' said Kopy-Keck.

'If it should, she would yet die in doing our duty,' retorted Hum-Drum.

But their Majesties had too much tenderness for their volatile offspring to subject her to either of the schemes of the equally unscrupulous philosophers. Indeed, the most complete knowledge of the laws of nature would have been unserviceable in her case; for it was impossible to classify her. She was a fifth imponderable body, sharing all the other properties of the ponderable.

8. *Try a Drop of Water*

Perhaps the best thing for the princess would have been to fall in love. But how a princess who had no gravity could fall into anything is a difficulty—perhaps *the* difficulty.

As for her own feelings on the subject, she did not even know that there was such a beehive of honey and stings to be fallen into. But now I come to mention another curious fact about her.

The palace was built on the shores of the loveliest lake in the world; and the princess loved this lake more than father or mother. The root of this preference no doubt, although the princess did not recognise it as such, was, that the moment she got into it, she recovered the natural right of which she had been so wickedly deprived—namely, gravity. Whether this was owing to the fact that water had been employed as the means of conveying the injury, I do not know. But it is certain that she could swim and dive like the duck that her old nurse said she was. The manner in which this alleviation of her misfortune was discovered was as follows.

One summer evening, during the carnival of the country, she had been taken upon the lake by the king and queen, in the royal barge. They were accompanied by many of the courtiers in a fleet of little boats. In the middle of the lake she wanted to get into the lord chancellor's barge, for his daughter, who was a great favourite with her, was in it with her father. Now though the old king rarely condescended to make light of his misfortune, yet, happening on this occasion to be in a particularly good humour, as the barges approached each other, he caught up the princess to throw her into the chancellor's barge. He lost his balance, however, and, dropping into the bottom of the barge, lost his hold of his daughter; not, however, before imparting to her the downward tendency of his own person, though in a somewhat different direction; for, as the king fell into the boat, she fell into the water. With a burst of delighted laughter she disappeared in the lake. A cry of horror ascended from the boats. They had never seen the princess go down before. Half the men were under water in a moment; but they had all, one after another, come up to the surface again for breath, when—tinkle, tinkle, babble, and gush! came the princess's laugh over the water from far away. There she was, swimming like a swan. Nor would she come out for king or queen, chancellor or daughter. She was perfectly obstinate.

But at the same time she seemed more sedate than usual. Perhaps that was because a great pleasure spoils laughing. At all events, after this, the passion of her life was to get into the water, and she was always the better behaved and the more beautiful the more she had of it. Summer and winter it was quite the same; only she could not stay so long in the water when they had to break the ice to let her in. Any day, from morning till evening in summer, she might be descried—a streak of white in the blue water—lying as still as the shadow of a cloud, or shooting along like a dolphin; disappearing, and coming up again far off, just where one did not expect her. She would have been in the lake of a night, too, if she could have had her way; for the balcony of her window overhung a deep pool in it; and through a shallow reedy passage she could have swum out into the wide wet water, and no one would have been any the wiser. Indeed, when she happened to wake in the moonlight she could hardly resist the temptation. But there was the sad difficulty of getting into it. She had as great a dread of the air as some children have of the water. For the slightest gust of wind would blow her away; and a gust might arise in the stillest moment. And if she gave herself a push towards the water and just failed of reaching it, her situation would be dreadfully awkward, irrespective of the wind; for at best there she would have to remain, suspended in her night-gown, till she was seen and angled for by someone from the window.

'Oh! if I had my gravity,' thought she, contemplating the water, 'I would flash off this balcony like a long white sea-bird, headlong into the darling wetness. Heigh-ho!'

This was the only consideration that made her wish to be like other people.

Another reason for her being fond of the water was that in it alone she enjoyed any freedom. For she could not walk out without a *cortége*, consisting in part of a troop of light horse, for fear of the liberties which the wind might take with her. And the king grew more apprehensive with increasing years, till at last he would not allow her to walk abroad at all without some twenty silken cords fastened to as many parts of her dress, and held by twenty noblemen. Of course horseback was out of the question. But she bade good-by to all this ceremony when she got into the water.

And so remarkable were its effects upon her, especially in restoring her for the time to the ordinary human gravity, that Hum-Drum and Kopy-Keck agreed in recommending the king to bury her alive for three years; in the hope that, as the water did her so much good, the earth would do her yet more. But the king had some vulgar prejudices against the experiment, and would not give his consent. Foiled in this, they yet agreed in another recommendation; which, seeing that one imported his opinions from China and the other from Thibet, was very remarkable indeed. They argued that, if water of external origin and application could be so efficacious, water from a deeper source might work a perfect cure; in short, that if the poor afflicted princess could by any means be made to cry, she might recover her lost gravity.

But how was this to be brought about? Therein lay all the difficulty—to meet which the philosophers were not wise enough. To make the princess cry was as impossible as to make her weigh. They sent for a professional beggar; commanded him to prepare his most touching oracle of woe; helped him out of the court charade box, to whatever he wanted for dressing up, and promised great rewards in the event of his success. But it was all in vain. She listened to the mendicant artist's story, and gazed at his marvellous make up, till she could contain herself no longer, and went into the most undignified contortions for relief, shrieking, positively screeching with laughter.

When she had a little recovered herself, she ordered her attendants to drive him away, and not give him a single copper; whereupon his look of mortified discomfiture wrought her punishment and his revenge, for it sent her into violent hysterics, from which she was with difficulty recovered.

But so anxious was the king that the suggestion should have a fair trial, that he put himself in a rage one day, and, rushing up to her room, gave her an awful whipping. Yet not a tear would flow.

She looked grave, and her laughing sounded uncommonly like screaming—that was all. The good old tyrant, though he put on his best gold spectacles to look, could not discover the smallest cloud in the serene blue of her eyes.

9. *Put Me in Again*

It must have been about this time that the son of a king, who lived a thousand miles from Lagobel, set out to look for the daughter of a queen. He travelled far and wide, but as sure as he found a princess, he found some fault in her. Of course he could not marry a mere woman, however beautiful; and there was no princess to be found worthy of him. Whether the prince was so near perfection that he had a right to demand perfection itself, I cannot pretend to say. All I know is, that he was a fine, handsome, brave, generous, well-bred, and well-behaved youth, as all princes are.

In his wanderings he had come across some reports about our princess; but as everybody said she was bewitched, he never dreamed that she could bewitch him. For what indeed could a prince do with a princess that had lost her gravity? Who could tell what she might not lose next? She might lose her visibility, or her tangibility; or, in short, the power of making impressions upon the radical sensorium; so that he should never be able to tell whether she was dead or alive. Of course he made no further inquiries about her.

One day he lost sight of his retinue in a great forest. These forests are very useful in delivering princes from their courtiers, like a sieve that keeps back the bran. Then the princes get away to follow their fortunes. In this way they have the advantage of the princesses, who are forced to marry before they have had a bit of fun. I wish our princesses got lost in a forest sometimes.

One lovely evening, after wandering about for many days, he found that he was approaching the outskirts of this forest; for the trees had got so thin that he could see the sunset through them; and he soon came upon a kind of heath. Next he came upon signs of human neighbourhood; but by this time it was getting late, and there was nobody in the fields to direct him.

After travelling for another hour, his horse, quite worn out with long labour and lack of food, fell, and was unable to rise again. So he continued his journey on foot. At length he entered another wood—not a wild forest, but a civilized wood, through which a footpath led him to the side of a lake. Along this path the prince pursued his way through the gathering darkness. Suddenly he paused, and listened. Strange sounds came across the water. It was, in fact, the princess laughing. Now there was something odd in her laugh, as I have already hinted; for the hatching of a real hearty laugh requires the incubation of gravity; and perhaps this was how the prince mistook the laughter for screaming. Looking over the lake, he saw something white in the water; and, in an instant, he had torn off his tunic, kicked off his sandals, and plunged in. He soon reached the white object, and found that it was a woman. There was not light enough to show that she was a princess, but quite enough to show that she was a lady, for it does not want much light to see that.

Now I cannot tell how it came about,—whether she pretended to be drowning, or whether he frightened her, or caught her so as to embarrass her,—but certainly he brought her to shore in a fashion ignominious to a swimmer, and more nearly drowned than she had ever expected to be; for the water had got into her throat as often as she had tried to speak.

At the place to which he bore her, the bank was only a foot or two above the water; so he gave her a strong lift out of the water, to lay her on the bank. But, her gravitation ceasing the moment she left the water, away she went up into the air, scolding and screaming.

'You naughty, *naughty*, NAUGHTY, NAUGHTY man!' she cried.

No one had ever succeeded in putting her into a passion before.—When the prince saw her ascend, he thought he must have been bewitched, and have mistaken a great swan for a lady. But the princess caught hold of the topmost cone upon a lofty fir. This came off; but she caught at another; and, in fact, stopped herself by gathering cones, dropping them as the stalks gave way. The prince, meantime, stood in the water, staring, and forgetting to get out. But the princess disappearing, he scrambled on shore, and went in the direction of the tree. There he found her climbing down one of the branches towards the stem. But in the darkness of the wood, the prince continued in some bewilderment as to what the phenomenon could be; until, reaching the ground, and seeing him standing there, she caught hold of him, and said,—

'I'll tell papa.'

'Oh no, you won't!' returned the prince.

'Yes, I will,' she persisted. 'What business had you to pull me down out of the water, and throw me to the bottom of the air? I never did you any harm.'

'Pardon me. I did not mean to hurt you.'

'I don't believe you have any brains; and that is a worse loss than your wretched gravity. I pity you.'

The prince now saw that he had come upon the bewitched princess, and had already offended her. But before he could think what to say next, she burst out angrily, giving a stamp with her foot that would have sent her aloft again but for the hold she had of his arm,—

'Put me up directly.'

'Put you up where, you beauty?' asked the prince.

He had fallen in love with her almost, already; for her anger made her more charming than any one else had ever beheld her; and, as far as he could see, which certainly was not far, she had not a single fault about her, except, of course, that she had not any gravity. No prince, however, would judge of a princess by weight. The loveliness of her foot he would hardly estimate by the depth of the impression it could make in mud.

'Put you up where, you beauty?' asked the prince.

'In the water, you stupid!' answered the princess.

'Come, then,' said the prince.

The condition of her dress, increasing her usual difficulty in walking, compelled her to cling to him; and he could hardly persuade himself that he was not in a delightful dream, notwithstanding the torrent of musical abuse with which she overwhelmed him. The prince being therefore in no hurry, they came upon the lake at quite another part, where the bank was twenty-five feet high at least; and when they had reached the edge, he turned towards the princess, and said,—

'How am I to put you in?'

'That is your business,' she answered, quite snappishly. 'You took me out—put me in again.'

'Very well,' said the prince; and, catching her up in his arms, he sprang with her from the rock. The princess had just time to give one delighted shriek of laughter before the water closed over them. When they came to the surface, she found that, for a moment or two, she could not even laugh, for she had gone down with such a rush, that it was with difficulty she recovered her breath. The instant they reached the surface—

'How do you like falling in?' said the prince.

After some effort the princess panted out,—

'Is that what you call *falling in*?'

'Yes,' answered the prince, 'I should think it a very tolerable specimen.'

'It seemed to me like going up,' rejoined she.

'My feeling was certainly one of elevation too,' the prince conceded.

The princess did not appear to understand him, for she retorted his question:—

'How do *you* like falling in?' said the princess.

'Beyond everything,' answered he, 'for I have fallen in with the only perfect creature I ever saw.'

'No more of that: I am tired of it,' said the princess.

Perhaps she shared her father's aversion to punning.

'Don't you like falling in then?' said the prince.

'It is the most delightful fun I ever had in my life,' answered she. 'I never fell before. I wish I could learn. To think I am the only person in my father's kingdom that can't fall!'

Here the poor princess looked almost sad.

'I shall be most happy to fall in with you any time you like,' said the prince, devotedly.

'Thank you. I don't know. Perhaps it would not be proper. But I don't care. At all events, as we have fallen in, let us have a swim together.'

'With all my heart,' responded the prince.

And away they went, swimming, and diving, and floating, until at last they heard cries along the shore, and saw lights glancing in all directions. It was now quite late, and there was no moon.

'I must go home,' said the princess. 'I am very sorry, for this is delightful.'

'So am I,' returned the prince. 'But I am glad I haven't a home to go to—at least, I don't exactly know where it is.'

'I wish I hadn't one either,' rejoined the princess; 'it is so stupid! I have a great mind,' she continued, 'to play them all a trick. Why couldn't they leave me alone? They won't trust me in the lake for a single night!—You see where that green light is burning? That is the window of my room. Now if you would just swim there with me very quietly, and when we are all but under the balcony, give me such a push—*up* you call it—as you did a little while ago, I should be able to catch hold of the balcony, and get in at the window; and then they may look for me till to-morrow morning!'

'With more obedience than pleasure,' said the prince, gallantly; and away they swam, very gently.

'Will you be in the lake to-morrow night?' the prince ventured to ask.

'To be sure I will. I don't think so. Perhaps,' was the princess's somewhat strange answer.

But the prince was intelligent enough not to press her further; and merely whispered, as he gave her the parting lift, 'Don't tell.' The only answer the princess returned was a roguish look. She was already a yard above his head. The look seemed to say, 'Never fear. It is too good fun to spoil that way.'

So perfectly like other people had she been in the water, that even yet the prince could scarcely believe his eyes when he saw her ascend slowly, grasp the balcony, and disappear through the window. He turned, almost expecting to see her still by his side. But he was alone in the water. So he swam away quietly, and watched the lights roving about the shore for hours after the princess was safe in her chamber. As soon as they disappeared, he landed in search of his tunic and sword, and, after some trouble, found them again. Then he made the best of his way round the lake to the other side. There the wood was wilder, and the shore steeper—rising more immediately towards the mountains which surrounded the lake on all sides, and kept sending it messages of silvery streams from morning to night, and all night long. He soon found a spot whence he could see the green light in the princess's room, and where, even in the broad daylight, he would be in no danger of being discovered from the opposite shore. It was a sort of cave in the rock, where he provided himself a bed of withered leaves, and lay down too tired for hunger to keep him awake. All night long he dreamed that he was swimming with the princess.

10. *Look at the Moon*

Early the next morning the prince set out to look for something to eat, which he soon found at a forester's hut, where for many following days he was supplied with all that a brave prince could consider necessary. And having plenty to keep him alive for the present, he would not think of wants not yet in existence. Whenever Care intruded, this prince always bowed him out in the most princely manner.

When he returned from his breakfast to his watch-cave, he saw the princess already floating about in the lake, attended by the king and queen—whom he knew by their crowns—and a great company in lovely little boats, with canopies of all the colours of the rainbow, and flags and streamers of a great many more. It was a very bright day, and soon the prince, burned up with the heat, began to long for the cold water and the cool princess. But he had to endure till twilight; for the boats had provisions on board, and it was not till the sun went down that the gay party began to vanish. Boat after boat drew away to the shore, following that of the king and queen, till only one, apparently the princess's own boat, remained. But she did not want to go home even yet, and the prince thought he saw her order the boat to the shore without her. At all events, it rowed away; and now, of all the radiant company, only one white speck remained. Then the prince began to sing.

And this is what he sung:—

'Lady fair,
Swan-white,
Lift thine eyes,
Banish night
By the might
Of thine eyes.

Snowy arms,
Oars of snow,
Oar her hither,
Plashing low.
Soft and slow,
Oar her hither.

Stream behind her
O'er the lake,
Radiant whiteness!
In her wake
Following, following for her sake
Radiant whiteness!

Cling about her,
Waters blue;

Part not from her,
But renew
Cold and true
Kisses round her.

Lap me round,
Waters sad
That have left her
Make me glad,
For ye had
Kissed her ere ye left her.'

Before he had finished his song, the princess was just under the place where he sat, and looking up to find him. Her ears had led her truly.

'Would you like a fall, princess?' said the prince, looking down.

'Ah! there you are! Yes, if you please, prince,' said the princess, looking up.

'How do you know I am a prince, princess?' said the prince.

'Because you are a very nice young man, prince,' said the princess.

'Come up then, princess.'

'Fetch me, prince.'

The prince took off his scarf, then his sword-belt, then his tunic, and tied them all together, and let them down. But the line was far too short. He unwound his turban, and added it to the rest, when it was all but long enough; and his purse completed it. The princess just managed to lay hold of the knot of money, and was beside him in a moment. This rock was much higher than the other, and the splash and the dive were tremendous. The princess was in ecstasies of delight, and their swim was delicious.

Night after night they met, and swam about in the dark clear lake; where such was the prince's gladness, that (whether the princess's way of looking at things infected him, or he was actually getting light-headed) he often fancied that he was swimming in the sky instead of the lake. But when he talked about being in heaven, the princess laughed at him dreadfully.

When the moon came, she brought them fresh pleasure. Everything looked strange and new in her light, with an old, withered, yet unfading newness. When the moon was nearly full, one of their great delights was, to dive deep in the water, and then, turning round, look up through it at the great blot of light close above them, shimmering and trembling and wavering, spreading and contracting, seeming to melt away, and again grow solid. Then they would shoot up through the blot; and lo! there was the moon, far off, clear and steady and cold, and very lovely, at the bottom of a deeper and bluer lake than theirs, as the princess said.

The prince soon found out that while in the water the princess was very like other people. And besides this, she was not so forward in her questions or pert in her replies at sea as on shore. Neither did she laugh so much; and when she did laugh, it was more gently. She seemed altogether more modest and maidenly in the water than out of it. But when the prince, who had really fallen in love when he fell in the lake, began to talk to her about love, she always turned her head towards him and laughed. After a while she began to look puzzled, as if she were trying to understand what he meant, but could not—revealing a notion that he meant something. But as soon as ever she left the lake, she was so altered, that the prince said to himself, 'If I marry her, I see no help for it: we must turn merman and mermaid, and go out to sea at once.'

11. *Hiss!*

The princess's pleasure in the lake had grown to a passion, and she could scarcely bear to be out of it for an hour. Imagine then her consternation, when, diving with the prince one night, a sudden suspicion seized her that the lake was not so deep as it used to be. The prince could not imagine what had happened. She shot to the surface, and, without a word, swam at full speed towards the higher side of the lake. He followed, begging to know if she was ill, or what was the matter. She never turned her head, or took the smallest notice of his question. Arrived at the shore, she coasted the rocks with minute inspection. But she was not able to come to a conclusion, for the moon was very small, and so she could not see well. She turned therefore and swam home, without saying a word to explain her conduct to the prince, of whose presence she seemed no longer conscious. He withdrew to his cave, in great perplexity and distress.

Next day she made many observations, which, alas! strengthened her fears. She saw that the banks were too dry; and that the grass on the shore, and the trailing plants on the rocks, were withering away. She caused marks to be made along the borders, and examined them, day after day, in all directions of the wind; till at last the horrible idea became a certain fact—that the surface of the lake was slowly sinking.

The poor princess nearly went out of the little mind she had. It was awful to her to see the lake, which she loved more than any living thing, lie dying before her eyes. It sank away, slowly vanishing. The tops of rocks that had never been seen till now, began to appear far down in the clear water. Before long they were dry in the sun. It was fearful to think of the mud that would soon lie there baking and festering, full of lovely creatures dying, and ugly creatures coming to life, like the unmaking of a world. And how hot the sun would be without any lake! She could not bear to swim in it any more, and began to pine away. Her life seemed bound up with it; and ever as the lake sank, she pined. People said she would not live an hour after the lake was gone.

But she never cried.

Proclamation was made to all the kingdom, that whosoever should discover the cause of the lake's decrease, would be rewarded after a princely fashion. Hum-Drum and Kopy-Keck applied themselves to their physics and metaphysics; but in vain. Not even they could suggest a cause.

Now the fact was that the old princess was at the root of the mischief. When she heard that her niece found more pleasure in the water than any one else out of it, she went into a rage, and cursed herself for her want of foresight.

'But,' said she, 'I will soon set all right. The king and the people shall die of thirst; their brains shall boil and frizzle in their skulls before I will lose my revenge.'

And she laughed a ferocious laugh, that made the hairs on the back of her black cat stand erect with terror.

Then she went to an old chest in the room, and opening it, took out what looked like a piece of dried seaweed. This she threw into a tub of water. Then she threw some powder into the water, and stirred it with her bare arm, muttering over it words of hideous sound, and yet more hideous import. Then she set the tub aside, and took from the chest a huge bunch of a hundred rusty keys, that clattered in her shaking hands. Then she sat down and proceeded to oil them all. Before she had finished, out from the tub, the water of which had kept on a slow motion ever since she had ceased stirring it, came the head and half the body of a huge gray snake. But the witch did not look round. It grew out of the tub, waving itself backwards and forwards with a slow horizontal motion, till it reached the princess, when it laid its head upon her shoulder, and gave a low hiss in her ear. She started—but with joy; and seeing the head resting on her shoulder, drew it towards her and kissed it. Then she drew it all out of the tub, and wound it round her body. It was one of those dreadful creatures which few have ever beheld—the White Snakes of Darkness.

Then she took the keys and went down to her cellar; and as she unlocked the door she said to herself,—

'This *is* worth living for!'

Locking the door behind her, she descended a few steps into the cellar, and crossing it, unlocked another door into a dark, narrow passage. She locked this also behind her, and descended a few more steps. If any one had followed the witch-princess, he would have heard her unlock exactly one hundred doors, and descend a few steps after unlocking each. When she had unlocked the last, she entered a vast cave, the roof of which was supported by huge natural pillars of rock. Now this roof was the under side of the bottom of the lake.

She then untwined the snake from her body, and held it by the tail high above her. The hideous creature stretched up its head towards the roof of the cavern, which it was just able to reach. It then began to move its head backwards and forwards, with a slow oscillating motion, as if looking for something. At the same moment the witch began to walk round and round the cavern, coming nearer to the centre every circuit; while the head of the snake described the same path over the roof that she did over the floor, for she kept holding it up. And still it kept slowly oscillating. Round and round the cavern they went, ever lessening the circuit, till at last the snake made a sudden dart, and clung to the roof with its mouth.

'That's right, my beauty!' cried the princess; 'drain it dry.'

She let it go, left it hanging, and sat down on a great stone, with her black cat, which had followed her all round the cave, by her side. Then she began to knit and mutter awful words. The snake hung like a huge leech, sucking at the stone; the cat stood with his back arched, and his tail like a piece of cable, looking up at the snake; and the old woman sat and knitted and muttered. Seven days and seven nights they remained thus; when suddenly the serpent dropped from the roof as if exhausted, and shrivelled up till it was again like a piece of dried seaweed. The witch started to her feet, picked it up, put it in her pocket, and looked up at the roof. One drop of water was trembling on the spot where the snake had been sucking. As soon as she saw that, she turned and fled, followed by her cat. Shutting the door in a terrible hurry, she locked it, and having muttered some frightful words, sped to the next, which also she locked and muttered over; and so with all the hundred doors, till she arrived in her own cellar. Then she sat down on the floor ready to faint, but listening with malicious delight to the rushing of the water, which she could hear distinctly through all the hundred doors.

But this was not enough. Now that she had tasted revenge, she lost her patience. Without further measures, the lake would be too long in disappearing. So the next night, with the last shred of the dying old moon rising, she took some of the water in which she had revived the snake, put it in a bottle, and set out, accompanied by her cat. Before morning she had made the entire circuit of the lake, muttering fearful words as she crossed every stream, and casting into it some of the water out of her bottle. When she had finished the circuit she muttered yet again, and flung a handful of water towards the moon. Thereupon every spring in the country ceased to throb and bubble, dying away like the pulse of a dying man. The next day there was no sound of falling water to be heard along the borders of the lake. The very courses were dry; and the mountains showed no silvery streaks down their dark sides. And not alone had the fountains of mother Earth ceased to flow; for all the babies throughout the country were crying dreadfully—only without tears.

12. *Where Is the Prince?*

Never since the night when the princess left him so abruptly had the prince had a single interview with her. He had seen her once or twice in the lake; but as far as he could discover, she had not been in it any more at night. He had sat and sung, and looked in vain for his Nereid; while she, like a true Nereid, was wasting away with her lake, sinking as it sank, withering as it dried. When at length he discovered the change that was taking place in the level of the water, he was in great alarm and perplexity. He could not tell whether the lake was dying because the lady had forsaken it; or whether the lady would not come because the lake had begun to sink. But he resolved to know so much at least.

He disguised himself, and, going to the palace, requested to see the lord chamberlain. His appearance at once gained his request; and the lord chamberlain, being a man of some insight, perceived that there was more in the prince's solicitation than met the ear. He felt likewise that no one could tell whence a solution of the present difficulties might arise. So he granted the prince's prayer to be made shoeblack to the princess. It was rather cunning in the prince to request such an easy post, for the princess could not possibly soil as many shoes as other princesses.

He soon learned all that could be told about the princess. He went nearly distracted; but after roaming about the lake for days, and diving in every depth that remained, all that he could do was to put an extra polish on the dainty pair of boots that was never called for.

For the princess kept her room, with the curtains drawn to shut out the dying lake. But she could not shut it out of her mind for a moment. It haunted her imagination so that she felt as if the lake were her soul, drying up within her, first to mud, then to madness and death. She thus brooded over the change, with all its dreadful accompaniments, till she was nearly distracted. As for the prince, she had forgotten him. However much she had enjoyed his company in the water, she did not care for him without it. But she seemed to have forgotten her father and mother too.

The lake went on sinking. Small slimy spots began to appear, which glittered steadily amidst the changeful shine of the water. These grew to broad patches of mud, which widened and spread, with rocks here and there, and floundering fishes and crawling eels swarming. The people went everywhere catching these, and looking for anything that might have dropped from the royal boats.

At length the lake was all but gone, only a few of the deepest pools remaining unexhausted.

It happened one day that a party of youngsters found themselves on the brink of one of these pools in the very centre of the lake. It was a rocky basin of considerable depth. Looking in, they saw at the bottom something that shone yellow in the sun. A little boy jumped in and dived for it. It was a plate of gold covered with writing. They carried it to the king. On one side of it stood these words:—

> 'Death alone from death can save.
> Love is death, and so is brave—
> Love can fill the deepest grave.
> Love loves on beneath the wave.'

Now this was enigmatical enough to the king and courtiers. But the reverse of the plate explained it a little. Its writing amounted to this:—

'If the lake should disappear, they must find the hole through which the water ran. But it would be useless to try to stop it by any ordinary means. There was but one effectual mode.—The body of a living man could alone stanch the flow. The man must give himself of his own will; and the lake must take his life as it filled. Otherwise the offering would be of no avail. If the nation could not provide one hero, it was time it should perish.'

13. *Here I Am*

This was a very disheartening revelation to the king—not that he was unwilling to sacrifice a subject, but that he was hopeless of finding a man willing to sacrifice himself. No time was to be lost, however, for the princess was lying motionless on her bed, and taking no nourishment but lake-water, which was now none of the best. Therefore the king caused the contents of the wonderful plate of gold to be published throughout the country.

No one, however, came forward.

The prince, having gone several days' journey into the forest, to consult a hermit whom he had met there on his way to Lagobel, knew nothing of the oracle till his return.

When he had acquainted himself with all the particulars, he sat down and thought,—

'She will die if I don't do it, and life would be nothing to me without her; so I shall lose nothing by doing it. And life will be as pleasant to her as ever, for she will soon forget me. And there will be so much more beauty and happiness in the world!—To be sure, I shall not see it.' (Here the poor prince gave a sigh.) 'How lovely the lake will be in the moonlight, with that glorious creature sporting in it like a wild goddess!—It is rather hard to be drowned by inches, though. Let me see—that will be seventy inches of me to drown.' (Here he tried to laugh, but could not.) 'The longer the better, however,' he resumed: 'for can I not bargain that the princess shall be beside me all the time? So I shall see her once more, kiss her perhaps,—who knows? and die looking in her eyes. It will be no death. At least, I shall not feel it. And to see the lake filling for the beauty again!—All right! I am ready.'

He kissed the princess's boot, laid it down, and hurried to the king's apartment. But feeling, as he went, that anything sentimental would be disagreeable, he resolved to carry off the whole affair with nonchalance. So he knocked at the door of the king's counting-house, where it was all but a capital crime to disturb him.

When the king heard the knock he started up, and opened the door in a rage. Seeing only the shoeblack, he drew his sword. This, I am sorry to say, was his usual mode of asserting his regality when he thought his dignity was in danger. But the prince was not in the least alarmed.

'Please your Majesty, I'm your butler,' said he.

'My butler! you lying rascal! What do you mean?'

'I mean, I will cork your big bottle.'

'Is the fellow mad?' bawled the king, raising the point of his sword.

'I will put a stopper—plug—what you call it, in your leaky lake, grand monarch,' said the prince.

The king was in such a rage that before he could speak he had time to cool, and to reflect that it would be great waste to kill the only man who was willing to be useful in the present emergency, seeing that in the end the insolent fellow would be as dead as if he had died by his Majesty's own hand.

'Oh!' said he at last, putting up his sword with difficulty, it was so long; 'I am obliged to you, you young fool! Take a glass of wine?'

'No, thank you,' replied the prince.

'Very well,' said the king. 'Would you like to run and see your parents before you make your experiment?'

'No, thank you,' said the prince.

'Then we will go and look for the hole at once,' said his Majesty, and proceeded to call some attendants.

'Stop, please your Majesty; I have a condition to make,' interposed the prince.

'What!' exclaimed the king, 'a condition! and with me! How dare you?'

'As you please,' returned the prince, coolly. 'I wish your Majesty a good morning.'

'You wretch! I will have you put in a sack, and stuck in the hole.'

'Very well, your Majesty,' replied the prince, becoming a little more respectful, lest the wrath of the king should deprive him of the pleasure of dying for the princess. 'But what good will that do your Majesty? Please to remember that the oracle says the victim must offer himself.'

'Well, you *have* offered yourself,' retorted the king.

'Yes, upon one condition.'

'Condition again!' roared the king, once more drawing his sword. 'Begone! Somebody else will be glad enough to take the honour off your shoulders.'

'Your Majesty knows it will not be easy to get another to take my place.'

'Well, what is your condition?' growled the king, feeling that the prince was right.

'Only this,' replied the prince: 'that, as I must on no account die before I am fairly drowned, and the waiting will be rather wearisome, the princess, your daughter, shall go with me, feed me with her own hands, and look at me now and then to comfort me; for you must confess it *is* rather hard. As soon as the water is up to my eyes, she may go and be happy, and forget her poor shoeblack.'

Here the prince's voice faltered, and he very nearly grew sentimental, in spite of his resolution.

'Why didn't you tell me before what your condition was? Such a fuss about nothing!' exclaimed the king.

'Do you grant it?' persisted the prince.

'Of course I do,' replied the king.

'Very well. I am ready.'

'Go and have some dinner, then, while I set my people to find the place.'

The king ordered out his guards, and gave directions to the officers to find the hole in the lake at once. So the bed of the lake was marked out in divisions and thoroughly examined, and in an hour or so the hole was discovered. It was in the middle of a stone, near the centre of the lake, in the very pool where the golden plate had been found. It was a three-cornered hole of no great size. There was water all round the stone, but very little was flowing through the hole.

14. *This Is Very Kind of You*

The prince went to dress for the occasion, for he was resolved to die like a prince.

When the princess heard that a man had offered to die for her, she was so transported that she jumped off the bed, feeble as she was, and danced about the room for joy. She did not care who the man was; that was nothing to her. The hole wanted stopping; and if only a man would do, why, take one. In an hour or two more everything was ready. Her maid dressed her in haste, and they carried her to the side of the lake. When she saw it she shrieked, and covered her face with her hands. They bore her across to the stone where they had already placed a little boat for her.

The water was not deep enough to float it, but they hoped it would be, before long. They laid her on cushions, placed in the boat wines and fruits and other nice things, and stretched a canopy over all.

In a few minutes the prince appeared. The princess recognized him at once, but did not think it worth while to acknowledge him.

'Here I am,' said the prince. 'Put me in.'

'They told me it was a shoeblack,' said the princess.

'So I am,' said the prince. 'I blacked your little boots three times a day, because they were all I could get of you. Put me in.'

The courtiers did not resent his bluntness, except by saying to each other that he was taking it out in impudence.

But how was he to be put in? The golden plate contained no instructions on this point. The prince looked at the hole, and saw but one way. He put both his legs into it, sitting on the stone, and, stooping forward, covered the corner that remained open with his two hands. In this uncomfortable position he resolved to abide his fate, and turning to the people, said,—

'Now you can go.'

The king had already gone home to dinner.

'Now you can go,' repeated the princess after him, like a parrot.

The people obeyed her and went.

Presently a little wave flowed over the stone, and wetted one of the prince's knees. But he did not mind it much. He began to sing, and the song he sang was this:—

'As a world that has no well,
Darkly bright in forest dell;
As a world without the gleam
Of the downward-going stream;
As a world without the glance
Of the ocean's fair expanse;
As a world where never rain
Glittered on the sunny plain;—
Such, my heart, thy world would be,
If no love did flow in thee.

As a world without the sound
Of the rivulets underground;
Or the bubbling of the spring
Out of darkness wandering;
Or the mighty rush and flowing
Of the river's downward going;
Or the music-showers that drop
On the outspread beech's top;
Or the ocean's mighty voice,
When his lifted waves rejoice;—
Such, my soul, thy world would be,
If no love did sing in thee.

Lady, keep thy world's delight;
Keep the waters in thy sight.
Love hath made me strong to go,
For thy sake, to realms below,
Where the water's shine and hum
Through the darkness never come;
Let, I pray, one thought of me
Spring, a little well, in thee;
Lest thy loveless soul be found
Like a dry and thirsty ground.'

'Sing again, prince. It makes it less tedious,' said the princess.

But the prince was too much overcome to sing any more, and a long pause followed.

'This is very kind of you, prince,' said the princess at last, quite coolly, as she lay in the boat with her eyes shut.

'I am sorry I can't return the compliment,' thought the prince; 'but you are worth dying for, after all.'

Again a wavelet, and another, and another flowed over the stone, and wetted both the prince's knees; but he did not speak or move. Two—three—four hours passed in this way, the princess apparently asleep, and the prince very patient. But he was much disappointed in his position, for he had none of the consolation he had hoped for.

As last he could bear it no longer.

'Princess!' said he.

But at the moment up started the princess, crying,—

'I'm afloat! I'm afloat!'

And the little boat bumped against the stone.

'Princess!' repeated the prince, encouraged by seeing her wide awake and looking eagerly at the water.

'Well?' said she, without looking round.

'Your papa promised that you should look at me, and you haven't looked at me once.'

'Did he? Then I suppose I must. But I am so sleepy!'

'Sleep then, darling, and don't mind me,' said the poor prince.

'Really, you are very good,' replied the princess. 'I think I will go to sleep again.'

'Just give me a glass of wine and a biscuit first,' said the prince, very humbly.

'With all my heart,' said the princess, and gaped as she said it.

She got the wine and the biscuit, however, and leaning over the side of the boat towards him, was compelled to look at him.

'Why, prince,' she said, 'you don't look well! Are you sure you don't mind it?'

'Not a bit,' answered he, feeling very faint indeed. 'Only I shall die before it is of any use to you, unless I have something to eat.'

'There, then,' said she, holding out the wine to him.

'Ah! you must feed me. I dare not move my hands. The water would run away directly.'

'Good gracious!' said the princess; and she began at once to feed him with bits of biscuit and sips of wine.

As she fed him, he contrived to kiss the tips of her fingers now and then. She did not seem to mind it, one way or the other. But the prince felt better.

'Now for your own sake, princess,' said he, 'I cannot let you go to sleep. You must sit and look at me, else I shall not be able to keep up.'

'Well, I will do anything I can to oblige you,' answered she, with condescension; and, sitting down, she did look at him, and kept looking at him with wonderful steadiness, considering all things.

The sun went down, and the moon rose, and, gush after gush, the waters were rising up the prince's body. They were up to his waist now.

'Why can't we go and have a swim?' said the princess. 'There seems to be water enough just about here.'

'I shall never swim more,' said the prince.

'Oh, I forgot,' said the princess, and was silent.

So the water grew and grew, and rose up and up on the prince. And the princess sat and looked at him. She fed him now and then. The night wore on. The waters rose and rose. The moon rose likewise higher and higher, and shone full on the face of the dying prince. The water was up to his neck.

'Will you kiss me, princess?' said he, feebly. The nonchalance was all gone now.

'Yes, I will,' answered the princess, and kissed him with a long, sweet, cold kiss.

'Now,' said he, with a sigh of content, 'I die happy.'

He did not speak again. The princess gave him some wine for the last time: he was past eating. Then she sat down again, and looked at him. The water rose and rose. It touched his chin. It touched his lower lip. It touched between his lips. He shut them hard to keep it out. The princess began to feel strange. It touched his upper lip. He breathed through his nostrils. The princess looked wild. It covered his nostrils. Her eyes looked scared, and shone strange in the moonlight. His head fell back; the water closed over it, and the bubbles of his last breath bubbled up through the water. The princess gave a shriek, and sprang into the lake.

She laid hold first of one leg, and then of the other, and pulled and tugged, but she could not move either. She stopped to take breath, and that made her think that he could not get any breath. She was frantic. She got hold of him, and held his head above the water, which was possible now his hands were no longer on the hole. But it was of no use, for he was past breathing.

Love and water brought back all her strength. She got under the water, and pulled and pulled with her whole might, till at last she got one leg out. The other easily followed. How she got him into the boat she never could tell; but when she did, she fainted away. Coming to herself, she seized the oars, kept herself steady as best she could, and rowed and rowed, though she had never rowed before. Round rocks, and over shallows, and through mud she rowed, till she got to the landing-stairs of the palace. By this time her people were on the shore, for they had heard her shriek. She made them carry the prince to her own room, and lay him in her bed, and light a fire, and send for the doctors.

'But the lake, your Highness!' said the chamberlain, who, roused by the noise, came in, in his nightcap.

'Go and drown yourself in it!' she said.

This was the last rudeness of which the princess was ever guilty; and one must allow that she had good cause to feel provoked with the lord chamberlain.

Had it been the king himself, he would have fared no better. But both he and the queen were fast asleep. And the chamberlain went back to his bed. Somehow, the doctors never came. So the princess and her old nurse were left with the prince. But the old nurse was a wise woman, and knew what to do.

They tried everything for a long time without success. The princess was nearly distracted between hope and fear, but she tried on and on, one thing after another, and everything over and over again.

At last, when they had all but given it up, just as the sun rose, the prince opened his eyes.

15. *Look at the Rain!*

The princess burst into a passion of tears, and *fell* on the floor. There she lay for an hour, and her tears never ceased. All the pent-up crying of her life was spent now. And a rain came on, such as had never been seen in that country. The sun shone all the time, and the great drops, which fell straight to the earth, shone likewise. The palace was in the heart of a rainbow. It was a rain of rubies, and sapphires, and emeralds, and topazes. The torrents poured from the mountains like molten gold; and if it had not been for its subterraneous outlet, the lake would have overflowed and inundated the country. It was full from shore to shore.

But the princess did not heed the lake. She lay on the floor and wept, and this rain within doors was far more wonderful than the rain out of doors. For when it abated a little, and she proceeded to rise, she found, to her astonishment, that she could not. At length, after many efforts, she succeeded in getting upon her feet. But she tumbled down again directly. Hearing her fall, her old nurse uttered a yell of delight, and ran to her, screaming,—

'My darling child! she's found her gravity!'

'Oh, that's it! is it?' said the princess, rubbing her shoulder and her knee alternately. 'I consider it very unpleasant. I feel as if I should be crushed to pieces.'

'Hurrah!' cried the prince from the bed. 'If you've come round, princess, so have I. How's the lake?'

'Brimful,' answered the nurse.

'Then we're all happy.'

'That we are indeed!' answered the princess, sobbing.

And there was rejoicing all over the country that rainy day. Even the babies forgot their past troubles, and danced and crowed amazingly. And the king told stories, and the queen listened to them. And he divided the money in his box, and she the honey in her pot, among all the children. And there was such jubilation as was never heard of before.

Of course the prince and princess were betrothed at once. But the princess had to learn to walk, before they could be married with any propriety. And this was not so easy at her time of life, for she could walk no more than a baby. She was always falling down and hurting herself.

'Is this the gravity you used to make so much of?' said she one day to the prince, as he raised her from the floor. 'For my part, I was a great deal more comfortable without it.'

'No, no, that's not it. This is it,' replied the prince, as he took her up, and carried her about like a baby, kissing her all the time. 'This is gravity.'

'That's better,' said she. 'I don't mind that so much.'

And she smiled the sweetest, loveliest smile in the prince's face. And she gave him one little kiss in return for all his; and he thought them overpaid, for he was beside himself with delight. I fear she complained of her gravity more than once after this, notwithstanding.

It was a long time before she got reconciled to walking. But the pain of learning it was quite counterbalanced by two things, either of which would have been sufficient consolation. The first was, that the prince himself was her teacher; and the second, that she could tumble into the lake as often as she pleased. Still, she preferred to have the prince jump in with her; and the splash they made before was nothing to the splash they made now.

The lake never sank again. In process of time, it wore the roof of the cavern quite through, and was twice as deep as before.

The only revenge the princess took upon her aunt was to tread pretty hard on her gouty toe the next time she saw her. But she was sorry for it the very next day, when she heard that the water had undermined her house, and that it had fallen in the night, burying her in its ruins; whence no one ever ventured to dig up her body. There she lies to this day.

So the prince and princess lived and were happy; and had crowns of gold, and clothes of cloth, and shoes of leather, and children of boys and girls, not one of whom was ever known, on the most critical occasion, to lose the smallest atom of his or her due proportion of gravity.

1864

CHARLES
DICKENS

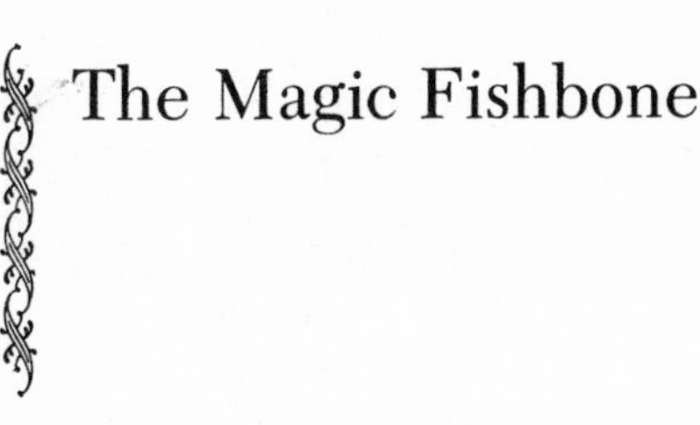

The Magic Fishbone

THERE was once a King, and he had a Queen, and he was the manliest of his sex, and she was the loveliest of hers. The King was, in his private profession, Under Government. The Queen's father had been a medical man out of town.

They had nineteen children, and were always having more. Seventeen of these children took care of the baby, and Alicia, the eldest, took care of them all. Their ages varied from seven years to seven months.

Let us now resume our story.

One day the King was going to the Office, when he stopped at the fishmonger's to buy a pound and a half of salmon not too near the tail, which the Queen (who was a careful housekeeper) had requested him to send home. Mr Pickles, the fishmonger, said, 'Certainly, sir, is there any other article, good morning.'

The King went on towards the Office in a melancholy mood, for Quarter Day was such a long way off, and several of the dear children were growing out of their clothes. He had not proceeded far, when Mr Pickles's errand-boy came running after him, and said, 'Sir, you didn't notice the old lady in our shop.'

'What old lady?' inquired the King. 'I saw none.'

Now, the King had not seen any old lady, because this old lady had been invisible to him, though visible to Mr Pickles's boy. Probably because he messed and splashed the water about to that degree, and flopped the pairs of soles down in that violent manner, that, if she had not been visible to him, he would have spoilt her clothes.

Just then the old lady came trotting up. She was dressed in shot-silk of the richest quality, smelling of dried lavender.

'King Watkins the First, I believe?' said the old lady.

'Watkins,' replied the King, 'is my name.'

'Papa, if I am not mistaken, of the beautiful Princess Alicia?' said the old lady.

'And of eighteen other darlings,' replied the King.

'Listen. You are going to the Office,' said the old lady.

It instantly flashed upon the King that she must be a Fairy, or how could she know that?

'You are right,' said the old lady, answering his thoughts, 'I am the Good Fairy Grandmarina. Attend. When you return home to dinner, politely invite the Princess Alicia to have some of the salmon you bought just now.'

'It may disagree with her,' said the King.

The old lady became so very angry at this absurd idea, that the King was quite alarmed, and humbly begged her pardon.

'We hear a great deal too much about this thing disagreeing and that thing disagreeing,' said the old lady, with the greatest contempt it was possible to express. 'Don't be greedy. I think you want it all yourself.'

The King hung his head under this reproof, and said he wouldn't talk about things disagreeing, any more.

'Be good then,' said the Fairy Grandmarina, 'and don't! When the beautiful Princess Alicia consents to partake of the salmon—as I think she will—you will find she will leave a fish-bone on her plate. Tell her to dry it, and to rub it, and to polish it till it shines like mother-of-pearl, and to take care of it as a present from me.'

'Is that all?' asked the King.

'Don't be impatient, sir,' returned the Fairy Grandmarina, scolding him severely. 'Don't catch people short, before they have done speaking. Just the way with you grown-up persons. You are always doing it.'

The King again hung his head, and said he wouldn't do so any more.

'Be good then,' said the Fairy Grandmarina, 'and don't! Tell the Princess Alicia, with my love, that the fish-bone is a magic present which can only be used once; but that it will bring her, that once, whatever she wishes for, PROVIDED SHE WISHES FOR IT AT THE RIGHT TIME. That is the message. Take care of it.'

The King was beginning, 'Might I ask the reason—?' When the Fairy became absolutely furious.

'*Will* you be good, sir?' she exclaimed, stamping her foot on the ground. 'The reason for this, and the reason for that, indeed! You are always wanting the reason. No reason. There! Hoity toity me! I am sick of your grown-up reasons.'

The King was extremely frightened by the old lady's flying into such a passion, and said he was very sorry to have offended her, and he wouldn't ask for reasons any more.

'Be good then,' said the old lady, 'and don't!'

With those words, Grandmarina vanished, and the King went on and on and on, till he came to the Office. There he wrote and wrote and wrote, till it was time to go home again. Then he politely invited the Princess Alicia, as the Fairy had directed him, to partake of the salmon. And when she had enjoyed it very much, he saw the fish-bone on her plate, as the Fairy had told him he would, and he delivered the Fairy's message, and the Princess Alicia took care to dry the bone, and to rub it, and to polish it till it shone like mother-of-pearl.

And so when the Queen was going to get up in the morning, she said, 'O dear me, dear me, my head, my head!' And then she fainted away.

The Princess Alicia, who happened to be looking in at the chamber door, asking about breakfast, was very much alarmed when she saw her Royal Mamma in this state, and she rang the bell for Peggy,—which was the name of the Lord Chamberlain. But remembering where the smelling-bottle was, she climbed on a chair and got it, and after that she climbed on another chair by the bedside and held the smelling-bottle to the Queen's nose, and after that she jumped down and got some water, and after that she jumped up again and wetted the Queen's forehead, and, in short, when the Lord Chamberlain came in, that dear old woman said to the little Princess, 'What a Trot you are! I couldn't have done it better myself!'

But that was not the worst of the good Queen's illness. O no! She was very ill indeed, for a long time. The Princess Alicia kept the seventeen young Princes and Princesses quiet, and dressed and undressed and danced the baby, and made the kettle boil, and heated the soup, and swept the hearth, and poured out the medicine, and nursed the Queen, and did all that ever she could, and was as busy busy busy, as busy could be. For there were not many servants at that Palace, for three reasons; because the King was short of money, because a rise in his office never seemed to come, and because quarter-day was so far off that it looked almost as far off and as little as one of the stars.

But on the morning when the Queen fainted away, where was the magic fish-bone? Why, there it was in the Princess Alicia's pocket. She had almost taken it out to bring the Queen to life again, when she put it back, and looked for the smelling-bottle.

After the Queen had come out of her swoon that morning, and was dozing, the Princess Alicia hurried up stairs to tell a most particular secret to a most particularly confidential friend of hers, who was a Duchess. People did suppose her to be a Doll, but she was really a Duchess, though nobody knew it except the Princess.

This most particular secret was the secret about the magic fish-bone, the history of which was well known to the Duchess, because the Princess told her everything. The Princess kneeled down by the bed on which the Duchess was lying, full dressed and wide-awake, and whispered the secret to her. The Duchess smiled and nodded. People might have supposed that she never smiled and nodded, but she often did, though nobody knew it except the Princess.

Then the Princess Alicia hurried down stairs again, to keep watch in the Queen's room. She often kept watch by herself in the Queen's room; but every evening, while the illness lasted, she sat there watching with the King. And every evening the King sat looking at her with a cross look, wondering why she never brought out the magic fish-bone. As often as she noticed this, she ran up stairs, whispered the secret to the Duchess over again, and said to the Duchess besides, 'They think we children never have a reason or a meaning!' And the Duchess, though the most fashionable Duchess that ever was heard of, winked her eye.

'Alicia,' said the King, one evening when she wished him Good Night.

'Yes, Papa.'

'What is become of the magic fish-bone?'

'In my pocket, Papa.'

'I thought you had lost it?'

'O no, Papa!'

'Or forgotten it?'

'No, indeed, Papa!'

And so another time the dreadful little snapping pug-dog next door made a rush at one of the young Princes as he stood on the steps coming home from school, and terrified him out of his wits, and he put his hand through a pane of glass, and bled bled bled. When the seventeen other young Princes and Princesses saw him bleed bleed bleed, they were terrified out of their wits too, and screamed themselves black in their seventeen faces all at once. But the Princess Alicia put her hands over all their seventeen mouths, one after another, and persuaded them to be quiet because of the sick Queen. And then she put the wounded Prince's hand in a basin of fresh cold water, while they stared with their twice seventeen are thirty-four put down four and carry three eyes, and then she looked in the hand for bits of glass, and there were fortunately no bits of glass there. And then she said to two chubby-legged Princes who were sturdy though small, 'Bring me in the Royal rag-bag; I must snip and stitch and cut and contrive.' So those two young Princes tugged at the Royal rag-bag and lugged it in, and the Princess Alicia sat down on the floor with a large pair of scissors and a needle and thread, and snipped and stitched and cut and contrived, and made a bandage and put it on, and it fitted beautifully, and so when it was all done she saw the King her Papa looking on by the door.

'Alicia.'

'Yes, Papa.'

'What have you been doing?'

'Snipping stitching cutting and contriving, Papa.'

'Where is the magic fish-bone?'

'In my pocket, Papa.'

'I thought you had lost it?'

'O no, Papa!'

'Or forgotten it?'

'No, indeed, Papa!'

After that, she ran up stairs to the Duchess and told her what had passed, and told her the secret over again, and the Duchess shook her flaxen curls and laughed with her rosy lips.

Well! and so another time the baby fell under the grate. The seventeen young Princes and Princesses were used to it, for they were almost always falling under the grate or down the stairs, but the baby was not used to it yet, and it gave him a swelled face and a black eye. The way the poor little darling came to tumble was, that he slid out of the Princess Alicia's lap just as she was sitting, in a great coarse apron that quite smothered her, in front of the kitchen fire, beginning to peel the turnips for the broth for dinner; and the way she came to be doing that was, that the King's cook had run away that morning with her own true love, who was a very tall but very tipsy soldier. Then, the seventeen young Princes and Princesses, who cried at everything that happened, cried and roared. But the Princess Alicia (who couldn't help crying a little herself) quietly called to them to be still, on account of not throwing back the Queen up stairs, who was fast getting well, and said, 'Hold your tongues you wicked little monkeys, every one of you, while I examine baby!' Then she examined baby, and found that he hadn't broken anything, and she held cold iron to his poor dear eye, and smoothed his poor dear face, and he presently fell asleep in her arms. Then she said to the seventeen Princes and Princesses, 'I am afraid to lay him down yet, lest he should wake and feel pain, be good and you shall all be cooks.' They jumped for joy when they heard that, and began making themselves cooks' caps out of old newspapers. So to one she gave the salt-box, and to one she gave the barley, and to one she gave the herbs, and to one she gave the turnips, and to one she gave the carrots, and to one she gave the onions, and to one she gave the spice-box, till they were all cooks, and all running about at work, she sitting in the middle, smothered in the great coarse apron, nursing baby. By and by the broth was done, and the baby woke up, smiling like an angel, and was trusted to the sedatest Princess to hold, while the other Princes and Princesses were squeezed into a far-off corner to look at the Princess Alicia turning out the saucepan-full of broth, for fear (as they were always getting into trouble) they should get splashed and scalded. When the broth came tumbling out, steaming beautifully, and smelling like a nosegay good to eat, they clapped their hands. That made the baby clap his hands; and that, and his looking as if he had a comic toothache, made all the Princes and Princesses laugh. So the Princess Alicia said, 'Laugh and be good, and after dinner we will make him a nest on the floor in a corner, and he shall sit in his nest and see a dance of eighteen cooks.' That delighted the young Princes and Princesses, and they ate up all the broth, and washed up all the plates and dishes, and cleared away, and pushed the table into a corner, and then they in their cooks' caps, and the Princess Alicia in the smothering coarse apron that belonged to the cook that had run away with her own true love that was the very tall but very tipsy soldier, danced a dance of eighteen cooks before the angelic baby, who forgot his swelled face and his black eye, and crowed with joy.

And so then, once more the Princess Alicia saw King Watkins the First, her father, standing in the doorway looking on, and he said: 'What have you been doing, Alicia?'

'Cooking and contriving, Papa.'

'What else have you been doing, Alicia?'

'Keeping the children light-hearted, Papa.'

'Where is the magic fish-bone, Alicia?'

'In my pocket, Papa.'

'I thought you had lost it?'

'O no, Papa.'

'Or forgotten it?'

'No, indeed, Papa.'

The King then sighed so heavily, and seemed so low-spirited, and sat down so miserably, leaning his head upon his hand, and his elbow upon the kitchen table pushed away in the corner, that the seventeen Princes and Princesses crept softly out of the kitchen, and left him alone with the Princess Alicia and the angelic baby.

'What is the matter, Papa?'

'I am dreadfully poor, my child.'

'Have you no money at all, Papa?'

'None, my child.'

'Is there no way left of getting any, Papa?'

'No way,' said the King. 'I have tried very hard, and I have tried all ways.'

When she heard those last words, the Princess Alicia began to put her hand into the pocket where she kept the magic fish-bone.

'Papa,' said she, 'when we have tried very hard, and tried all ways, we must have done our very very best?'

'No doubt, Alicia.'

'When we have done our very very best, Papa, and that is not enough, then I think the right time must have come for asking help of others.' This was the very secret connected with the magic fish-bone, which she had found out for herself from the good fairy Grandmarina's words, and which she had so often whispered to her beautiful and fashionable friend the Duchess.

So she took out of her pocket the magic fish-bone that had been dried and rubbed and polished till it shone like mother-of-pearl, and she gave it one little kiss and wished it was quarter-day. And immediately it *was* Quarter-Day, and the King's quarter's salary came rattling down the chimney, and bounced into the middle of the floor.

But this was not half of what happened, no not a quarter, for immediately afterwards the good fairy Grandmarina came riding in, in a carriage and four (Peacocks), with Mr Pickles's boy up behind, dressed in silver and gold, with a cocked-hat, powdered hair, pink silk stockings, a jewelled cane, and a nosegay. Down jumped Mr Pickles's boy with his cocked-hat in his hand and wonderfully polite (being entirely changed by enchantment), and handed Grandmarina out, and there she stood, in her rich shot-silk smelling of dried lavender, fanning herself with a sparkling fan.

'Alicia, my dear,' said this charming old Fairy, 'how do you do, I hope I see you pretty well, give me a kiss.'

The Princess Alicia embraced her, and then Grandmarina turned to the King, and said rather sharply: 'Are you good?'

The King said he hoped so.

'I suppose you know the reason, *now*, why my god-Daughter here,' kissing the Princess again, 'did not apply to the fish-bone sooner?' said the Fairy.

The King made her a shy bow.

'Ah! But you didn't *then!*' said the Fairy.

The King made her a shyer bow.

'Any more reasons to ask for?' said the Fairy.

The King said no, and he was very sorry.

'Be good then,' said the Fairy, 'and live happy ever afterwards.'

Then, Grandmarina waved her fan, and the Queen came in most splendidly dressed, and the seventeen young Princes and Princesses, no longer grown out of their clothes, came in, newly fitted out from top to toe, with tucks in everything to admit of its being let out. After that, the Fairy tapped the Princess Alicia with her fan, and the smothering coarse apron flew away, and she appeared exquisitely dressed, like a little Bride, with a wreath of orange-flowers, and a silver veil. After that, the kitchen dresser changed of itself into a wardrobe, made of beautiful woods and gold and looking-glass, which was full of dresses of all sorts, all for her and all exactly fitting her. After that, the angelic baby came in, running alone, with his face and eye not a bit the worse but much the better. Then, Grandmarina begged to be introduced to the

Duchess, and when the Duchess was brought down many compliments passed between them.

A little whispering took place between the Fairy and the Duchess, and then the Fairy said out loud, 'Yes. I thought she would have told you.' Grandmarina then turned to the King and Queen, and said, 'We are going in search of Prince Certainpersonio. The pleasure of your company is requested at church in half an hour precisely.' So she and the Princess Alicia got into the carriage, and Mr Pickles's boy handed in the Duchess who sat by herself on the opposite seat, and then Mr Pickles's boy put up the steps and got up behind, and the Peacocks flew away with their tails spread.

Prince Certainpersonio was sitting by himself, eating barley-sugar and waiting to be ninety. When he saw the Peacocks followed by the carriage coming in at the window, it immediately occurred to him that something uncommon was going to happen.

'Prince,' said Grandmarina, 'I bring you your Bride.'

The moment the Fairy said those words, Prince Certainpersonio's face left off being sticky, and his jacket and corduroys changed to peach-bloom velvet, and his hair curled, and a cap and feather flew in like a bird and settled on his head. He got into the carriage by the Fairy's invitation, and there he renewed his acquaintance with the Duchess whom he had seen before.

In the church were the Prince's relations and friends, and the Princess Alicia's relations and friends, and the seventeen Princes and Princesses, and the baby, and a crowd of the neighbors. The marriage was beautiful beyond expression. The Duchess was bridesmaid, and beheld the ceremony from the pulpit where she was supported by the cushion of the desk.

Grandmarina gave a magnificent wedding feast afterwards, in which there was everything and more to eat, and everything and more to drink. The wedding cake was delicately ornamented with white satin ribbons, frosted silver and white lilies, and was forty-two yards round.

When Grandmarina had drunk her love to the young couple, and Prince Certainpersonio had made a speech, and everybody had cried Hip Hip Hip Hurrah! Grandmarina announced to the King and Queen that in future there would be eight Quarter-Days in every year, except in leap-year, when there would be ten. She then turned to Certainpersonio and Alicia, and said, 'My dears, you will have thirty-five children, and they will all be good and beautiful. Seventeen of your children will be boys, and eighteen will be girls. The hair of the whole of your children will curl naturally. They will never have the measles, and will have recovered from the whooping-cough before being born.'

On hearing such good news, everybody cried out 'Hip Hip Hip Hurrah!' again.

'It only remains,' said Grandmarina in conclusion, 'to make an end of the fish-bone.'

So she took it from the hand of the Princess Alicia, and it instantly flew down the throat of the dreadful little snapping pug-dog next door and choked him, and he expired in convulsions.

1868

MARY DE MORGAN

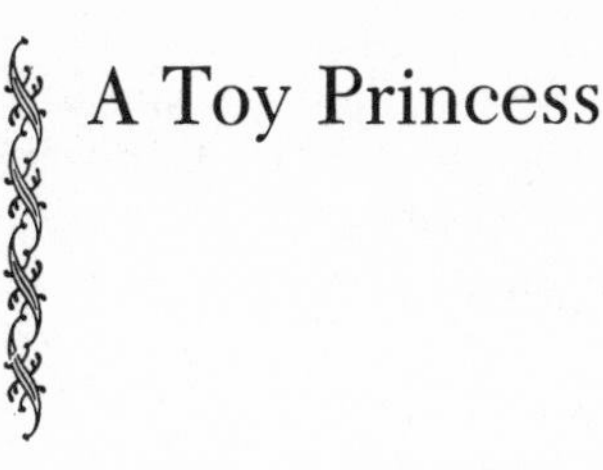

A Toy Princess

More than a thousand years ago, in a country quite on the other side of the world, it fell out that the people all grew so very polite that they hardly ever spoke to each other. And they never said more than was quite necessary, as 'Just so,' 'Yes indeed,' 'Thank you,' and 'If you please.' And it was thought to be the rudest thing in the world for any one to say they liked or disliked, or loved or hated, or were happy or miserable. No one ever laughed aloud, and if any one had been seen to cry they would at once have been avoided by their friends.

The King of this country married a Princess from a neighbouring land, who was very good and beautiful, but the people in her own home were as unlike her husband's people as it was possible to be. They laughed, and talked, and were noisy and merry when they were happy, and cried and lamented if they were sad. In fact, whatever they felt they showed at once, and the Princess was just like them.

So when she came to her new home, she could not at all understand her subjects, or make out why there was no shouting and cheering to welcome her, and why every one was so distant and formal. After a time, when she found they never changed, but were always the same, just as stiff and quiet, she wept, and began to pine for her own old home.

Every day she grew thinner and paler. The courtiers were much too polite to notice how ill their young Queen looked; but she knew it herself, and believed she was going to die.

Now she had a fairy godmother, named Taboret, whom she loved very dearly, and who was always kind to her. When she knew her end was drawing near she sent for her godmother, and when she came had a long talk with her quite alone.

No one knew what was said, and soon afterwards a little Princess was born, and the Queen died. Of course all the courtiers were sorry for the poor Queen's death, but it would have been thought rude to say so. So, although there was a grand funeral, and the court put on mourning, everything else went on much as it had done before.

The little baby was christened Ursula, and given to some court ladies to be taken charge of. Poor little Princess! *She* cried hard enough, and nothing could stop her.

All her ladies were frightened, and said that they had not heard such a dreadful noise for a long time. But, till she was about two years old, nothing could stop her crying when she was cold or hungry, or crowing when she was pleased.

After that she began to understand a little what was meant when her nurses told her, in cold, polite tones, that she was being naughty, and she grew much quieter.

She was a pretty little girl, with a round baby face and big merry blue eyes; but as she grew older, her eyes grew less and less merry and bright, and her fat little face grew thin and pale. She was not allowed to play with any other children, lest she might learn bad manners; and she was not taught any games or given any toys. So she passed most of her time, when she was not at her lessons, looking out of the window at the birds flying against the clear blue sky; and sometimes she would give a sad little sigh when her ladies were not listening.

One day the old fairy Taboret made herself invisible, and flew over to the King's palace to see how things were going on there. She went straight up to the nursery, where she found poor little Ursula sitting by the window, with her head leaning on her hand.

It was a very grand room, but there were no toys or dolls about, and when the fairy saw this, she frowned to herself and shook her head.

'Your Royal Highness's dinner is now ready,' said the head nurse to Ursula.

'I don't want any dinner,' said Ursula, without turning her head.

'I think I have told your Royal Highness before that it is not polite to say you don't want anything, or that you don't like it,' said the nurse. 'We are waiting for your Royal Highness.'

So the Princess got up and went to the dinner-table, and Taboret watched them all the time. When she saw how pale little Ursula was, and how little she ate, and that there was no talking or laughing allowed, she sighed and frowned even more than before, and then she flew back to her fairy home, where she sat for some hours in deep thought.

At last she rose, and went out to pay a visit to the largest shop in Fairyland.

It was a queer sort of shop. It was neither a grocer's, nor a draper's, nor a hatter's. Yet it contained sugar, and dresses, and hats. But the sugar was magic sugar, which transformed any liquid into which it was put; the dresses each had some special charm, and the hats were wishing-caps. It was, in fact, a shop where every sort of spell or charm was sold.

Into this shop Taboret flew; and as she was well known there as a good customer, the master of the shop came forward to meet her at once, and bowing, begged to know what he could get for her.

'I want,' said Taboret, 'a Princess.'

'A Princess!' said the shopman, who was in reality an old wizard. 'What size do you want it? I have one or two in stock.'

'It must look now about six years old. But it must grow.'

'I can make you one,' said the wizard, 'but it'll come rather expensive.'

'I don't mind that,' said Taboret. 'See! I want it to look exactly like this,' and so saying she took a portrait of Ursula out of her bosom and gave it to the old man, who examined it carefully.

'I'll get it for you,' he said. 'When will you want it?'

'As soon as possible,' said Taboret. 'By to-morrow evening if possible. How much will it cost?'

'It'll come to a good deal,' said the wizard, thoughtfully. 'I have such difficulty in getting these things properly made in these days. What sort of a voice is it to have?'

'It need not be at all talkative,' said Taboret, 'so that won't add much to the price. It need only say, "If you please," "No, thank you," "Certainly," and "Just so."'

'Well, under those circumstances,' said the wizard, 'I will do it for four cats' footfalls, two fish's screams, and two swans' songs.'

'It is too much,' cried Taboret. 'I'll give you the footfalls and the screams, but to ask for swans' songs!'

She did not really think it dear, but she always made a point of trying to beat tradesmen down.

'I can't do it for less,' said the wizard, 'and if you think it too much, you'd better try another shop.'

'As I am really in a hurry for it, and cannot spend time in searching about, I suppose I must have it,' said Taboret; 'but I consider the price very high. When will it be ready?'

'By to-morrow evening.'

'Very well, then, be sure it is ready for me by the time I call for it, and whatever you do, don't make it at all noisy or rough in its ways'; and Taboret swept out of the shop and returned to her home.

Next evening she returned and asked if her job was done.

'I will fetch it, and I am sure you will like it,' said the wizard, leaving the shop as he spoke. Presently he came back, leading by the hand a pretty little girl of about six years old—a little girl so like the Princess Ursula that no one could have told them apart.

'Well,' said Taboret, 'it looks well enough. But are you sure that it's a good piece of workmanship, and won't give way anywhere?'

'It's as good a piece of work as ever was done,' said the wizard, proudly, striking the child on the back as he spoke. 'Look at it! Examine it all over, and see if you find a flaw anywhere. There's not one fairy in twenty who could tell it from the real thing, and no mortal could.'

'It seems to be fairly made,' said Taboret, approvingly, as she turned the little girl round. 'Now I'll pay you, and then will be off'; with which she raised her wand in the air and waved it three times, and there arose a series of strange sounds.

The first was a low tramping, the second shrill and piercing screams, the third voices of wonderful beauty, singing a very sorrowful song.

The wizard caught all the sounds and pocketed them at once, and Taboret, without ceremony, picked up the child, took her head downwards under her arm, and flew away.

At court that night the little Princess had been naughty, and had refused to go to bed. It was a long time before her ladies could get her into her crib, and when she was there, she did not really go to sleep, only lay still and pretended, till every one went away; then she got up and stole noiselessly to the window, and sat down on the window-seat all curled up in a little bunch, while she looked out wistfully at the moon. She was such a pretty soft little thing, with all her warm bright hair falling over her shoulders, that it would have been hard for most people to be angry with her. She leaned her chin on her tiny white hands, and as she gazed out, the tears rose to her great blue eyes; but remembering that her ladies would call this naughty, she wiped them hastily away with her nightgown sleeve.

'Ah moon, pretty bright moon!' she said to herself, 'I wonder if they let you cry when you want to. I think I'd like to go up there and live with you; I'm sure it would be nicer than being here.'

'Would you like to go away with me?' said a voice close beside her; and looking up she saw a funny old woman in a red cloak, standing near to her. She was not frightened, for the old woman had a kind smile and bright black eyes, though her nose was hooked and her chin long.

'Where would you take me?' said the little Princess, sucking her thumb, and staring with all her might.

'I'd take you to the sea-shore, where you'd be able to play about on the sands, and where you'd have some little boys and girls to play with, and no one to tell you not to make a noise.'

'I'll go,' cried Ursula, springing up at once.

'Come along,' said the old woman, taking her tenderly in her arms and folding her in her warm red cloak. Then they rose up in the air, and flew out of the window, right away over the tops of the houses.

The night air was sharp, and Ursula soon fell asleep; but still they kept flying on, on, over hill and dale, for miles and miles, away from the palace, towards the sea.

Far away from the court and the palace, in a tiny fishing village, on the sea, was a little hut where a fisherman named Mark lived with his wife and three children. He was a poor man, and lived on the fish he caught in his little boat. The children, Oliver, Philip, and little Bell, were rosy-cheeked and bright-eyed. They played all day long on the shore, and shouted till they were hoarse. To this village the fairy bore the still sleeping Ursula, and gently placed her on the door-step of Mark's cottage; then she kissed her cheeks, and with one gust blew the door open, and disappeared before any one could come to see who it was.

The fisherman and his wife were sitting quietly within. She was making the children clothes, and he was mending his net, when without any noise the door opened and the cold night air blew in.

'Wife,' said the fisherman, 'just see who's at the door.'

The wife got up and went to the door, and there lay Ursula, still sleeping soundly, in her little white nightdress.

The woman gave a little scream at sight of the child, and called to her husband.

'Husband, see, here's a little girl!' and so saying she lifted her in her arms, and carried her into the cottage. When she was brought into the warmth and light, Ursula awoke, and sitting up, stared about her in fright. She did not cry, as another child might have done, but she trembled very much, and was almost too frightened to speak.

Oddly enough, she had forgotten all about her strange flight through the air, and could remember nothing to tell the fisherman and his wife, but that she was the Princess Ursula; and, on hearing this, the good man and woman thought the poor little girl must be a trifle mad. However, when they examined her little nightdress, made of white fine linen and embroidery, with a crown worked in one corner, they agreed that she must belong to very grand people. They said it would be cruel to send the poor little thing away on such a cold night, and they must of course keep her till she was claimed. So the woman gave her some warm bread-and-milk, and put her to bed with their own little girl.

In the morning, when the court ladies came to wake Princess Ursula, they found her sleeping as usual in her little bed, and little did they think it was not she, but a toy Princess placed there in her stead. Indeed the ladies were much pleased; for when they said, 'It is time for your Royal Highness to arise,' she only answered, 'Certainly,' and let herself be dressed without another word. And as the time passed, and she was never naughty, and scarcely ever spoke, all said she was vastly improved, and she grew to be a great favourite.

The ladies all said that the young Princess bid fair to have the most elegant manners in the country, and the King smiled and noticed her with pleasure.

In the meantime, in the fisherman's cottage far away, the real Ursula grew tall and straight as an alder, and merry and light-hearted as a bird.

No one came to claim her, so the good fisherman and his wife kept her and brought her up among their own little ones. She played with them on the beach, and learned her lessons with them at school, and her old life had become like a dream she barely remembered.

But sometimes the mother would take out the little embroidered nightgown and show it to her, and wonder whence she came, and to whom she belonged.

'I don't care who I belong to,' said Ursula; 'they won't come and take me from you, and that's all I care about.' So she grew tall and fair, and as she grew, the toy Princess, in her place at the court, grew too, and always was just like her, only that whereas Ursula's face was sunburnt and her cheeks red, the face of the toy Princess was pale, with only a very slight tint in her cheeks.

Years passed, and Ursula at the cottage was a tall young woman, and Ursula at the court was thought to be the most beautiful there, and every one admired her manners, though she never said anything but 'If you please,' 'No, thank you,' 'Certainly,' and 'Just so.'

The King was now an old man, and the fisherman Mark and his wife were grey-headed. Most of their fishing was now done by their eldest son, Oliver, who was their great pride. Ursula waited on them, and cleaned the house, and did the needlework, and was so useful that they could not have done without her. The fairy Taboret had come to the cottage from time to time, unseen by any one, to see Ursula, and always finding her healthy and merry, was pleased to think of how she had saved her from a dreadful life. But one evening when she paid them a visit, not having been there for some time, she saw something which made her pause and consider. Oliver and Ursula were standing together watching the waves, and Taboret stopped to hear what they said,—

'When we are married,' said Oliver, softly, 'we will live in that little cottage yonder, so that we can come and see them every day. But that will not be till little Bell is old enough to take your place, for how would my mother do without you?'

'And we had better not tell them,' said Ursula, 'that we mean to marry, or else the thought that they are preventing us will make them unhappy.'

When Taboret heard this she became grave, and pondered for a long time. At last she flew back to the court to see how things were going on there. She found the King in the middle of a state council. On seeing this, she at once made herself visible, when the King begged her to be seated near him, as he was always glad of her help and advice.

'You find us,' said his Majesty, 'just about to resign our sceptre into younger and more vigorous hands; in fact, we think we are growing too old to reign, and mean to abdicate in favour of our dear daughter, who will reign in our stead.'

'Before you do any such thing,' said Taboret, 'just let me have a little private conversation with you'; and she led the King into a corner, much to his surprise and alarm.

In about half an hour he returned to the council, looking very white, and with a dreadful expression on his face, whilst he held a handkerchief to his eyes.

'My lords,' he faltered, 'pray pardon our apparently extraordinary behaviour. We have just received a dreadful blow; we hear on authority, which we cannot doubt, that our dear, dear daughter'—here sobs choked his voice, and he was almost unable to proceed—'is—is—in fact, not our daughter at all, and only a *sham*.' Here the King sank back in his chair, overpowered with grief, and the fairy Taboret, stepping to the front, told the courtiers the whole story; how she had stolen the real Princess, because she feared they were spoiling her, and how she had placed a toy Princess in her place. The courtiers looked from one to another in surprise, but it was evident they did not believe her.

'The Princess is a truly charming young lady,' said the Prime Minister.

'Has your Majesty any reason to complain of her Royal Highness's conduct?' asked the old Chancellor.

'None whatever,' sobbed the King; 'she was ever an excellent daughter.'

'Then I don't see,' said the Chancellor, 'what reason your Majesty can have for paying any attention to what this—this person says.'

'If you don't believe me, you old idiots,' cried Taboret, 'call the Princess here, and I'll soon prove my words.'

'By all means,' cried they.

So the King commanded that her Royal Highness should be summoned.

In a few minutes she came, attended by her ladies. She said nothing, but then she never did speak till she was spoken to. So she entered, and stood in the middle of the room silently.

'We have desired that your presence be requested,' the King was beginning, but Taboret without any ceremony advanced towards her, and struck her lightly on the head with her wand. In a moment the head rolled on the floor, leaving the body standing motionless as before, and showing that it was but an empty shell. 'Just so,' said the head, as it rolled towards the King, and he and the courtiers nearly swooned with fear.

When they were a little recovered, the King spoke again. 'The fairy tells me,' he said, 'that there is somewhere a real Princess whom she wishes us to adopt as our daughter. And in the meantime let her Royal Highness be carefully placed in a cupboard, and a general mourning be proclaimed for this dire event.'

So saying he glanced tenderly at the body and head, and turned weeping away.

So it was settled that Taboret was to fetch Princess Ursula, and the King and council were to be assembled to meet her.

That evening the fairy flew to Mark's cottage, and told them the whole truth about Ursula, and that they must part from her:

Loud were their lamentations, and great their grief, when they heard she must leave them. Poor Ursula herself sobbed bitterly.

'Never mind,' she cried after a time, 'if I am really a great Princess, I will have you all to live with me. I am sure the King, my father, will wish it, when he hears how good you have all been to me.'

On the appointed day, Taboret came for Ursula in a grand coach and four, and drove her away to the court. It was a long, long drive; and she stopped on the way and had the Princess dressed in a splendid white silk dress trimmed with gold, and put pearls round her neck and in her hair, that she might appear properly at court.

The King and all the council were assembled with great pomp, to greet their new Princess, and all looked grave and anxious. At last the door opened, and Taboret appeared, leading the young girl by the hand.

'That is your father!' said she to Ursula, pointing to the King; and on this, Ursula, needing no other bidding, ran at once to him, and putting her arms round his neck, gave him a sounding kiss.

His Majesty almost swooned, and all the courtiers shut their eyes and shivered.

'This is really!' said one.

'This is truly!' said another.

'What have I done?' cried Ursula, looking from one to another, and seeing that something was wrong, but not knowing what.

'Have I kissed the *wrong person*?' On hearing which every one groaned.

'Come now,' cried Taboret, 'if you don't like her, I shall take her away to those who do. I'll give you a week, and then I'll come back and see how you're treating her. She's a great deal too good for any of you.' So saying she flew away on her wand, leaving Ursula to get on with her new friends as best she might. But Ursula could not get on with them at all, as she soon began to see.

If she spoke or moved they looked shocked, and at last she was so frightened and troubled by them that she burst into tears, at which they were more shocked still.

'This is indeed a change after our sweet Princess,' said one lady to another.

'Yes, indeed,' was the answer, 'when one remembers how even after her head was struck off she behaved so beautifully, and only said, "Just so."'

And all the ladies disliked poor Ursula, and soon showed her their dislike. Before the end of the week, when Taboret was to return, she had grown quite thin and pale, and seemed afraid of speaking above a whisper.

'Why, what is wrong?' cried Taboret, when she returned and saw how much poor Ursula had changed. 'Don't you like being here? Aren't they kind to you?'

'Take me back, dear Taboret,' cried Ursula, weeping. 'Take me back to Oliver, and Philip, and Bell. As for these people, I *hate* them.'

And she wept again.

Taboret only smiled and patted her head, and then went into the King and courtiers.

'Now, how is it,' she cried, 'I find the Princess Ursula in tears? and I am sure you are making her unhappy. When you had that bit of wood-and-leather Princess, you could behave well enough to it, but now that you have a real flesh-and-blood woman, you none of you care for her.'

'Our late dear daughter—' began the King, when the fairy interrupted him.

'I do believe,' she said, 'that you would like to have the doll back again. Now I will give you your choice. Which will you have—my Princess Ursula, the real one, or your Princess Ursula, the sham?'

The King sank back into his chair. 'I am not equal to this,' he said: 'summon the council, and let them settle it by vote.' So the council were summoned, and the fairy explained to them why they were wanted.

'Let both Princesses be fetched,' she said; and the toy Princess was brought in with great care from her cupboard, and her head stood on the table beside her, and the real Princess came in with her eyes still red from crying and her bosom heaving.

'I should think there could be no doubt which one would prefer,' said the Prime Minister to the Chancellor.

'I should think not either,' answered the Chancellor.

'Then vote,' said Taboret; and they all voted, and every vote was for the sham Ursula, and not one for the real one. Taboret only laughed.

'You are a pack of sillies and idiots,' she said, 'but you shall have what you want'; and she picked up the head, and with a wave of her wand stuck it on to the body, and it moved round slowly and said, 'Certainly,' just in its old voice; and on hearing this, all the courtiers gave something as like a cheer as they thought polite, whilst the old King could not speak for joy.

'We will,' he cried, 'at once make our arrangements for abdicating and leaving the government in the hands of our dear daughter'; and on hearing this the courtiers all applauded again.

But Taboret laughed scornfully, and taking up the real Ursula in her arms, flew back with her to Mark's cottage.

In the evening the city was illuminated, and there were great rejoicings at the recovery of the Princess, but Ursula remained in the cottage and married Oliver, and lived happily with him for the rest of her life.

1877

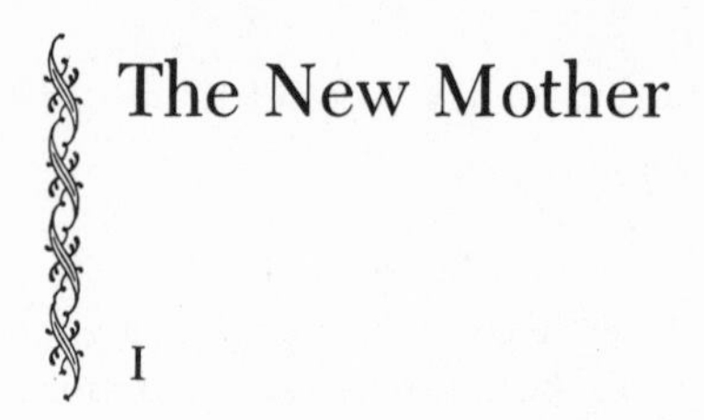

LUCY LANE CLIFFORD

The New Mother

I

THE children were always called Blue-Eyes and the Turkey, and they came by the names in this manner. The elder one was like her dear father who was far away at sea, and when the mother looked up she would often say, 'Child, you have taken the pattern of your father's eyes'; for the father had the bluest of blue eyes, and so gradually his little girl came to be called after them. The younger one had once, while she was still almost a baby, cried bitterly because a turkey that lived near to the cottage, and sometimes wandered into the forest, suddenly vanished in the middle of the winter; and to console her she had been called by its name.

Now the mother and Blue-Eyes and the Turkey and the baby all lived in a lonely cottage on the edge of the forest. The forest was so near that the garden at the back seemed a part of it, and the tall fir-trees were so close that their big black arms stretched over the little thatched roof, and when the moon shone upon them their tangled shadows were all over the white-washed walls.

It was a long way to the village, nearly a mile and a half, and the mother had to work hard and had not time to go often herself to see if there was a letter at the post-office from the dear father, and so very often in the afternoon she used to send the two children. They were very proud of being able to go alone, and often ran half the way to the post-office. When they came back tired with the long walk, there would be the mother waiting and watching for them, and the tea would be ready, and the baby crowing with delight; and if by any chance there was a letter from the sea, then they were happy indeed. The cottage room was so cosy: the walls were as white as snow inside as well as out, and against them hung the cake-tin and the baking-dish, and the lid of a large saucepan that had been worn out long before the children could remember, and the fish-slice, all polished and shining as bright as silver. On one side of the fireplace, above the bellows hung the almanac, and on the other the clock that always struck the wrong hour and was always running down too soon, but it was a good clock, with a little picture on its face and sometimes ticked away for nearly a week without stopping. The baby's high chair stood in one corner, and in another there was a cupboard hung up high against the wall, in which the mother kept all manner of little surprises. The children often wondered how the things that came out of that cupboard had got into it, for they seldom saw them put there.

'Dear children,' the mother said one afternoon late in the autumn, 'it is very chilly for you to go to the village, but you must walk quickly, and who knows but what you may bring back a letter saying that dear father is already on his way to England.' Then Blue-Eyes and the Turkey made haste and were soon ready to go. 'Don't be long,' the mother said, as she always did before they started. 'Go the nearest way and don't look at any strangers you meet, and be sure you do not talk with them.'

'No, mother,' they answered; and then she kissed them and called them dear good children, and they joyfully started on their way.

The village was gayer than usual, for there had been a fair the day before, and the people who had made merry still hung about the street as if reluctant to own that their holiday was over.

'I wish we had come yesterday,' Blue-Eyes said to the Turkey; 'then we might have seen something.'

'Look there,' said the Turkey, and she pointed to a stall covered with gingerbread; but the children had no money. At the end of the street, close to the Blue Lion where the coaches stopped, an old man sat on the ground with his back resting against the wall of a house, and by him, with smart collars round their necks, were two dogs. Evidently they were dancing dogs, the children thought, and longed to see them perform, but they seemed as tired as their master, and sat quite still beside him, looking as if they had not even a single wag left in their tails.

'Oh, I *do* wish we had been here yesterday,' Blue-Eyes said again as they went on to the grocer's, which was also the post-office. The post-mistress was very busy weighing out half-pounds of coffee, and when she had time to attend to the children she only just said 'No letter for you to-day,' and went on with what she was doing. Then Blue-Eyes and the Turkey turned away to go home. They went back slowly down the village street, past the man with the dogs again. One dog had roused himself and sat up rather crookedly with his head a good deal on one side, looking very melancholy and rather ridiculous; but on the children went towards the bridge and the fields that led to the forest.

They had left the village and walked some way, and then, just before they reached the bridge, they noticed, resting against a pile of stones by the wayside, a strange dark figure. At first they thought it was some one asleep, then they thought it was a poor woman ill and hungry, and then they saw that it was a strange wild-looking girl, who seemed very unhappy, and they felt sure that something was the matter. So they went and looked at her, and thought they would ask her if they could do anything to help her, for they were kind children and sorry indeed for any one in distress.

The girl seemed to be tall, and was about fifteen years old. She was dressed in very ragged clothes. Round her shoulders there was an old brown shawl, which was torn at the corner that hung down the middle of her back. She wore no bonnet, and an old yellow handkerchief which she had tied round her head had fallen backwards and was all huddled up round her neck. Her hair was coal black and hung down uncombed and unfastened, just anyhow. It was not very long, but it was very shiny, and it seemed to match her bright black eyes and dark freckled skin. On her feet were coarse gray stockings and thick shabby boots, which she had evidently forgotten to lace up. She had something hidden away under her shawl, but the children did not know what it was. At first they thought it was a baby, but when, on seeing them coming towards her, she carefully put it under her and sat upon it, they thought they must be mistaken. She sat watching the children approach, and did not move or stir till they were within a yard of her; then she wiped her eyes just as if she had been crying bitterly, and looked up.

The children stood still in front of her for a moment, staring at her and wondering what they ought to do.

'Are you crying?' they asked shyly.

To their surprise she said in a most cheerful voice,

'Oh dear, no! quite the contrary. Are you?'

They thought it rather rude of her to reply in this way, for any one could see that they were not crying. They felt half in mind to walk away; but the girl looked at them so hard with her big black eyes, they did not like to do so till they had said something else.

'Perhaps you have lost yourself?' they said gently.

But the girl answered promptly, 'Certainly not. Why, you have just found me. Besides,' she added, 'I live in the village.'

The children were surprised at this, for they had never seen her before, and yet they thought they knew all the village folk by sight.

'We often go to the village,' they said, thinking it might interest her.

'Indeed,' she answered. That was all; and again they wondered what to do.

Then the Turkey, who had an inquiring mind, put a good straightforward question. 'What are you sitting on?' she asked.

'On a peardrum,' the girl answered, still speaking in a most cheerful voice, at which the children wondered, for she looked very cold and uncomfortable.

'What is a peardrum?' they asked.

'I am surprised at your not knowing,' the girl answered. 'Most people in good society have one.' And then she pulled it out and showed it to them. It was a curious instrument, a good deal like a guitar in shape; it had three strings, but only two pegs by which to tune them. The third string was never tuned at all, and thus added to the singular effect produced by the village girl's music. And yet, oddly, the peardrum was not played by touching its strings, but by turning a little handle cunningly hidden on one side.

But the strange thing about the peardrum was not the music it made, or the strings, or the handle, but a little square box attached to one side. The box had a little flat lid that appeared to open by a spring. That was all the children could make out at first. They were most anxious to see inside the box, or to know what it contained, but they thought it might look curious to say so.

'It really is a most beautiful thing, is a peardrum,' the girl said, looking at it, and speaking in a voice that was almost affectionate.

'Where did you get it?' the children asked.

'I bought it,' the girl answered.

'Didn't it cost a great deal of money?' they asked.

'Yes,' answered the girl slowly, nodding her head, 'it cost a great deal of money. I am very rich,' she added.

And this the children thought a really remarkable statement, for they had not supposed that rich people dressed in old clothes, or went about without bonnets. She might at least have done her hair, they thought, but they did not like to say so.

'You don't look rich,' they said slowly, and in as polite a voice as possible.

'Perhaps not,' the girl answered cheerfully.

At this the children gathered courage, and ventured to remark, 'You look rather shabby'—they did not like to say ragged.

'Indeed?' said the girl in the voice of one who had heard a pleasant but surprising statement. 'A little shabbiness is very respectable,' she added in a satisfied voice. 'I must really tell them this,' she continued. And the children wondered what she meant. She opened the little box by the side of the peardrum, and said, just as if she were speaking to some one who could hear her, 'They say I look rather shabby; it is quite lucky, isn't it?'

'Why, you are not speaking to any one!' they said, more surprised than ever.

'Oh dear, yes! I am speaking to them both.'

'Both?' they said, wondering.

'Yes, I have here a little man dressed as a peasant, and wearing a wide slouch hat with a large feather, and a little woman to match, dressed in a red petticoat, and a white handkerchief pinned across her bosom. I put them on the lid of the box, and when I play they dance most beautifully. The little man takes off his hat and waves it in the air, and the little woman holds up her petticoat a little bit on one side with one hand, and with the other sends forward a kiss.'

'Oh! let us see; do let us see!' the children cried, both at once.

Then the village girl looked at them doubtfully.

'Let you see!' she said slowly. 'Well, I am not sure that I can. Tell me, are you good?'

'Yes, yes,' they answered eagerly, 'we are very good!'

'Then it's quite impossible,' she answered, and resolutely closed the lid of the box.

They stared at her in astonishment.

'But we are good,' they cried, thinking she must have misunderstood them. 'We are very good. Mother always says we are.'

'So you remarked before,' the girl said, speaking in a tone of decision.

Still the children did not understand.

'Then can't you let us see the little man and woman?' they asked.

'Oh dear, no!' the girl answered. 'I only show them to naughty children.'

'To naughty children!' they exclaimed.

'Yes, to naughty children,' she answered; 'and the worse the children the better do the man and woman dance.'

She put the peardrum carefully under her ragged cloak, and prepared to go on her way.

'I really could not have believed that you were good,' she said, reproachfully, as if they had accused themselves of some great crime. 'Well, good day.'

'Oh, but do show us the little man and woman,' they cried.

'Certainly not. Good day,' she said again.

'Oh, but we will be naughty,' they said in despair.

'I am afraid you couldn't,' she answered, shaking her head. 'It requires a great deal of skill, especially to be naughty well. Good day,' she said for the third time. 'Perhaps I shall see you in the village to-morrow.'

And swiftly she walked away, while the children felt their eyes fill with tears, and their hearts ache with disappointment.

'If we had only been naughty,' they said, 'we should have seen them dance; we should have seen the little woman holding her red petticoat in her hand, and the little man waving his hat. Oh, what shall we do to make her let us see them?'

'Suppose,' said the Turkey, 'we try to be naughty to-day; perhaps she would let us see them to-morrow.'

'But, oh!' said Blue-Eyes, 'I don't know how to be naughty; no one ever taught me.'

The Turkey thought for a few minutes in silence. 'I think I can be naughty if I try,' she said. 'I'll try to-night.'

And then poor Blue-Eyes burst into tears.

'Oh, don't be naughty without me!' she cried. 'It would be so unkind of you. You know I want to see the little man and woman just as much as you do. You are very, very unkind.' And she sobbed bitterly.

And so, quarrelling and crying, they reached their home.

Now, when their mother saw them, she was greatly astonished, and, fearing they were hurt, ran to meet them.

'Oh, my children, oh, my dear, dear children,' she said; 'what is the matter?'

But they did not dare tell their mother about the village girl and the little man and woman, so they answered, 'Nothing is the matter; nothing at all is the matter,' and cried all the more.

'But why are you crying?' she asked in surprise.

'Surely we may cry if we like,' they sobbed. 'We are very fond of crying.'

'Poor children!' the mother said to herself. 'They are tired, and perhaps they are hungry; after tea they will be better.' And she went back to the cottage, and made the fire blaze, until its reflection danced about on the tin lids upon the wall; and she put the kettle on to boil, and set the tea-things on the table, and opened the window to let in the sweet fresh air, and made all things look bright. Then she went to the little cupboard, hung up high against the wall, and took out some bread and put it on the table, and said in a loving voice, 'Dear little children, come and have your tea; it is all quite ready for you. And see, there is the baby waking up from her sleep; we will put her in the high chair, and she will crow at us while we eat.'

But the children made no answer to the dear mother; they only stood still by the window and said nothing.

'Come, children,' the mother said again. 'Come, Blue-Eyes, and come, my Turkey; here is nice sweet bread for tea.'

Then Blue-Eyes and the Turkey looked round, and when they saw the tall loaf, baked crisp and brown, and the cups all in a row, and the jug of milk, all waiting for them, they went to the table and sat down and felt a little happier; and the mother did not put the baby in the high chair after all, but took it on her knee, and danced it up and down, and sang little snatches of songs to it, and laughed, and looked content, and thought of the father far away at sea, and wondered what he would say to them all when he came home again. Then suddenly she looked up and saw that the Turkey's eyes were full of tears.

'Turkey!' she exclaimed, 'my dear little Turkey! what is the matter? Come to mother, my sweet; come to own mother.' And putting the baby down on the rug, she held out her arms, and the Turkey, getting up from her chair, ran swiftly into them.

'Oh, mother,' she sobbed, 'oh, dear mother! I do so want to be naughty.'

'My dear child!' the mother exclaimed.

'Yes, mother,' the child sobbed, more and more bitterly. 'I do so want to be very, very naughty.'

And then Blue-Eyes left her chair also, and, rubbing her face against the mother's shoulder, cried sadly. 'And so do I, mother. Oh, I'd give anything to be very, very naughty.'

'But, my dear children,' said the mother, in astonishment, 'why do you want to be naughty?'

'Because we do; oh, what shall we do?' they cried together.

'I should be very angry if you were naughty. But you could not be, for you love me,' the mother answered.

'Why couldn't we be naughty because we love you?' they asked.

'Because it would make me very unhappy; and if you love me you couldn't make me unhappy.'

'Why couldn't we?' they asked.

Then the mother thought a while before she answered; and when she did so they hardly understood, perhaps because she seemed to be speaking rather to herself than to them.

'Because if one loves well,' she said gently, 'one's love is stronger than all bad feelings in one, and conquers them. And this is the test whether love be real or false, unkindness and wickedness have no power over it.'

'We don't know what you mean,' they cried; 'and we do love you; but we want to be naughty.'

'Then I should know you did not love me,' the mother said.

'And what should you do?' asked Blue-Eyes.

'I cannot tell. I should try to make you better.'

'But if you couldn't? If we were very, very, very naughty, and wouldn't be good, what then?'

'Then,' said the mother sadly—and while she spoke her eyes filled with tears, and a sob almost choked her—'then,' she said, 'I should have to go away and leave you, and to send home a new mother, with glass eyes and wooden tail.'

'You couldn't,' they cried.

'Yes, I could,' she answered in a low voice; 'but it would make me very unhappy, and I will never do it unless you are very, very naughty, and I am obliged.'

'We won't be naughty,' they cried; 'we will be good. We should hate a new mother; and she shall never come here.' And they clung to their own mother, and kissed her fondly.

But when they went to bed they sobbed bitterly, for they remembered the little man and woman, and longed more than ever to see them; but how could they bear to let their own mother go away, and a new one take her place?

II

'Good-day,' said the village girl, when she saw Blue-Eyes and the Turkey approach. She was again sitting by the heap of stones, and under her shawl the peardrum was hidden. She looked just as if she had not moved since the day before. 'Good-day,' she said, in the same cheerful voice in which she had spoken yesterday; 'the weather is really charming.'

'Are the little man and woman there?' the children asked, taking no notice of her remark.

'Yes; thank you for inquiring after them,' the girl answered; 'they are both here and quite well. The little man is learning how to rattle the money in his pocket, and the little woman has heard a secret—she tells it while she dances.'

'Oh, do let us see,' they entreated.

'Quite impossible, I assure you,' the girl answered promptly. 'You see, you are good.'

'Oh!' said Blue-Eyes, sadly; 'but mother says if we are naughty she will go away and send home a new mother, with glass eyes and a wooden tail.'

'Indeed,' said the girl, still speaking in the same unconcerned voice, 'that is what they all say.'

'What do you mean?' asked the Turkey.

'They all threaten that kind of thing. Of course really there are no mothers with glass eyes and wooden tails; they would be much too expensive to make.' And the common sense of this remark the children, especially the Turkey, saw at once, but they merely said, half crying—

'We think you might let us see the little man and woman dance.'

'The kind of thing you would think,' remarked the village girl.

'But will you if we are naughty?' they asked in despair.

'I fear you could not be naughty—that is, really—even if you tried,' she said scornfully.

'Oh, but we will try; we will indeed,' they cried; 'so do show them to us.'

'Certainly not beforehand,' answered the girl, getting up and preparing to walk away.

'But if we are very naughty to-night, will you let us see them to-morrow?'

'Questions asked to-day are always best answered to-morrow,' the girl said, and turned round as if to walk on. 'Good-day,' she said blithely; 'I must really go and play a little to myself; good-day,' she repeated, and then suddenly she began to sing—

'Oh, sweet and fair's the lady-bird,
  And so's the bumble-bee,
But I myself have long preferred
  The gentle chimpanzee,
  The gentle chimpanzee-e-e,
  The gentle chim——'

'I beg your pardon,' she said, stopping, and looking over her shoulder; 'it's very rude to sing without leave before company. I won't do it again.'

'Oh, do go on,' the children said.

'I'm going,' she said, and walked away.

'No, we meant go on singing,' they explained, 'and do let us just hear you play,' they entreated, remembering that as yet they had not heard a single sound from the peardrum.

'Quite impossible,' she called out as she went along. 'You are good, as I remarked before. The pleasure of goodness centres in itself; the pleasures of naughtiness are many and varied. Good-day,' she shouted, for she was almost out of hearing.

For a few minutes the children stood still looking after her, then they broke down and cried.

'She might have let us see them,' they sobbed.

The Turkey was the first to wipe away her tears.

'Let us go home and be very naughty,' she said; 'then perhaps she will let us see them to-morrow.'

'But what shall we do?' asked Blue-Eyes, looking up. Then together all the way home they planned how to begin being naughty. And that afternoon the dear mother was sorely distressed, for, instead of sitting at their tea as usual with smiling happy faces, and then helping her to clear away and doing all she told them, they broke their mugs and threw their bread and butter on the floor, and when the mother told them to do one thing they carefully went and did another, and as for helping her to put away, they left her to do it all by herself, and only stamped their feet with rage when she told them to go upstairs until they were good.

'We won't be good,' they cried. 'We hate being good, and we always mean to be naughty. We like being naughty very much.'

'Do you remember what I told you I should do if you were very very naughty?' she asked sadly.

'Yes, we know, but it isn't true,' they cried. 'There is no mother with a wooden tail and glass eyes, and if there were we should just stick pins into her and send her away; but there is none.'

Then the mother became really angry at last, and sent them off to bed, but instead of crying and being sorry at her anger they laughed for joy, and when they were in bed they sat up and sang merry songs at the top of their voices.

The next morning quite early, without asking leave from the mother, the children got up and ran off as fast as they could over the fields towards the bridge to look for the village girl. She was sitting as usual by the heap of stones with the peardrum under her shawl.

'Now please show us the little man and woman,' they cried, 'and let us hear the peardrum. We were very naughty last night.' But the girl kept the peardrum carefully hidden. 'We were very naughty,' the children cried again.

'Indeed,' she said in precisely the same tone in which she had spoken yesterday.

'But we were,' they repeated; 'we were indeed.'

'So you say,' she answered. 'You were not half naughty enough.'

'Why, we were sent to bed!'

'Just so,' said the girl, putting the other corner of the shawl over the peardrum. 'If you had been really naughty you wouldn't have gone; but you can't help it, you see. As I remarked before, it requires a great deal of skill to be naughty well.'

'But we broke our mugs, we threw our bread and butter on the floor, we did everything we could to be tiresome.'

'Mere trifles,' answered the village girl scornfully. 'Did you throw cold water on the fire, did you break the clock, did you pull all the tins down from the walls, and throw them on the floor?'

'No!' exclaimed the children, aghast, 'we did not do that.'

'I thought not,' the girl answered. 'So many people mistake a little noise and foolishness for real naughtiness; but, as I remarked before, it wants skill to do the thing properly. Well, good-day,' and before they could say another word she had vanished.

'We'll be much worse,' the children cried, in despair. 'We'll go and do all the things she says'; and then they went home and did all these things. They threw water on the fire; they pulled down the baking-dish and the cake-tin, the fish-slice and the lid of the saucepan they had never seen, and banged them on the floor; they broke the clock and danced on the butter; they turned everything upside down; and then they sat still and wondered if they were naughty enough. And when the mother saw all that they had done she did not scold them as she had the day before or send them to bed, but she just broke down and cried, and then she looked at the children and said sadly—

'Unless you are good to-morrow, my poor Blue-Eyes and Turkey, I shall indeed have to go away and come back no more, and the new mother I told you of will come to you.'

They did not believe her; yet their hearts ached when they saw how unhappy she looked, and they thought within themselves that when they once had seen the little man and woman dance, they would be good to the dear mother for ever afterwards; but they could not be good now till they had heard the sound of the peardrum, seen the little man and woman dance, and heard the secret told—then they would be satisfied.

The next morning, before the birds were stirring, before the sun had climbed high enough to look in at their bedroom window, or the flowers had wiped their eyes ready for the day, the children got up and crept out of the cottage and ran across the fields. They did not think the village girl would be up so very early, but their hearts had ached so much at the sight of the mother's sad face that they had not been able to sleep, and they longed to know if they had been naughty enough, and if they might just once hear the peardrum and see the little man and woman, and then go home and be good for ever.

To their surprise they found the village girl sitting by the heap of stones, just as if it were her natural home. They ran fast when they saw her, and they noticed that the box containing the little man and woman was open, but she closed it quickly when she saw them, and they heard the clicking of the spring that kept it fast.

'We have been very naughty,' they cried. 'We have done all the things you told us; now will you show us the little man and woman?' The girl looked at them curiously, then drew the yellow silk handkerchief she sometimes wore round her head out of her pocket, and began to smooth out the creases in it with her hands.

'You really seem quite excited,' she said in her usual voice. 'You should be calm; calmness gathers in and hides things like a big cloak, or like my shawl does here; for instance'; and she looked down at the ragged covering that hid the peardrum.

'We have done all the things you told us,' the children cried again, 'and we do so long to hear the secret'; but the girl only went on smoothing out her handkerchief.

'I am so very particular about my dress,' she said. They could hardly listen to her in their excitement.

'But do tell if we may see the little man and woman,' they entreated again. 'We have been so very naughty, and mother says she will go away to-day and send home a new mother if we are not good.'

'Indeed,' said the girl, beginning to be interested and amused. 'The things that people say are most singular and amusing. There is an endless variety in language.' But the children did not understand, only entreated once more to see the little man and woman.

'Well, let me see,' the girl said at last, just as if she were relenting. 'When did your mother say she would go?'

'But if she goes what shall we do?' they cried in despair. 'We don't want her to go; we love her very much. Oh! what shall we do if she goes?'

'People go and people come; first they go and then they come. Perhaps she will go before she comes; she couldn't come before she goes. You had better go back and be good,' the girl added suddenly; 'you are really not clever enough to be anything else; and the little woman's secret is very important; she never tells it for make-believe naughtiness.'

'But we did do all the things you told us,' the children cried, despairingly.

'You didn't throw the looking-glass out of the window, or stand the baby on its head.'

'No, we didn't do that,' the children gasped.

'I thought not,' the girl said triumphantly. 'Well, good-day. I shall not be here to-morrow. Good-day.'

'Oh, but don't go away,' they cried. 'We are so unhappy; do let us see them just once.'

'Well, I shall go past your cottage at eleven o'clock this morning,' the girl said. 'Perhaps I shall play the peardrum as I go by.'

'And will you show us the man and woman?' they asked.

'Quite impossible, unless you have really deserved it; make-believe naughtiness is only spoilt goodness. Now if you break the looking-glass and do the things that are desired——'

'Oh, we will,' they cried. 'We will be very naughty till we hear you coming.'

'It's waste of time, I fear,' the girl said politely; 'but of course I should not like to interfere with you. You see the little man and woman, being used to the best society, are very particular. Good-day,' she said, just as she always said, and then quickly turned away, but she looked back and called out, 'Eleven o'clock, I shall be quite punctual; I am very particular about my engagements.'

Then again the children went home, and were naughty, oh, so very very naughty that the dear mother's heart ached, and her eyes filled with tears, and at last she went upstairs and slowly put on her best gown and her new sun-bonnet, and she dressed the baby all in its Sunday clothes, and then she came down and stood before Blue-Eyes and the Turkey, and just as she did so the Turkey threw the looking-glass out of the window, and it fell with a loud crash upon the ground.

'Good-bye, my children,' the mother said sadly, kissing them. 'Good-bye, my Blue-Eyes; good-bye, my Turkey; the new mother will be home presently. Oh, my poor children!' and then weeping bitterly the mother took the baby in her arms and turned to leave the house.

'But, mother,' the children cried, 'we are——' and then suddenly the broken clock struck half-past ten, and they knew that in half an hour the village girl would come by playing on the peardrum. 'But, mother, we will be good at half-past eleven, come back at half-past eleven,' they cried, 'and we'll both be good, we will indeed; we must be naughty till eleven o'clock.' But the mother only picked up the little bundle in which she had tied up her cotton apron and a pair of old shoes, and went slowly out at the door. It seemed as if the children were spellbound, and they could not follow her. They opened the window wide, and called after her—

'Mother! mother! oh, dear mother, come back again! We will be good, we will be good now, we will be good for evermore if you will come back.' But the mother only looked round and shook her head, and they could see the tears falling down her cheeks.

'Come back, dear mother!' cried Blue-Eyes; but still the mother went on across the fields.

'Come back, come back!' cried the Turkey; but still the mother went on. Just by the corner of the field she stopped and turned, and waved her handkerchief, all wet with tears, to the children at the window; she made the baby kiss its hand; and in a moment mother and baby had vanished from their sight.

Then the children felt their hearts ache with sorrow, and they cried bitterly just as the mother had done, and yet they could not believe that she had gone. Surely she would come back, they thought; she would not leave them altogether; but, oh, if she did—if she did—if she did. And then the broken clock struck eleven, and suddenly there was a sound—a quick, clanging, jangling sound, with a strange discordant one at intervals; and they looked at each other, while their hearts stood still, for they knew it was the peardrum. They rushed to the open window, and there they saw the village girl coming towards them from the fields, dancing along and playing as she did so. Behind her, walking slowly, and yet ever keeping the same distance from her, was the man with the dogs whom they had seen asleep by the Blue Lion, on the day they first saw the girl with the peardrum. He was playing on a flute that had a strange shrill sound; they could hear it plainly above the jangling of the peardrum. After the man followed the two dogs, slowly waltzing round and round on their hind legs.

'We have done all you told us,' the children called, when they had recovered from their astonishment. 'Come and see; and now show us the little man and woman.'

The girl did not cease her playing or her dancing, but she called out in a voice that was half speaking half singing, and seemed to keep time to the strange music of the peardrum.

'You did it all badly. You threw the water on the wrong side of the fire, the tin things were not quite in the middle of the room, the clock was not broken enough, you did not stand the baby on its head.'

Then the children, still standing spellbound by the window, cried out, entreating and wringing their hands, 'Oh, but we have done everything you told us, and mother has gone away. Show us the little man and woman now, and let us hear the secret.'

As they said this the girl was just in front of the cottage, but she did not stop playing. The sound of the strings seemed to go through their hearts. She did not stop dancing; she was already passing the cottage by. She did not stop singing, and all she said sounded like part of a terrible song. And still the man followed her, always at the same distance, playing shrilly on his flute; and still the two dogs waltzed round and round after him—their tails motionless, their legs straight, their collars clear and white and stiff. On they went, all of them together.

'Oh, stop!' the children cried, 'and show us the little man and woman now.'

But the girl sang out loud and clear, while the string that was out of tune twanged above her voice.

'The little man and woman are far away. See, their box is empty.'

And then for the first time the children saw that the lid of the box was raised and hanging back, and that no little man and woman were in it.

'I am going to my own land,' the girl sang, 'to the land where I was born.' And she went on towards the long straight road that led to the city many many miles away.

'But our mother is gone,' the children cried; 'our dear mother, will she ever come back?'

'No,' sang the girl; 'she'll never come back, she'll never come back. I saw her by the bridge: she took a boat upon the river; she is sailing to the sea; she will meet your father once again, and they will go sailing on, sailing on to the countries far away.'

And when they heard this, the children cried out, but could say no more, for their hearts seemed to be breaking.

Then the girl, her voice getting fainter and fainter in the distance, called out once more to them. But for the dread that sharpened their ears they would hardly have heard her, so far was she away, and so discordant was the music.

'Your new mother is coming. She is already on her way; but she only walks slowly, for her tail is rather long, and her spectacles are left behind; but she is coming, she is coming—coming—coming.'

The last word died away; it was the last one they ever heard the village girl utter. On she went, dancing on; and on followed the man, they could see that he was still playing, but they could no longer hear the sound of his flute; and on went the dogs round and round and round. On they all went, farther and farther away, till they were separate things no more, till they were just a confused mass of faded colour, till they were a dark misty object that nothing could define, till they had vanished altogether,—altogether and for ever.

Then the children turned, and looked at each other and at the little cottage home, that only a week before had been so bright and happy, so cosy and so spotless. The fire was out, and the water was still among the cinders; the baking-dish and cake-tin, the fish-slice and the saucepan lid, which the dear mother used to spend so much time in rubbing, were all pulled down from the nails on which they had hung so long, and were lying on the floor. And there was the clock all broken and spoilt, the little picture upon its face could be seen no more; and though it sometimes struck a stray hour, it was with the tone of a clock whose hours are numbered. And there was the baby's high chair, but no little baby to sit in it; there was the cupboard on the wall, and never a sweet loaf on its shelf; and there were the broken mugs, and the bits of bread tossed about, and the greasy boards which the mother had knelt down to scrub until they were white as snow. In the midst of all stood the children, looking at the wreck they had made, their hearts aching, their eyes blinded with tears, and their poor little hands clasped together in their misery.

'Oh, what shall we do?' cried Blue-Eyes. 'I wish we had never seen the village girl and the nasty, nasty peardrum.'

'Surely mother will come back,' sobbed the Turkey. 'I am sure we shall die if she doesn't come back.'

'I don't know what we shall do if the new mother comes,' cried Blue-Eyes. 'I shall never, never like any other mother. I don't know what we shall do if that dreadful mother comes.'

'We won't let her in,' said the Turkey.

'But perhaps she'll walk in,' sobbed Blue-Eyes.

Then Turkey stopped crying for a minute, to think what should be done.

'We will bolt the door,' she said, 'and shut the window; and we won't take any notice when she knocks.'

So they bolted the door, and shut the window, and fastened it. And then, in spite of all they had said, they felt naughty again, and longed after the little man and woman they had never seen, far more than after the mother who had loved them all their lives. But then they did not really believe that their own mother would not come back, or that any new mother would take her place.

When it was dinner-time, they were very hungry, but they could only find some stale bread, and they had to be content with it.

'Oh, I wish we had heard the little woman's secret,' cried the Turkey; 'I wouldn't have cared then.'

All through the afternoon they sat watching and listening for fear of the new mother; but they saw and heard nothing of her, and gradually they became less and less afraid lest she should come. Then they thought that perhaps when it was dark their own dear mother would come home; and perhaps if they asked her to forgive them she would. And then Blue-Eyes thought that if their mother did come she would be very cold, so they crept out at the back door and gathered in some wood, and at last, for the grate was wet, and it was a great deal of trouble to manage it, they made a fire. When they saw the bright fire burning, and the little flames leaping and playing among the wood and coal, they began to be happy again, and to feel certain that their own mother would return; and the sight of the pleasant fire reminded them of all the times she had waited for them to come from the post-office, and of how she had welcomed them, and comforted them, and given them nice warm tea and sweet bread, and talked to them. Oh, how sorry they were they had been naughty, and all for that nasty village girl! They did not care a bit about the little man and woman now, or want to hear the secret.

They fetched a pail of water and washed the floor; they found some rag, and rubbed the tins till they looked bright again, and, putting a footstool on a chair, they got up on it very carefully and hung up the things in their places; and then they picked up the broken mugs and made the room as neat as they could, till it looked more and more as if the dear mother's hands had been busy about it. They felt more and more certain she would return, she and the dear little baby together, and they thought they would set the tea-things for her, just as she had so often set them for her naughty children. They took down the tea-tray, and got out the cups, and put the kettle on the fire to boil, and made everything look as home-like as they could. There was no sweet loaf to put on the table, but perhaps the mother would bring something from the village, they thought. At last all was ready, and Blue-Eyes and the

Turkey washed their faces and their hands, and then sat and waited, for of course they did not believe what the village girl had said about their mother sailing away.

Suddenly, while they were sitting by the fire, they heard a sound as of something heavy being dragged along the ground outside, and then there was a loud and terrible knocking at the door. The children felt their hearts stand still. They knew it could not be their own mother, for she would have turned the handle and tried to come in without any knocking at all.

'Oh, Turkey!' whispered Blue-Eyes, 'if it should be the new mother, what shall we do?'

'We won't let her in,' whispered the Turkey, for she was afraid to speak aloud, and again there came a long and loud and terrible knocking at the door.

'What shall we do? oh, what shall we do?' cried the children, in despair. 'Oh, go away!' they called out. 'Go away; we won't let you in; we will never be naughty any more; go away, go away!'

But again there came a loud and terrible knocking.

'She'll break the door if she knocks so hard,' cried Blue-Eyes.

'Go and put your back to it,' whispered the Turkey, 'and I'll peep out of the window and try to see if it is really the new mother.'

So in fear and trembling Blue-Eyes put her back against the door, and the Turkey went to the window, and, pressing her face against one side of the frame, peeped out. She could just see a black satin poke bonnet with a frill round the edge, and a long bony arm carrying a black leather bag. From beneath the bonnet there flashed a strange bright light, and Turkey's heart sank and her cheeks turned pale, for she knew it was the flashing of two glass eyes. She crept up to Blue-Eyes. 'It is—it is—it is!' she whispered, her voice shaking with fear, 'it is the new mother! She has come, and brought her luggage in a black leather bag that is hanging on her arm!'

'Oh, what shall we do?' wept Blue-Eyes; and again there was the terrible knocking.

'Come and put your back against the door too, Turkey,' cried Blue-Eyes; 'I am afraid it will break.'

So together they stood with their two little backs against the door. There was a long pause. They thought perhaps the new mother had made up her mind that there was no one at home to let her in, and would go away, but presently the two children heard through the thin wooden door the new mother move a little, and then say to herself—'I must break open the door with my tail.'

For one terrible moment all was still, but in it the children could almost hear her lift up her tail, and then, with a fearful blow, the little painted door was cracked and splintered.

With a shriek the children darted from the spot and fled through the cottage, and out at the back door into the forest beyond. All night long they stayed in the darkness and the cold, and all the next day and the next, and all through the cold, dreary days and the long dark nights that followed.

They are there still, my children. All through the long weeks and months have they been there, with only green rushes for their pillows and only the brown dead leaves to cover them, feeding on the wild strawberries in the summer, or on the nuts when they hang green; on the blackberries when they are no longer sour in the autumn, and in the winter on the little red berries that ripen in the snow. They wander about among the tall dark firs or beneath the great trees beyond. Sometimes they stay to rest beside the little pool near the copse where the ferns grow thickest, and they long and long, with a longing that is greater than words can say, to see their own dear mother again, just once again, to tell her that they'll be good for evermore—just once again.

And still the new mother stays in the little cottage, but the windows are closed and the doors are shut, and no one knows what the inside looks like. Now and then, when the darkness has fallen and the night is still, hand in hand Blue-Eyes and the Turkey creep up near to the home in which they once were so happy, and with beating hearts they watch and listen; sometimes a blinding flash comes through the window, and they know it is the light from the new mother's glass eyes, or they hear a strange muffled noise, and they know it is the sound of her wooden tail as she drags it along the floor.

1882

JULIANA HORATIA EWING

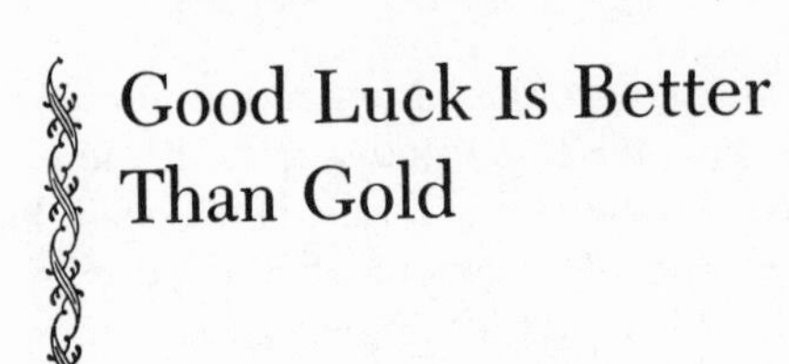

Good Luck Is Better Than Gold

THERE was once upon a time a child who had Good Luck for his godfather.

'I am not Fortune,' said Good Luck to the parents; 'I have no gifts to bestow, but whenever he needs help I will be at hand.'

'Nothing could be better,' said the old couple. They were delighted. But what pleases the father often fails to satisfy the son: moreover, every man thinks that he deserves just a little more than he has got, and does not reckon it to the purpose if his father had less.

Many a one would be thankful to have as good reasons for contentment as he who had Good Luck for his godfather.

If he fell, Good Luck popped something soft in the way to break his fall; if he fought, Good Luck directed his blows, or tripped up his adversary; if he got into a scrape, Good Luck helped him out of it; and if ever Misfortune met him, Good Luck contrived to hustle her on the pathway till his godson got safely by.

In games of hazard the godfather played over his shoulder. In matters of choice he chose for him. And when the lad began to work on his father's farm the farmer began to get rich. For no bird or field-mouse touched a seed that his son had sown, and every plant he planted throve when Good Luck smiled on it.

The boy was not fond of work, but when he did go into the fields, Good Luck followed him.

'Your christening-day was a blessed day for us all,' said the old farmer.

'He has never given me so much as a lucky sixpence,' muttered Good Luck's godson.

'I am not Fortune—I make no presents,' said the godfather.

When we are discontented it is oftener to please our neighbours than ourselves. It was because the other boys had said—'Simon, the shoemaker's son, has an alderman for his godfather. He gave him a silver spoon with the Apostle Peter for the handle; but thy godfather is more powerful than any alderman'—that Good Luck's godson complained, 'He has never given me so much as a bent sixpence.'

By and by the old farmer died, and his son grew up, and had the largest farm in the country. The other boys grew up also, and as they looked over the farmer's boundary-wall, they would say:

'Good-morning, Neighbour. That is certainly a fine farm of yours. Your cattle thrive without loss. Your crops grow in the rain and are reaped with the sunshine. Mischance never comes your road. What you have worked for you enjoy. Such success would turn the heads of poor folk like us. At the same time one would think a man need hardly work for his living at all who has Good Luck for his godfather.'

'That is very true,' thought the farmer. 'Many a man is prosperous, and reaps what he sows, who had no more than the clerk and the sexton for gossips at his christening.'

'What is the matter, Godson?' asked Good Luck, who was with him in the field.

'I want to be rich,' said the farmer.

'You will not have to wait long,' replied the godfather. 'In every field you sow, in every flock you rear, there is increase without abatement. Your wealth is already tenfold greater than your father's.'

'Aye, aye,' replied the farmer. 'Good wages for good work. But many a young man has gold at his command who need never turn a sod, and none of the Good People came to *his* christening. Fortunatus's Purse now, or even a sack or two of gold—'

'Peace!' cried the godfather; 'I have said that I give no gifts.'

Though he had not Fortunatus's Purse, the farmer had now money and to spare, and when the harvest was gathered in, he bought a fine suit of clothes, and took his best horse and went to the royal city to see the sights.

The pomp and splendour, the festivities and fine clothes dazzled him.

'This is a gay life which these young courtiers lead,' said he. 'A man has nothing to do but to enjoy himself.'

'If he has plenty of gold in his pocket,' said a bystander.

By and by the Princess passed in her carriage. She was the

King's only daughter. She had hair made of sunshine, and her eyes were stars.

'What an exquisite creature!' cried the farmer. 'What would not one give to possess her?'

'She has as many suitors as hairs on her head,' replied the bystander. 'She wants to marry the Prince of Moonshine, but he only dresses in silver, and the King thinks he might find a richer son-in-law. The Princess will go to the highest bidder.'

'And I have Good Luck for my godfather, and am not even at court!' cried the farmer; and he put spurs to his horse, and rode home.

Good Luck was taking care of the farm.

'Listen, Godfather!' cried the young man. 'I am in love with the King's daughter, and want her to wife.'

'It is not an easy matter,' replied Good Luck, 'but I will do what I can for you. Say that by good luck you saved the Princess's life, or perhaps better the King's—for they say he is selfish—'

'Tush!' cried the farmer. 'The King is covetous, and wants a rich son-in-law.'

'A wise man may bring wealth to a kingdom with his head, if not with his hands,' said Good Luck, 'and I can show you a district where the earth only wants mining to be flooded with wealth. Besides, there are a thousand opportunities that can be turned to account and influence. By wits and work, and with Good Luck to help him, many a poorer man than you has risen to greatness.'

'Wits and work!' cried the indignant godson. 'You speak well—truly! A hillman would have made a better godfather. Give me as much gold as will fill three meal-bins, and you may keep the rest of your help for those who want it.'

Now at this moment by Good Luck stood Dame Fortune. She likes handsome young men, and there was some little jealousy between her and the godfather; so she smiled at the quarrel.

'You would rather have had me for your gossip?' said she.

'If you would give me three wishes, I would,' replied the farmer boldly, 'and I would trouble you no more.'

'Will you make him over to me?' said Dame Fortune to the godfather.

'If he wishes it,' replied Good Luck. 'But if he accepts your gifts he has no further claim on me.'

'Nor on me either,' said the Dame. 'Hark ye, young man, you

mortals are apt to make a hobble of your three wishes, and you may end with a sausage at your nose, like your betters.'

'I have thought of it too often,' replied the farmer, 'and I know what I want. For my first wish I desire imperishable beauty.'

'It is yours,' said Dame Fortune, smiling as she looked at him.

'The face of a prince and the manners of a clown are poor partners,' said the farmer. 'My second wish is for suitable learning and courtly manners, which cannot be gained at the plough-tail.'

'You have them in perfection,' said the Dame, as the young man thanked her by a graceful bow.

'Thirdly,' said he, 'I demand a store of gold that I can never exhaust.'

'I will lead you to it,' said Dame Fortune; and the young man was so eager to follow her that he did not even look back to bid farewell to his godfather.

He was soon at court. He lived in the utmost pomp. He had a suit of armour made for himself out of beaten gold. No metal less precious might come near his person, except for the blade of his sword. This was obliged to be made of steel, for gold is not always strong enough to defend one's life or his honour. But the Princess still loved the Prince of Moonshine.

'Stuff and nonsense!' said the King. 'I shall give you to the Prince of Gold.'

'I wish I had the good luck to please her,' muttered the young Prince. But he had not, for all his beauty and his wealth. However, she was to marry him, and that was something.

The preparations for the wedding were magnificent.

'It is a great expense,' sighed the King, 'but then I get the Prince of Gold for a son-in-law.'

The Prince and his bride drove round the city in a triumphal procession. Her hair fell over her like sunshine, but the starlight of her eyes was cold.

In the train rode the Prince of Moonshine, dressed in silver, and with no colour in his face.

As the bridal chariot approached one of the city gates, two black ravens hovered over it, and then flew away, and settled on a tree.

Good Luck was sitting under the tree to see his godson's triumph, and he heard the birds talking above him.

'Has the Prince of Gold no friend who can tell him that there is a loose stone above the archway that is tottering to fall?' said they.

And Good Luck covered his face with his mantle as the Prince drove through.

Just as they were passing out of the gateway the stone fell on to the Prince's head. He wore a casque of pure gold, but his neck was broken.

'We can't have all this expense for nothing,' said the King: so he married his daughter to the Prince of Moonshine. If one can't get gold one must be content with silver.

'Will you come to the funeral?' asked Dame Fortune of the godfather.

'Not I,' replied Good Luck. 'I had no hand in *this* matter.'

The rain came down in torrents. The black feathers on the ravens' backs looked as if they had been oiled.

'Caw! caw!' said they. 'It was an unlucky end.'

However, the funeral was a very magnificent one, for there was no stint of gold.

1882

HOWARD PYLE

The Apple of Contentment

I

THERE was a woman once, and she had three daughters. The first daughter squinted with both eyes, yet the woman loved her as she loved salt, for she herself squinted with both eyes. The second daughter had one shoulder higher than the other, and eyebrows as black as soot in the chimney, yet the woman loved her as well as she loved the other, for she herself had black eyebrows and one shoulder higher than the other. The youngest daughter was as pretty as a ripe apple, and had hair as fine as silk and the color of pure gold, but the woman loved her not at all, for, as I have said, she herself was neither pretty, nor had she hair of the color of pure gold. Why all this was so, even Hans Pfifendrummel cannot tell, though he has read many books and one over.

The first sister and the second sister dressed in their Sunday clothes every day, and sat in the sun doing nothing, just as though they had been born ladies, both of them.

As for Christine—that was the name of the youngest girl—as for Christine, she dressed in nothing but rags, and had to drive the geese to the hills in the morning and home again in the evening, so that they might feed on the young grass all day and grow fat.

The first sister and the second sister had white bread (and butter beside) and as much fresh milk as they could drink; but Christine had to eat cheese-parings and bread-crusts, and had hardly enough of them to keep Goodman Hunger from whispering in her ear.

This was how the churn clacked in that house!

Well, one morning Christine started off to the hills with her flock of geese, and in her hands she carried her knitting, at which she worked to save time. So she went along the dusty road until, by-and-by, she came to a place where a bridge crossed the brook, and what should she see there but a little red cap, with a silver bell at the point of it, hanging from the alder branch. It was such a nice, pretty little red cap that Christine thought that she would take it home with her, for she had never seen the like of it in all of her life before.

So she put it in her pocket, and then off she went with her geese again. But she had hardly gone two-score of paces when she heard a voice calling her, 'Christine! Christine!'

She looked, and who should she see but a queer little gray man, with a great head as big as a cabbage and little legs as thin as young radishes.

'What do you want?' said Christine, when the little man had come to where she was.

Oh, the little man only wanted his cap again, for without it he could not go back home into the hill—that was where he belonged.

But how did the cap come to be hanging from the bush? Yes, Christine would like to know that before she gave it back again.

Well, the little hill-man was fishing by the brook over yonder when a puff of wind blew his cap into the water, and he just hung it up to dry. That was all that there was about it; and now would Christine please give it to him?

Christine did not know how about that; perhaps she would and perhaps she would not. It was a nice, pretty little cap; what would the little underground man give her for it? that was the question.

Oh, the little man would give her five thalers for it, and gladly.

No; five thalers was not enough for such a pretty little cap—see, there was a silver bell hanging to it too.

Well, the little man did not want to be hard at a bargain; he would give her a hundred thalers for it.

No; Christine did not care for money. What else would he give for this nice, dear little cap?

'See, Christine,' said the little man, 'I will give you this for the cap'; and he showed her something in his hand that looked just like a bean, only it was as black as a lump of coal.

'Yes, good; but what is that?' said Christine.

'That,' said the little man, 'is a seed from the apple of contentment. Plant it, and from it will grow a tree, and from the tree an apple. Everybody in the world that sees the apple will long for it, but nobody in the world can pluck it but you. It will always be meat and drink to you when you are hungry, and warm clothes to your back when you are cold. Moreover, as soon as you pluck it from the tree, another as good will grow in its place. *Now*, will you give me my hat?'

Oh yes; Christine would give the little man his cap for such a seed as that, and gladly enough. So the little man gave Christine the seed, and Christine gave the little man his cap again. He put the cap on his head, and—puff!—away he was gone, as suddenly as the light of a candle when you blow it out.

So Christine took the seed home with her, and planted it before the window of her room. The next morning when she looked out of the window she beheld a beautiful tree, and on the tree hung an apple that shone in the sun as though it were pure gold. Then she went to the tree and plucked the apple as easily as though it were a gooseberry, and as soon as she had plucked it another as good grew in its place. Being hungry she ate it, and thought that she had never eaten anything as good, for it tasted like pancake with honey and milk.

By-and-by the oldest sister came out of the house and looked around, but when she saw the beautiful tree with the golden apple hanging from it you can guess how she stared.

Presently she began to long and long for the apple as she had never longed for anything in her life. 'I will just pluck it,' said she, 'and no one will be the wiser for it.' But that was easier said than done. She reached and reached, but she might as well have reached for the moon; she climbed and climbed, but she might as well have climbed for the sun—for either one would have been as easy to get as that which she wanted. At last she had to give up trying for it, and her temper was none the sweeter for that, you may be sure.

After a while came the second sister, and when she saw the golden apple she wanted it just as much as the first had done. But to want and to get are very different things, as she soon found, for she was no more able to get it than the other had been.

Last of all came the mother, and she also strove to pluck the apple. But it was no use. She had no more luck of her trying than her daughters; all that the three could do was to stand under the tree and look at the apple, and wish for it and wish for it.

They are not the only ones who have done the like, with the apple of contentment hanging just above them.

As for Christine, she had nothing to do but to pluck an apple whenever she wanted it. Was she hungry? there was the apple hanging in the tree for her. Was she thirsty? there was the apple. Cold? there was the apple. So you see, she was the happiest girl betwixt all the seven hills that stand at the ends of the earth; for nobody in the world can have more than contentment, and that was what the apple brought her.

II

One day a king came riding along the road, and all of his people with him. He looked up and saw the apple hanging in the tree, and a great desire came upon him to have a taste of it. So he called one of the servants to him, and told him to go and ask whether it could be bought for a potful of gold.

So the servant went to the house, and knocked on the door—rap! tap! tap!

'What do you want?' said the mother of the three sisters, coming to the door.

Oh, nothing much; only a king was out there in the road, and wanted to know if she would sell the apple yonder for a potful of gold.

Yes, the woman would do that. Just pay her the pot of gold and he might go and pluck it and welcome.

So the servant gave her the pot of gold, and then he tried to pluck the apple. First he reached for it, and then he climbed for it, and then he shook the limb.

But it was no use for him to try; he could no more get it—well—than *I* could if I had been in his place.

At last the servant had to go back to the King. The apple was there, he said, and the woman had sold it, but try and try as he would he could no more get it than he could get the little stars in the sky.

Then the King told the steward to go and get it for him; but the steward, though he was a tall man and a strong man, could no more pluck the apple than the servant.

So he had to go back to the King with an empty fist. No; he could not gather it, either.

Then the King himself went. He knew that he could pluck it—of course he could! Well, he tried and tried; but nothing came of his trying, and he had to ride away at last without having had so much as a smell of the apple.

After the King came home, he talked and dreamed and thought of nothing but the apple; for the more he could not get it the more he wanted it—that is the way we are made in this world. At last he grew melancholy and sick for want of that which he could not get. Then he sent for one who was so wise that he had more in his head than ten men together. This wise man told him that the only one who could pluck the fruit of contentment for him was the one to whom the tree belonged. This was one of the daughters of the woman who had sold the apple to him for the pot of gold.

When the King heard this he was very glad; he had his horse saddled, and he and his court rode away, and so came at last to the cottage where Christine lived. There they found the mother and the elder sisters, for Christine was away on the hills with her geese.

The King took off his hat and made a fine bow.

The wise man at home had told him this and that; now to which one of her daughters did the apple-tree belong? so said the King.

'Oh, it is my oldest daughter who owns the tree,' said the woman.

So, good! Then if the oldest daughter would pluck the apple for him he would take her home and marry her and make a queen of her. Only let her get it for him without delay.

Prut! that would never do. What! was the girl to climb the apple-tree before the King and all of the court? No! no! Let the King go home, and she would bring the apple to him all in good time; that was what the woman said.

Well, the King would do that, only let her make haste, for he wanted it very much indeed.

As soon as the King had gone, the woman and her daughters sent for the goose-girl to the hills. Then they told her that the King wanted the apple yonder, and that she must pluck it for her sister to take to him; if she did not do as they said they would throw her into the well. So Christine had to pluck the fruit; and as soon as she had done so the oldest sister wrapped it up in a napkin and set off with it to the King's house, as pleased as pleased could be. Rap! tap! tap! she knocked at the door. Had she brought the apple for the King?

Oh yes, she had brought it. Here it was, all wrapped up in a fine napkin.

After that they did not let her stand outside the door till her toes were cold, I can tell you. As soon as she had come to the King she opened her napkin. Believe me or not as you please, all the same, I tell you that there was nothing in the napkin but a hard round stone. When the King saw only a stone he was so angry that he stamped like a rabbit and told them to put the girl out of the house. So they did, and she went home with a flea in her ear, I can tell you.

Then the King sent his steward to the house where Christine and her sisters lived.

He told the woman that he had come to find whether she had any other daughters.

Yes; the woman had another daughter, and, to tell the truth, it was she who owned the tree. Just let the steward go home again and the girl would fetch the apple in a little while.

As soon as the steward had gone, they sent to the hills for Christine again. Look! she must pluck the apple for the second sister to take to the King; if she did not do that they would throw her into the well.

So Christine had to pluck it, and gave it to the second sister, who wrapped it up in a napkin and set off for the King's house. But she fared no better than the other, for, when she opened the napkin, there was nothing in it but a lump of mud. So they packed her home again with her apron to her eyes.

After a while the King's steward came to the house again. Had the woman no other daughter than these two?

Well, yes, there was one, but she was a poor ragged thing, of no account, and fit for nothing in the world but to tend the geese.

Where was she?

Oh, she was up on the hills now tending her flock.

But could the steward see her?

Yes, he might see her, but she was nothing but a poor simpleton.

That was all very good, but the steward would like to see her, for that was what the King had sent him there for.

So there was nothing to do but to send to the hills for Christine.

After a while she came, and the steward asked her if she could pluck the apple yonder for the King.

Yes; Christine could do that easily enough. So she reached and picked it as though it had been nothing but a gooseberry on the bush. Then the steward took off his hat and made her a low bow in spite of her ragged dress, for he saw that she was the one for whom they had been looking all this time.

So Christine slipped the golden apple into her pocket, and then she and the steward set off to the King's house together.

When they had come there everybody began to titter and laugh behind the palms of their hands to see what a poor ragged goose-girl the steward had brought home with him. But for that the steward cared not a rap.

'Have you brought the apple?' said the King, as soon as Christine had come before him.

Yes; here it was; and Christine thrust her hand into her pocket and brought it forth. Then the King took a great bite of it, and as soon as he had done so he looked at Christine and thought that he had never seen such a pretty girl. As for her rags, he minded them no more than one minds the spots on a cherry; that was because he had eaten of the apple of contentment.

And were they married? Of course they were! and a grand wedding it was, I can tell you. It is a pity that you were not there; but though you were not, Christine's mother and sisters were, and, what is more, they danced with the others, though I believe they would rather have danced upon pins and needles.

'Never mind,' said they; 'we still have the apple of contentment at home, though we cannot taste of it.' But no; they had nothing of the kind. The next morning it stood before the young Queen Christine's window, just as it had at her old home, for it belonged to her and to no one else in all of the world. That was lucky for the King, for he needed a taste of it now and then as much as anybody else, and no one could pluck it for him but Christine.

Now, that is all of this story. What does it mean?
Can you not see? Prut! rub your
spectacles and look again!

1886

FRANK STOCKTON

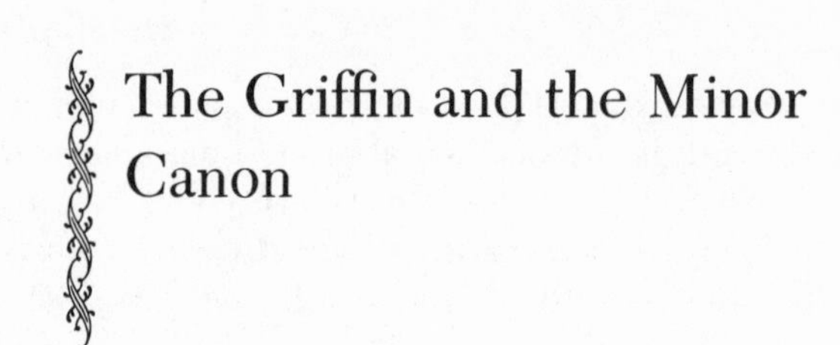

The Griffin and the Minor Canon

OVER the great door of an old, old church, which stood in a quiet town of a far-away land, there was carved in stone the figure of a large griffin. The old-time sculptor had done his work with great care, but the image he had made was not a pleasant one to look at. It had a large head, with enormous open mouth and savage teeth. From its back arose great wings, armed with sharp hooks and prongs. It had stout legs in front, with projecting claws, but there were no legs behind, the body running out into a long and powerful tail, finished off at the end with a barbed point. This tail was coiled up under him, the end sticking up just back of his wings.

The sculptor, or the people who had ordered this stone figure, had evidently been very much pleased with it, for little copies of it, also in stone, had been placed here and there along the sides of the church, not very far from the ground, so that people could easily look at them and ponder on their curious forms. There were a great many other sculptures on the outside of this church—saints, martyrs, grotesque heads of men, beasts, and birds, as well as those of other creatures which cannot be named, because nobody knows exactly what they were. But none were so curious and interesting as the great griffin over the door and the little griffins on the sides of the church.

A long, long distance from the town, in the midst of dreadful wilds scarcely known to man, there dwelt the Griffin whose image had been put up over the church door. In some way or other the old-time sculptor had seen him, and afterwards, to the best of his memory, had copied his figure in stone. The Griffin had never known this until, hundreds of years afterwards, he heard from a bird, from a wild animal, or in some manner which it is not easy to find out, that there was a likeness of him on the old church in the distant town.

Now, this Griffin had no idea whatever how he looked. He had never seen a mirror, and the streams where he lived were so turbulent and violent that a quiet piece of water, which would reflect the image of anything looking into it, could not be found. Being, as far as could be ascertained, the very last of his race, he had never seen another griffin. Therefore it was that, when he heard of this stone image of himself, he became very anxious to know what he looked like, and at last he determined to go to the old church and see for himself what manner of being he was. So he started off from the dreadful wilds, and flew on and on until he came to the countries inhabited by men, where his appearance in the air created great consternation. But he alighted nowhere, keeping up a steady flight until he reached the suburbs of the town which had his image on its church. Here, late in the afternoon, he alighted in a green meadow by the side of a brook, and stretched himself on the grass to rest. His great wings were tired, for he had not made such a long flight in a century or more.

The news of his coming spread quickly over the town, and the people, frightened nearly out of their wits by the arrival of so extraordinary a visitor, fled into their houses and shut themselves up. The Griffin called loudly for some one to come to him; but the more he called, the more afraid the people were to show themselves. At length he saw two laborers hurrying to their homes through the fields, and in a terrible voice he commanded them to stop. Not daring to disobey, the men stood, trembling.

'What is the matter with you all?' cried the Griffin. 'Is there not a man in your town who is brave enough to speak to me?'

'I think,' said one of the laborers, his voice shaking so that his words could hardly be understood, 'that—perhaps—the Minor Canon—would come.'

'Go, call him, then!' said the Griffin. 'I want to see him.'

The Minor Canon, who filled a subordinate position in the old church, had just finished the afternoon service, and was coming out of a side door, with three aged women who had formed the week-day congregation. He was a young man of a kind disposition, and very anxious to do good to the people of the town. Apart from his duties in the church, where he conducted services every week-day, he visited the sick and the poor; counselled and assisted persons who were in trouble, and taught a school composed entirely of the bad children in the town, with whom nobody else would have anything to do. Whenever the people wanted something difficult done for them, they always went to the Minor Canon. Thus it was that the laborer thought of the young priest when he found that some one must come and speak to the Griffin.

The Minor Canon had not heard of the strange event, which was known to the whole town except himself and the three old women, and when he was informed of it, and was told that the Griffin had asked to see him, he was greatly amazed and frightened.

'Me!' he exclaimed. 'He has never heard of me! What should he want with *me*?'

'Oh, you must go instantly!' cried the two men. 'He is very angry now because he has been kept waiting so long, and nobody knows what may happen if you don't hurry to him.'

The poor Minor Canon would rather have had his hand cut off than to go out to meet an angry griffin; but he felt that it was his duty to go, for it would be a woeful thing if injury should come to the people of the town because he was not brave enough to obey the summons of the Griffin; so, pale and frightened, he started off.

'Well,' said the Griffin, as soon as the young man came near, 'I am glad to see that there is some one who has the courage to come to me.'

The Minor Canon did not feel very courageous, but he bowed his head.

'Is this the town,' said the Griffin, 'where there is a church with a likeness of myself over one of the doors?'

The Minor Canon looked at the frightful creature before him, and saw that it was, without doubt, exactly like the stone image on the church. 'Yes,' he said, 'you are right.'

'Well, then,' said the Griffin, 'will you take me to it? I wish very much to see it.'

The Minor Canon instantly thought that if the Griffin entered the town without the people knowing what he came for, some of them would probably be frightened to death, and so he sought to gain time to prepare their minds.

'It is growing dark now,' he said, very much afraid, as he spoke, that his words might enrage the Griffin, 'and objects on the front of the church cannot be seen clearly. It will be better to wait until morning, if you wish to get a good view of the stone image of yourself.'

'That will suit me very well,' said the Griffin. 'I see you are a man of good sense. I am tired, and I will take a nap here on this soft grass, while I cool my tail in the little stream that runs near me. The end of my tail gets red-hot when I am angry or excited, and it is quite warm now. So you may go; but be sure and come early tomorrow morning, and show me the way to the church.'

The Minor Canon was glad enough to take his leave, and hurried into the town. In front of the church he found a great many people assembled to hear his report of his interview with the Griffin. When they found that he had not come to spread ruin and devastation, but simply to see his stony likeness on the church, they showed neither relief nor gratification, but began to upbraid the Minor Canon for consenting to conduct the creature into the town.

'What could I do?' cried the young man. 'If I should not bring him he would come himself, and perhaps end by setting fire to the town with his red-hot tail.'

Still the people were not satisfied, and a great many plans were proposed to prevent the Griffin from coming into the town. Some elderly persons urged that the young men should go out and kill him. But the young men scoffed at such a ridiculous idea. Then some one said that it would be a good thing to destroy the stone image, so that the Griffin would have no excuse for entering the town. This proposal was received with such favor that many of the people ran for hammers, chisels, and crowbars with which to tear down and break up the stone griffin. But the Minor Canon resisted this plan with all the strength of his mind and body. He assured the people that this action would enrage the Griffin beyond measure, for it would be impossible to conceal from him that his image had been destroyed during the night.

But they were so determined to break up the stone griffin that the Minor Canon saw that there was nothing for him to do but to stay there and protect it. All night he walked up and down in front of the church door, keeping away the men who brought ladders by which they might mount to the great stone griffin and knock it to pieces with their hammers and crowbars. After many hours the people were obliged to give up their attempts, and went home to sleep. But the Minor Canon remained at his post till early morning, and then he hurried away to the field where he had left the Griffin.

The monster had just awakened, and rising to his fore-legs and shaking himself, he said that he was ready to go into the town. The

Minor Canon, therefore, walked back, the Griffin flying slowly through the air at a short distance above the head of his guide. Not a person was to be seen in the streets, and they proceeded directly to the front of the church, where the Minor Canon pointed out the stone griffin.

The real Griffin settled down in the little square before the church and gazed earnestly at his sculptured likeness. For a long time he looked at it. First he put his head on one side, and then he put it on the other. Then he shut his right eye and gazed with his left, after which he shut his left eye and gazed with his right. Then he moved a little to one side and looked at the image, then he moved the other way. After a while he said to the Minor Canon, who had been standing by all this time:

'It is, it must be, an excellent likeness! That breadth between the eyes, that expansive forehead, those massive jaws! I feel that it must resemble me. If there is any fault to find with it, it is that the neck seems a little stiff. But that is nothing. It is an admirable likeness—admirable!'

The Griffin sat looking at his image all the morning and all the afternoon. The Minor Canon had been afraid to go away and leave him, and had hoped all through the day that he would soon be satisfied with his inspection and fly away home. But by evening the poor young man was utterly exhausted, and felt that he must eat and sleep. He frankly admitted this fact to the Griffin, and asked him if he would not like something to eat. He said this because he felt obliged in politeness to do so; but as soon as he had spoken the words, he was seized with dread lest the monster should demand half a dozen babies, or some tempting repast of that kind.

'Oh, no,' said the Griffin, 'I never eat between the equinoxes. At the vernal and at the autumnal equinox I take a good meal, and that lasts me for half a year. I am extremely regular in my habits, and do not think it healthful to eat at odd times. But if you need food, go and get it, and I will return to the soft grass where I slept last night, and take another nap.'

The next day the Griffin came again to the little square before the church, and remained there until evening steadfastly regarding the stone griffin over the door. The Minor Canon came once or twice to look at him, and the Griffin seemed very glad to see him. But the young clergyman could not stay as he had done before, for he had many duties to perform. Nobody went to the church, but the people came to the Minor Canon's house, and anxiously asked him how long the Griffin was going to stay.

'I do not know,' he answered, 'but I think he will soon be satisfied with looking at his stone likeness, and then he will go away.'

But the Griffin did not go away. Morning after morning he went to the church, but after a time he did not stay there all day. He seemed to have taken a great fancy to the Minor Canon, and followed him about as he pursued his various avocations. He would wait for him at the side door of the church, for the Minor Canon held services every day, morning and evening, though nobody came now. 'If any one should come,' he said to himself, 'I must be found at my post.' When the young man came out, the Griffin would accompany him in his visits to the sick and the poor, and would often look into the windows of the school-house where the Minor Canon was teaching his unruly scholars. All the other schools were closed, but the parents of the Minor Canon's scholars forced them to go to school, because they were so bad they could not endure them all day at home—griffin or no griffin. But it must be said they generally behaved very well when that great monster sat up on his tail and looked in at the school-room window.

When it was perceived that the Griffin showed no sign of going away, all the people who were able to do so, left the town. The canons and the higher officers of the church had fled away during the first day of the Griffin's visit, leaving behind only the Minor Canon and some of the men who opened the doors and swept the church. All the citizens who could afford it shut up their houses and travelled to distant parts, and only the working-people and the poor were left behind. After some days these ventured to go about and attend to their business, for if they did not work they would starve. They were getting a little used to seeing the Griffin, and having been told that he did not eat between equinoxes, they did not feel so much afraid of him as before.

Day by day the Griffin became more and more attached to the Minor Canon. He kept near him a great part of the time, and often spent the night in front of the little house where the young clergyman lived alone. This strange companionship was often burdensome to the Minor Canon. But, on the other hand, he could not deny that he derived a great deal of benefit and instruction from it.

The Griffin had lived for hundreds of years, and had seen much, and he told the Minor Canon many wonderful things.

'It is like reading an old book,' said the young clergyman to himself. 'But how many books I would have had to read before I would have found out what the Griffin has told me about the earth, the air, the water, about minerals, and metals, and growing things, and all the wonders of the world!'

Thus the summer went on, and drew toward its close. And now the people of the town began to be very much troubled again.

'It will not be long,' they said, 'before the autumnal equinox is here, and then that monster will want to eat. He will be dreadfully hungry, for he has taken so much exercise since his last meal. He will devour our children. Without doubt, he will eat them all. What is to be done?'

To this question no one could give an answer, but all agreed that the Griffin must not be allowed to remain until the approaching equinox. After talking over the matter a great deal, a crowd of the people went to the Minor Canon, at a time when the Griffin was not with him.

'It is all your fault,' they said, 'that that monster is among us. You brought him here, and you ought to see that he goes away. It is only on your account that he stays here at all, for, although he visits his image every day, he is with you the greater part of the time. If you were not here he would not stay. It is your duty to go away, and then he will follow you, and we shall be free from the dreadful danger which hangs over us.'

'Go away!' cried the Minor Canon, greatly grieved at being spoken to in such a way. 'Where shall I go? If I go to some other town, shall I not take this trouble there? Have I a right to do that?'

'No,' said the people, 'you must not go to any other town. There is no town far enough away. You must go to the dreadful wilds where the Griffin lives, and then he will follow you and stay there.'

They did not say whether or not they expected the Minor Canon to stay there also, and he did not ask them anything about it. He bowed his head, and went into his house to think. The more he thought, the more clear it became to his mind that it was his duty to go away, and thus free the town from the presence of the Griffin.

That evening he packed a leather bag full of bread and meat, and early the next morning he set out on his journey to the dreadful wilds. It was a long, weary, and doleful journey, especially after he had gone beyond the habitations of men; but the Minor Canon kept on bravely and never faltered. The way was longer than he had expected, and his provisions soon grew so scanty that he was obliged to eat but a little every day; but he kept up his courage, and pressed on, and after many days of toilsome travel he reached the dreadful wilds.

When the Griffin found that the Minor Canon had left the town, he seemed sorry, but showed no disposition to go and look for him. After a few days had passed, he became much annoyed, and asked some of the people where the Minor Canon had gone. But although the citizens had been so anxious that the young clergyman should go to the dreadful wilds, thinking that the Griffin would immediately follow him, they were now afraid to mention the Minor Canon's destination, for the monster seemed angry already, and if he should suspect their trick, he would doubtless become very much enraged. So every one said he did not know, and the Griffin wandered about disconsolate. One morning he looked into the Minor Canon's school-house, which was always empty now, and thought that it was a shame that everything should suffer on account of the young man's absence.

'It does not matter so much about the church,' he said, 'for nobody went there. But it is a pity about the school. I think I will teach it myself until he returns.'

It was the hour for opening the school, and the Griffin went inside and pulled the rope which rang the school bell. Some of the children who heard the bell ran in to see what was the matter, supposing it to be a joke of one of their companions. But when they saw the Griffin they stood astonished and scared.

'Go tell the other scholars,' said the monster, 'that school is about to open, and that if they are not all here in ten minutes I shall come after them.'

In seven minutes every scholar was in place.

Never was seen such an orderly school. Not a boy or girl moved or uttered a whisper. The Griffin climbed into the master's seat, his wide wings spread on each side of him, because he could not lean back in his chair while they stuck out behind, and his great tail coiled around in front of the desk, the barbed end sticking up, ready to tap any boy or girl who might misbehave. The Griffin now addressed the scholars, telling them that he intended to teach them while their master was away. In speaking he endeavored to imitate, as far as possible, the mild and gentle tones of the Minor Canon, but it must be admitted that in this he was not very successful. He had paid a good deal of attention to the studies of the school, and he determined not to attempt to teach them anything new, but to review them in what they had been studying. So he called up the various classes, and questioned them upon their previous lessons. The children racked their brains to remember what they had learned. They were so afraid of the Griffin's displeasure that they recited as they had never recited before. One of the boys, far down in his class, answered so well that the Griffin was astonished.

'I should think you would be at the head,' said he. 'I am sure you have never been in the habit of reciting so well. Why is this?'

'Because I did not choose to take the trouble,' said the boy, trembling in his boots. He felt obliged to speak the truth, for all the children thought that the great eyes of the Griffin could see right through them, and that he would know when they told a falsehood.

'You ought to be ashamed of yourself,' said the Griffin. 'Go down to the very tail of the class, and if you are not at the head in two days, I shall know the reason why.'

The next afternoon this boy was number one.

It was astonishing how much these children now learned of what they had been studying. It was as if they had been educated over again. The Griffin used no severity toward them, but there was a look about him which made them unwilling to go to bed until they were sure they knew their lessons for the next day.

The Griffin now thought that he ought to visit the sick and the poor, and he began to go about the town for this purpose. The effect upon the sick was miraculous. All except those who were very ill indeed, jumped from their beds when they heard he was coming, and declared themselves quite well. To those who could not get up he gave herbs and roots, which none of them had ever before thought of as medicines, but which the Griffin had seen used in various parts of the world, and most of them recovered. But, for all that, they afterwards said that no matter what happened to them, they hoped that they should never again have such a doctor coming to their bedsides, feeling their pulses and looking at their tongues.

As for the poor, they seemed to have utterly disappeared. All those who had depended upon charity for their daily bread were now at work in some way or other many of them offering to do odd jobs for their neighbors just for the sake of their meals—a thing which before had been seldom heard of in the town. The Griffin could find no one who needed his assistance.

The summer now passed, and the autumnal equinox was rapidly approaching. The citizens were in a state of great alarm and anxiety. The Griffin showed no signs of going away, but seemed to have settled himself permanently among them. In a short time the day for his semi-annual meal would arrive, and then what would happen? The monster would certainly be very hungry, and would devour all their children.

Now they greatly regretted and lamented that they had sent away the Minor Canon. He was the only one on whom they could have depended in this trouble, for he could talk freely with the Griffin, and so find out what could be done. But it would not do to be inactive. Some step must be taken immediately. A meeting of the citizens was called, and two old men were appointed to go and talk to the Griffin. They were instructed to offer to prepare a splendid dinner for him on equinox day—one which would entirely satisfy his hunger. They would offer him the fattest mutton, the most tender beef, fish and game of various sorts, and anything of the kind he might fancy. If none of these suited, they were to mention that there was an orphan asylum in the next town.

'Anything would be better,' said the citizens, 'than to have our dear children devoured.'

The old men went to the Griffin, but their propositions were not received with favor.

'From what I have seen of the people of this town,' said the monster, 'I do not think I could relish anything which was prepared by them. They appear to be all cowards, and, therefore, mean and selfish. As for eating one of them, old or young, I could not think of it for a moment. In fact, there was only one creature in the whole place for whom I could have had any appetite, and that is the Minor Canon, who has gone away. He was brave, and good, and honest, and I think I should have relished him.'

'Ah!' said one of the old men, very politely, 'in that case I wish we had not sent him to the dreadful wilds!'

'What!' cried the Griffin. 'What do you mean? Explain instantly what you are talking about!'

The old man, terribly frightened at what he had said, was obliged to tell how the Minor Canon had been sent away by the people, in the hope that the Griffin might be induced to follow him.

When the monster heard this he became furiously angry. He dashed away from the old men and, spreading his wings, flew backward and forward over the town. He was so much excited that his tail became red-hot, and glowed like a meteor against the evening sky. When at last he settled down in the little field where he usually rested, and thrust his tail into the brook, the steam arose like a cloud, and the water of the stream ran hot through the town. The citizens were greatly frightened, and bitterly blamed the old man for telling about the Minor Canon.

'It is plain,' they said, 'that the Griffin intended at last to go and look for him, and we should have been saved. Now who can tell what misery you have brought upon us?'

The Griffin did not remain long in the little field. As soon as his tail was cool he flew to the town hall and rang the bell. The citizens knew that they were expected to come there, and although they were afraid to go, they were still more afraid to stay away, and they crowded into the hall. The Griffin was on the platform at one end, flapping his wings and walking up and down, and the end of his tail was still so warm that it slightly scorched the boards as he dragged it after him.

When everybody who was able to come was there, the Griffin stood still and addressed the meeting.

'I have had a contemptible opinion of you,' he said, 'ever since I discovered what cowards you are, but I had no idea that you were so ungrateful, selfish, and cruel as I now find you to be. Here was your Minor Canon, who labored day and night for your good, and thought of nothing else but how he might benefit you and make you happy; and as soon as you imagine yourselves threatened with a danger,—for well I know you are dreadfully afraid of me,—you send him off, caring not whether he returns or perishes, hoping thereby to save yourselves. Now, I had conceived a great liking for that young man, and had intended, in a day or two, to go and look him up. But I have changed my mind about him. I shall go and find him, but I shall send him back here to live among you, and I intend that he shall enjoy the reward of his labor and his sacrifices. Go, some of you, to the officers of the church, who so cowardly ran away when I first came here, and tell them never to return to this town under penalty of death. And if, when your Minor Canon comes back to you, you do not bow yourselves before him, put him in the highest place among you, and serve and honor him all his life, beware of my terrible vengeance! There were only two good things in this town: the Minor Canon and the stone image of myself over your church door. One of these you have sent away, and the other I shall carry away myself.'

With these words he dismissed the meeting; and it was time, for the end of his tail had become so hot that there was danger of its setting fire to the building.

The next morning the Griffin came to the church, and tearing the stone image of himself from its fastenings over the great door, he grasped it with his powerful fore-legs and flew up into the air. Then, after hovering over the town for a moment, he gave his tail an angry shake, and took up his flight to the dreadful wilds. When he reached this desolate region, he set the stone griffin upon a ledge of a rock which rose in front of the dismal cave he called his home. There the image occupied a position somewhat similar to that it had had over the church door; and the Griffin, panting with the exertion of carrying such an enormous load to so great a distance, lay down upon the ground, and regarded it with much satisfaction. When he felt somewhat rested he went to look for the Minor Canon. He found the young man, weak and half starved, lying under the shadow of a rock. After picking him up and carrying him to his cave, the Griffin flew away to a distant marsh, where he procured some roots and herbs which he well knew were strengthening and beneficial to man, though he had never tasted them himself. After eating these the Minor Canon was greatly revived, and sat up and listened while the Griffin told him what had happened in the town.

'Do you know,' said the monster, when he had finished, 'that I have had, and still have, a great liking for you?'

'I am very glad to hear it,' said the Minor Canon, with his usual politeness.

'I am not at all sure that you would be,' said the Griffin, 'if you thoroughly understood the state of the case, but we will not consider that now. If some things were different, other things would be otherwise. I have been so enraged by discovering the manner in which you have been treated that I have determined that you shall at last enjoy the rewards and honors to which you are entitled. Lie down and have a good sleep, and then I will take you back to the town.'

As he heard these words, a look of trouble came over the young man's face.

'You need not give yourself any anxiety,' said the Griffin, 'about my return to the town. I shall not remain there. Now that I have that admirable likeness of myself in front of my cave, where I can sit at my leisure and gaze upon its noble features and magnificent proportions, I have no wish to see that abode of cowardly and selfish people.'

The Minor Canon, relieved from his fears, lay back, and dropped into a doze; and when he was sound asleep, the Griffin took him up and carried him back to the town. He arrived just before daybreak, and putting the young man gently on the grass in the little field where he himself used to rest, the monster, without having been seen by any of the people, flew back to his home.

When the Minor Canon made his appearance in the morning among the citizens, the enthusiasm and cordiality with which he was received were truly wonderful. He was taken to a house which had been occupied by one of the banished high officers of the place, and every one was anxious to do all that could be done for his health and comfort. The people crowded into the church when he held services, so that the three old women who used to be his week-day congregation could not get to the best seats, which they had always been in the habit of taking; and the parents of the bad children determined to reform them at home, in order that he might be spared the trouble of keeping up his former school. The Minor Canon was appointed to the highest office of the old church, and before he died he became a bishop.

During the first years after his return from the dreadful wilds, the people of the town looked up to him as a man to whom they were bound to do honor and reverence. But they often, also, looked up to the sky to see if there were any signs of the Griffin coming back. However, in the course of time they learned to honor and reverence their former Minor Canon without the fear of being punished if they did not do so.

But they need never have been afraid of the Griffin. The

autumnal equinox day came round, and the monster ate nothing. If he could not have the Minor Canon, he did not care for anything. So, lying down with his eyes fixed upon the great stone griffin, he gradually declined, and died. It was a good thing for some of the people of the town that they did not know this.

If you should ever visit the old town, you would still see the little griffins on the sides of the church, but the great stone griffin that was over the door is gone.

1887

OSCAR WILDE

The Selfish Giant

EVERY afternoon, as they were coming from school, the children used to go and play in the Giant's garden.

It was a large lovely garden, with soft green grass. Here and there over the grass stood beautiful flowers like stars, and there were twelve peach-trees that in the spring-time broke out into delicate blossoms of pink and pearl, and in the autumn bore rich fruit. The birds sat on the trees and sang so sweetly that the children used to stop their games in order to listen to them. 'How happy we are here!' they cried to each other.

One day the Giant came back. He had been to visit his friend the Cornish ogre, and had stayed with him for seven years. After the seven years were over he had said all that he had to say, for his conversation was limited, and he determined to return to his own castle. When he arrived he saw the children playing in the garden.

'What are you doing there?' he cried in a very gruff voice, and the children ran away.

'My own garden is my own garden,' said the Giant; 'any one can understand that, and I will allow nobody to play in it but myself.' So he built a high wall all round it, and put up a notice-board.

TRESPASSERS
WILL BE
PROSECUTED

He was a very selfish Giant.

The poor children had now nowhere to play. They tried to play on the road, but the road was very dusty and full of hard stones, and they did not like it. They used to wander round the high wall when their lessons were over, and talk about the beautiful garden inside. 'How happy we were there,' they said to each other.

Then the Spring came, and all over the country there were little blossoms and little birds. Only in the garden of the Selfish Giant it was still winter. The birds did not care to sing in it as there were no children, and the trees forgot to blossom. Once a beautiful flower put its head out from the grass, but when it saw the notice-board it was so sorry for the children that it slipped back into the ground again, and went off to sleep. The only people who were pleased were the Snow and the Frost. 'Spring has forgotten this garden,' they cried, 'so we will live here all the year round.' The Snow covered up the grass with her great white cloak, and the Frost painted all the trees silver. Then they invited the North Wind to stay with them, and he came. He was wrapped in furs, and he roared all day about the garden, and blew the chimney-pots down. 'This is a delightful spot,' he said, 'we must ask the Hail on a visit.' So the Hail came. Every day for three hours he rattled on the roof of the castle till he broke most of the slates, and then he ran round and round the garden as fast as he could go. He was dressed in grey, and his breath was like ice.

'I cannot understand why the Spring is so late in coming,' said the Selfish Giant, as he sat at the window and looked out at his cold white garden; 'I hope there will be a change in the weather.'

But the Spring never came, nor the Summer. The Autumn gave golden fruit to every garden, but to the Giant's garden she gave none. 'He is too selfish,' she said. So it was always Winter there, and the North Wind, and the Hail, and the Frost, and the Snow danced about through the trees.

One morning the Giant was lying awake in bed when he heard some lovely music. It sounded so sweet to his ears that he thought it must be the King's musicians passing by. It was really only a little linnet singing outside his window, but it was so long since he had heard a bird sing in his garden that it seemed to him to be the most beautiful music in the world. Then the Hail stopped dancing over his head, and the North Wind ceased roaring, and a delicious perfume came to him through the open casement. 'I believe the Spring has come at last,' said the Giant; and he jumped out of bed and looked out.

What did he see?

He saw a most wonderful sight. Through a little hole in the wall the children had crept in, and they were sitting in the branches of the trees. In every tree that he could see there was a little child.

And the trees were so glad to have the children back again that they had covered themselves with blossoms, and were waving their arms gently above the children's heads. The birds were flying about and twittering with delight, and the flowers were looking up through the green grass and laughing. It was a lovely scene, only in one corner it was still winter. It was the farthest corner of the garden, and in it was standing a little boy. He was so small that he could not reach up to the branches of the tree, and he was wandering all round it, crying bitterly. The poor tree was still quite covered with frost and snow, and the North Wind was blowing and roaring above it. 'Climb up! little boy,' said the Tree, and it bent its branches down as low as it could; but the boy was too tiny.

And the Giant's heart melted as he looked out. 'How selfish I have been!' he said; 'now I know why the Spring would not come here. I will put that poor little boy on the top of the tree, and then I will knock down the wall, and my garden shall be the children's playground for ever and ever.' He was really very sorry for what he had done.

So he crept downstairs and opened the front door quite softly, and went out into the garden. But when the children saw him they were so frightened that they all ran away, and the garden became winter again. Only the little boy did not run, for his eyes were so full of tears that he did not see the Giant coming. And the Giant stole up behind him and took him gently in his hand, and put him up into the tree. And the tree broke at once into blossom, and the birds came and sang on it, and the little boy stretched out his two arms and flung them round the Giant's neck, and kissed him. And the other children, when they saw that the Giant was not wicked any longer, came running back, and with them came the Spring. 'It is your garden now, little children,' said the Giant, and he took a great axe and knocked down the wall. And when the people were going to market at twelve o'clock they found the Giant playing with the children in the most beautiful garden they had ever seen.

All day long they played, and in the evening they came to the Giant to bid him good-bye.

'But where is your little companion?' he said: 'the boy I put into the tree.' The Giant loved him the best because he had kissed him.

'We don't know,' answered the children; 'he has gone away.'

'You must tell him to be sure and come here to-morrow,' said the Giant. But the children said that they did not know where he lived, and had never seen him before; and the Giant felt very sad.

Every afternoon, when school was over, the children came and played with the Giant. But the little boy whom the Giant loved was never seen again. The Giant was very kind to all the children, yet he longed for his first little friend, and often spoke of him. 'How I would like to see him!' he used to say.

Years went over, and the Giant grew very old and feeble. He could not play about any more, so he sat in a huge armchair, and watched the children at their games, and admired his garden. 'I have many beautiful flowers,' he said; 'but the children are the most beautiful flowers of all.'

One winter morning he looked out of his window as he was dressing. He did not hate the Winter now, for he knew that it was merely the Spring asleep, and that the flowers were resting.

Suddenly he rubbed his eyes in wonder, and looked and looked. It certainly was a marvellous sight. In the farthest corner of the garden was a tree quite covered with lovely white blossoms. Its branches were all golden, and silver fruit hung down from them, and underneath it stood the little boy he had loved.

Downstairs ran the Giant in great joy, and out into the garden. He hastened across the grass, and came near to the child. And when he came quite close his face grew red with anger, and he said, 'Who hath dared to wound thee?' For on the palms of the child's hands were the prints of two nails, and the prints of two nails were on the little feet.

'Who hath dared to wound thee?' cried the Giant; 'tell me, that I may take my big sword and slay him.'

'Nay!' answered the child; 'but these are the wounds of Love.'

'Who art thou?' said the Giant, and a strange awe fell on him, and he knelt before the little child.

And the child smiled on the Giant, and said to him, 'You let me play once in your garden, to-day you shall come with me to my garden, which is Paradise.'

And when the children ran in that afternoon, they found the Giant lying dead under the tree, all covered with white blossoms.

1888

LAURENCE HOUSMAN

The Rooted Lover

MORNING and evening a ploughboy went driving his team through a lane at the back of the palace garden. Over the hedge the wind came sweet with the scents of a thousand flowers, and through the hedge shot glimpses of all the colours of the rainbow, while now and then went the sheen of silver and gold tissue when the Princess herself paced by with her maidens. Also above all the crying and calling of the blackbirds and thrushes that filled the gardens with song, came now and then an airy exquisite voice flooding from bower to field; and that was the voice of the Princess Fleur-de-lis herself singing.

When she sang all the birds grew silent; new flowers came into bud to hear her and into blossom to look at her; apples and pears ripened and dropped down at her feet; her voice sang the bees home as if it were evening: and the ploughboy as he passed stuck his face into the thorny hedge, and feasted his eyes and ears with the sight and sound of her beauty.

He was a red-faced boy, red with the wind and the sun: over his face his hair rose like a fair flame, but his eyes were black and bold, and for love he had the heart of a true gentleman.

Yet he was but a ploughboy, rough-shod and poorly clad in coat of frieze, and great horses went at a word from him. But no word from him might move the heart of that great Princess; she never noticed the sound of his team as it jingled by, nor saw the dark eyes and the bronzed red face wedged into the thorn hedge for love of her.

'Ah! Princess,' sighed the ploughboy to himself, as the thorns pricked into his flesh, 'were it but a thorn-hedge which had to be trampled down, you should be my bride to-morrow!' But shut off by the thorns, he was not a whit further from winning her than if he had been kneeling at her feet.

He had no wealth in all the world, only a poor hut with poppies growing at the door; no mother or father, and his own living to get. To think at all of the Princess was the sign either of a knave or a fool.

No knave, but perhaps a fool, he thought himself to be. 'I will go,' he said at last, 'to the wise woman who tells fortunes and works strange cures, and ask her to help me.'

So he took all the money he had in the world and went to the wise woman in her house by the dark pool, and said, 'Show me how I may win Princess Fleur-de-lis to be my wife, and I will give you everything I possess.'

'That is a hard thing you ask,' said the wise woman; 'how much dare you risk for it?'

'Anything you can name,' said he.

'Your life?' said she.

'With all my heart,' he replied; 'for without her I shall but end by dying.'

'Then,' said the wise woman, 'give me your money, and you shall take your own risk.'

Then he gave her all.

'Now,' said she, 'you have but to choose any flower you like, and I will turn you into it; then, in the night I will take you and plant you in the palace garden; and if before you die the Princess touches you with her lips and lays you as a flower in her bosom, you shall become a man again and win her love; but if not, when the flower dies you will die too and be no more. So if that seem to you a good bargain, you have but to name your flower, and the thing is done.'

'Agreed, with all my heart!' cried the ploughboy. 'Only make me into some flower that is like me, for I would have the Princess to know what sort of a man I am, so that she shall not be deceived when she takes me to her bosom.'

He looked himself up and he looked himself down in the pool which was before the wise woman's home; at his rough frieze coat with its frayed edges, his long supple limbs, and his red face with its black eyes, and hair gleaming at the top.

'I am altogether like a poppy,' he said, 'what with my red head, and my rough coat, and my life among fields which the plough turns to furrow. Make a poppy of me, and put me in the palace garden, and I will be content.'

Then she stroked him down with her wand full couthly, and muttered her wise saws over him, for she was a wonderful witch-woman; and he turned before her very eyes into a great red poppy, and his coat of frieze became green and hairy all over him, and his feet ran down into the ground like roots.

The wise woman got a big flower-pot and a spade; and she dug him up out of the ground and planted him in the pot, and having watered him well, waited till it was quite dark.

As soon as the pole-star had hung out its light she got across her besom, tucked the flower-pot under her arm, and sailed away over hedge and ditch till she came to the palace garden.

There she dug a hole in a border by one of the walks, shook the ploughboy out of his flower-pot, and planted him with his feet deep down in the soil. Then giving a wink all round, and a wink up to the stars, she set her cap to the east, mounted her besom, and rode away into thin space.

But the poppy stood up where she had left him taking care of his petals, so as to be ready to show them off to the Princess the next morning. He did not go fast asleep, but just dozed the time away, and found it quite pleasant to be a flower, the night being warm. Now and then small insects ran up his stalks, or a mole passed under his roots, reminding him of the mice at home. But the poppy's chief thought was for the morning to return; for then would come the Princess walking straight to where he stood, and would reach out a hand and gather him, and lay her lips to his and his head upon her bosom, so that in the shaking of a breath he could turn again to his right shape, and her love would be won for ever.

Morning came, and gardeners with their brooms and barrows went all about, sweeping up the leaves, and polishing off the slugs from the gravel-paths. The head gardener came and looked at the poppy. 'Who has been putting this weed here?' he cried. And at that the poppy felt a shiver of red ruin go through him; for what if the gardener were to weed him up so that he could never see the Princess again?

All the other gardeners came and considered him, twisting wry faces at him. But they said, 'Perhaps it is a whim of the Princess's. It's none of our planting.' So after all they let him be.

The sun rose higher and higher, and the gardeners went carrying away their barrows and brooms; but the poppy stood waiting with his black eye turned to the way by which the Princess should come.

It was a long waiting, for princesses do not rise with the lark, and the poppy began to think his petals would be all shrivelled and old before she came. But at last he saw slim white feet under the green boughs and heard voices and shawm-like laughter and knew that it was the Princess coming to him.

Down the long walks he watched her go, pausing here and there to taste a fruit that fell or to look at a flower that opened. To him she would come shortly, and so bravely would he woo her with his red face, that she would at once bend down and press her lips to his, and lift him softly to her bosom. Yes, surely she would do this.

She came; she stopped full and began looking at him: he burned under her gaze. 'That is very beautiful!' she said at last. 'Why have I not seen that flower before? Is it so rare, then, that there is no other?' But, 'Oh, it is too common!' cried all her maids in a chorus; 'it is only a common poppy such as grows wild in the fields.'

'Yet it is very beautiful,' said the Princess; and she looked at it long before she passed on. She half bent to it. 'Surely now,' said the poppy, 'her lips to mine!'

'Has it a sweet smell?' she asked. But one of her maids said, 'No, only a poor little stuffy smell, not nice at all!' and the Princess drew back.

'Alas, alas,' murmured the poor poppy in his heart, as he watched her departing, 'why did I forget to choose a flower with a sweet smell? then surely at this moment she would have been mine.' He felt as if his one chance were gone, and death already overtaking him. But he remained brave: 'At least,' he said, 'I will die looking at her; I will not faint or wither, till I have no life left in me. And after all there is to-morrow.' So he went to sleep hoping much, and slept late into the morning of the next day.

Opening his eyes he was aware of a great blaze of red in a border to his right. Ears had been attentive to the words of Princess Fleur-de-lis, and a whole bed of poppies had been planted to gratify her latest fancy. There they were, in a thick mass, burning the air around them with their beauty. Alas! against their hundreds what chance had he?

And the Princess came and stood by them, lost in admiration, while the poppy turned to her his love-sick eye, trying to look braver than them all. And she being gracious, and not forgetful of what first had given her pleasure, came and looked at him also, but not very long; and as for her lips, there was no chance for him there now. Yet for the delight of those few moments he was almost contented with the fate he had chosen—to be a flower, and to die as a flower so soon as his petals fell.

Days came and went; they were all alike now, save that the Princess stayed less often to look at him or the other poppies which had stolen his last chance from him. He saw autumn changes coming over the garden: flowers sickened and fell, and were removed, and the nights began to get cold.

Beside him the other poppies were losing their leaves, and their flaming tops had grown scantier; but for a little while he would hold out still: so long as he had life his eye should stay open to look at the Princess as she passed by.

The sweet-smelling flowers were gone, but the loss of their fragrant rivalry gave him no greater hopes: one by one every gorgeous colour dropped away; only when a late evening primrose hung her lamp beside him in the dusk did he feel that there was anything left as bright as himself to the eye. And now death was taking hold of him, each night twisting and shrivelling his leaves; but still he held up his head, determined that, though but for one more day, his eye should be blessed by a sight of his Princess. If he could keep looking at her he believed he should dream of her when dead.

At length he could see that he was the very last of all the poppies, the only spot of flame in a garden that had gone grey. In the cold dewy mornings cobwebs hung their silvery hammocks about the leaves, and the sun came through mist, making them sparkle. And beautiful they were, but to him they looked like the winding-sheet of his dead hopes.

Now it happened just about this time that the Prince of a neighbouring country was coming to the Court to ask Princess Fleur-de-lis' hand in marriage. The fame of his manners and of his good looks had gone before him, and the Princess being bred to the understanding that princesses must marry for the good of nations according to the bidding of their parents, was willing, since the King her father wished it, to look upon his suit with favour. All that she looked for was to be wooed with sufficient ardour, and to be allowed time for a becoming hesitancy before yielding.

A great ball was prepared to welcome the Prince on his arrival; and when the day came, Princess Fleur-de-lis went into the garden to find some flower that she might wear as an adornment of her loveliness. But almost everything had died of frost, and the only flower that retained its full beauty was the poor bewitched poppy, kept alive for love of her.

'How wonderfully that red flower has lasted!' she said to one of her maidens. 'Gather it for me, and I will wear it with my dress to-night.'

The poppy, not knowing that he was about to meet a much more dangerous rival than any flower, thrilled and almost fainted for bliss as the maid picked him from the stalk and carried him in.

He lay upon Princess Fleur-de-lis' toilet-table and watched the putting on of her ballroom array. 'If she puts me in her breast,' he thought, 'she must some time touch me with her lips; and then!'

And then, when the maid was giving soft finishing touches to the Princess's hair, the beloved one herself took up the poppy and arranged it in the meshes of gold. 'Alas!' thought the poppy, even while he nestled blissfully in its warm depths, 'I shall never reach her lips from here; but I shall dream of her when dead; and for a ploughboy, that surely is enough of happiness.'

So he went down with her to the ball, and could feel the soft throbbing of her temples, for she had not yet seen this Prince who was to be her lover, and her head was full of gentle agitation and excitement to know what he would be like. Very soon he was presented to her in state. Certainly he was extremely passable: he was tall and fine and had a pair of splendid mustachios that stuck out under his nostrils like walrus-tusks, and curled themselves like ram's horns. Beyond a slight fear that these might sweep her away when he tried to kiss her, she favoured his looks sufficiently to be prepared to accept his hand when he offered it.

Then music called to them invitingly, and she was led away to the dance.

As they danced the Prince said: 'I cannot tell how it is, I feel as if some one were looking at me.'

'Half the world is looking at you,' said the Princess in slight mockery. 'Do you not know you are dancing with Princess Fleur-de-lis?'

'Beautiful Princess,' said the Prince, 'can I ever forget it? But it is not in that way I feel myself looked at. I could swear I have seen somewhere a man with a sunburnt face and a bold black eye looking at me.'

'There is no such here,' said the Princess; and they danced on.

When the dance was over the Prince led her to a seat screened from view by rich hangings of silken tapestry; and Princess Fleur-de-lis knew that the time for the wooing was come.

She looked at him; quite clearly she meant to say 'yes.' Without being glad, she was not sorry. If he wooed well she would have him.

'It is strange,' said the Prince, 'I certainly feel that I am being looked at.'

The Princess was offended. 'I am not looking at you in the least,' she said slightingly.

'Ah!' replied the other, 'if you did, I should lose at once any less pleasant sensation; for when your eyes are upon me I know only that I love you—you, Princess, who are the most beautiful, the most radiant, the most accomplished, the most charming of your sex! Why should I waste time in laying my heart bare before you? It is here; it is yours. Take it!'

'Truly,' thought the Princess, 'this is very pretty wooing, and by no means ill done.' She bent down her head, and she toyed and she coyed, but she would not say 'yes' yet.

But the poppy, when he heard the Prince's words, first went all of a tremble, and then giving a great jump fell down at the Princess's feet. And she, toying and coying, and not wishing to say 'yes' yet, bent down and taking up the poppy from where it had fallen, brushed it gently to and fro over her lips to conceal her smiles, and then tucking her chin down into the dimples of her neck began to arrange the flower in the bosom of her gown.

As she did so, all of a sudden a startled look came over her face. 'Oh! I am afraid!' she cried. 'The man, the man with the red face, and the strong black eyes!'

'What is the matter?' demanded the Prince, bending over her in the greatest concern.

'No, no!' she cried, 'go away! Don't touch me! I can't and I won't marry you! Oh, dear! Oh, dear! what is going to become of me?' And she jumped up and ran right away out of the ballroom, and up the great staircase, where she let the poppy fall, and right into her own room, where she barred and bolted herself in.

In the palace there was the greatest confusion: everybody was running about and shaking heads at everybody else. 'Heads and tails! has it come to this?' cried the King, as he saw a party of serving men turning out a ploughboy who by some unheard-of means had found his way into the palace. Then he went up to interview his daughter as to her strange and sudden refusal of the Prince.

The Princess wrung her hands and cried: she didn't know why, but she couldn't help herself: nothing on earth should induce her to marry him.

Then the King was full of wrath, and declared that if she were not ready to obey him in three days' time, she should be turned out into the world like a beggar to find a living for herself.

So for three days the Princess was locked up and kept on nothing but bread and water; and every day she cried less, and was more determined than ever not to marry the Prince.

'Whom do you suppose you are going to marry then?' demanded the King in a fury.

'I don't know,' said the Princess, 'I only know he is a dear; and has got a beautiful tanned face and bold black eyes.'

The King felt inclined to have all the tanned faces and bold black eyes in his kingdom put to death: but as the Princess's obstinacy showed no signs of abating he ended by venting all his anger upon her. So on the third day she was clothed in rags, and had all her jewelry taken off her, and was turned out of the palace to find her way through the world alone.

And as she went on and on, crying and wondering what would become of her, she suddenly saw by the side of the road a charming cottage with poppies growing at the door. And in the doorway stood a beautiful man, with a tanned face and bold black eyes, looking as like a poppy as it was possible for a man to look.

Then he opened his arms: and the Princess opened her arms: and he ran, and she ran. And they ran and they ran and they ran, till they were locked in each other's arms, and lived happily ever after.

1894

ROBERT LOUIS STEVENSON

The Song of the Morrow

THE King of Duntrine had a daughter when he was old, and she was the fairest King's daughter between two seas; her hair was like spun gold, and her eyes like pools in a river; and the King gave her a castle upon the sea beach, with a terrace, and a court of the hewn stone, and four towers at the four corners. Here she dwelt and grew up, and had no care for the morrow, and no power upon the hour, after the manner of simple men.

It befell that she walked one day by the beach of the sea, when it was autumn, and the wind blew from the place of rains; and upon the one hand of her the sea beat, and upon the other the dead leaves ran. This was the loneliest beach between two seas, and strange things had been done there in the ancient ages. Now the King's daughter was aware of a crone that sat upon the beach. The sea foam ran to her feet, and the dead leaves swarmed about her back, and the rags blew about her face in the blowing of the wind.

'Now,' said the King's daughter, and she named a holy name, 'this is the most unhappy old crone between two seas.'

'Daughter of a King,' said the crone, 'you dwell in a stone house, and your hair is like the gold: but what is your profit? Life is not long, nor lives strong; and you live after the way of simple men, and have no thought for the morrow and no power upon the hour.'

'Thought for the morrow, that I have,' said the King's daughter; 'but power upon the hour, that have I not.' And she mused with herself.

Then the crone smote her lean hands one within the other, and laughed like a sea-gull. 'Home!' cried she. 'O daughter of a King, home to your stone house; for the longing is come upon you now, nor can you live any more after the manner of simple men. Home, and toil and suffer, till the gift come that will make you bare, and till the man come that will bring you care.'

The King's daughter made no more ado, but she turned about and went home to her house in silence. And when she was come into her chamber she called for her nurse.

'Nurse,' said the King's daughter, 'thought is come upon me for the morrow, so that I can live no more after the manner of simple men. Tell me what I must do that I may have power upon the hour.'

Then the nurse moaned like a snow wind. 'Alas!' said she, 'that this thing should be; but the thought is gone into your marrow, nor is there any cure against the thought. Be it so, then, even as you will; though power is less than weakness, power shall you have; and though the thought is colder than winter, yet shall you think it to an end.'

So the King's daughter sat in her vaulted chamber in the masoned house, and she thought upon the thought. Nine years she sat; and the sea beat upon the terrace, and the gulls cried about the turrets, and wind crooned in the chimneys of the house. Nine years she came not abroad, nor tasted the clean air, neither saw God's sky. Nine years she sat and looked neither to the right nor to the left, nor heard speech of any one, but thought upon the thought of the morrow. And her nurse fed her in silence, and she took of the food with her left hand, and ate it without grace.

Now when the nine years were out, it fell dusk in the autumn, and there came a sound in the wind like a sound of piping. At that the nurse lifted up her finger in the vaulted house.

'I hear a sound in the wind,' said she, 'that is like the sound of piping.'

'It is but a little sound,' said the King's daughter, 'but yet is it sound enough for me.'

So they went down in the dusk to the doors of the house, and along the beach of the sea. And the waves beat upon the one hand, and upon the other the dead leaves ran; and the clouds raced in the sky, and the gulls flew widdershins. And when they came to that part of the beach where strange things had been done in the ancient ages, lo! there was the crone, and she was dancing widdershins.

'What makes you dance widdershins, old crone?' said the King's daughter; 'here upon the bleak beach, between the waves and the dead leaves?'

'I hear a sound in the wind that is like a sound of piping,' quoth she. 'And it is for that that I dance widdershins. For the gift comes that will make you bare, and the man comes that must bring you care. But for me the morrow is come that I have thought upon, and the hour of my power.'

'How comes it, crone,' said the King's daughter, 'that you waver like a rag, and pale like a dead leaf before my eyes?'

'Because the morrow has come that I have thought upon, and the hour of my power,' said the crone; and she fell on the beach, and, lo! she was but stalks of the sea tangle, and dust of the sea sand, and the sand lice hopped upon the place of her.

'This is the strangest thing that befell between two seas,' said the King's daughter of Duntrine.

But the nurse broke out and moaned like an autumn gale. 'I am weary of the wind,' quoth she; and she bewailed her day.

The King's daughter was aware of a man upon the beach; he went hooded so that none might perceive his face, and a pipe was underneath his arm. The sound of his pipe was like singing wasps, and like the wind that sings in windlestraw; and it took hold upon men's ears like the crying of gulls.

'Are you the comer?' quoth the King's daughter of Duntrine.

'I am the comer,' said he, 'and these are the pipes that a man may hear, and I have power upon the hour, and this is the song of the morrow.' And he piped the song of the morrow, and it was as long as years; and the nurse wept out aloud at the hearing of it.

'This is true,' said the King's daughter, 'that you pipe the song of the morrow; but that ye have power upon the hour, how may I know that? Show me a marvel here upon the beach, between the waves and the dead leaves.'

And the man said, 'Upon whom?'

'Here is my nurse,' quoth the King's daughter. 'She is weary of the wind. Show me a good marvel upon her.'

And, lo! the nurse fell upon the beach as it were two handfuls of dead leaves, and the wind whirled them widdershins, and the sand lice hopped between.

'It is true,' said the King's daughter of Duntrine; 'you are the comer, and you have power upon the hour. Come with me to my stone house.'

So they went by the sea margin, and the man piped the song of the morrow, and the leaves followed behind them as they went. Then they sat down together; and the sea beat on the terrace, and the gulls cried about the towers, and the wind crooned in the chimneys of the house. Nine years they sat, and every year when it fell autumn, the man said, 'This is the hour, and I have power in it'; and the daughter of the King said, 'Nay, but pipe me the song of the morrow.' And he piped it, and it was long like years.

Now when the nine years were gone, the King's daughter of Duntrine got her to her feet, like one that remembers; and she looked about her in the masoned house; and all her servants were gone; only the man that piped sat upon the terrace with the hood upon his face; and as he piped the leaves ran about the terrace and the sea beat along the wall. Then she cried to him with a great voice, 'This is the hour, and let me see the power in it.' And with that the wind blew off the hood from the man's face, and, lo! there was no man there, only the clothes and the hood and the pipes tumbled one upon another in a corner of the terrace, and the dead leaves ran over them.

And the King's daughter of Duntrine got her to that part of the beach where strange things had been done in the ancient ages; and there she sat her down. The sea foam ran to her feet, and the dead leaves swarmed about her back, and the veil blew about her face in the blowing of the wind. And when she lifted up her eyes, there was the daughter of a King come walking on the beach. Her hair was like the spun gold, and her eyes like pools in a river, and she had no thought for the morrow and no power upon the hour, after the manner of simple men.

1894

KENNETH GRAHAME

The Reluctant Dragon

LONG ago—might have been hundreds of years ago—in a cottage half-way between this village and yonder shoulder of the Downs up there, a shepherd lived with his wife and their little son. Now the shepherd spent his days—and at certain times of the year his nights too—up on the wide ocean-bosom of the Downs, with only the sun and the stars and the sheep for company, and the friendly chattering world of men and women far out of sight and hearing. But his little son, when he wasn't helping his father, and often when he was as well, spent much of his time buried in big volumes that he borrowed from the affable gentry and interested parsons of the country round about. And his parents were very fond of him, and rather proud of him too, though they didn't let on in his hearing, so he was left to go his own way and read as much as he liked; and instead of frequently getting a cuff on the side of the head, as might very well have happened to him, he was treated more or less as an equal by his parents, who sensibly thought it a very fair division of labour that they should supply the practical knowledge, and he the book-learning. They knew that book-learning often came in useful at a pinch, in spite of what their neighbours said. What the Boy chiefly dabbled in was natural history and fairy-tales, and he just took them as they came, in a sandwichy sort of way, without making any distinctions; and really his course of reading strikes one as rather sensible.

One evening the shepherd, who for some nights past had been disturbed and preoccupied, and off his usual mental balance, came home all of a tremble, and, sitting down at the table where his wife and son were peacefully employed, she with her seam, he in following out the adventures of the Giant with no Heart in his Body, exclaimed with much agitation:

'It's all up with me, Maria! Never no more can I go up on them there Downs, was it ever so!'

'Now don't you take on like that,' said his wife, who was a *very* sensible woman: 'but tell us all about it first, whatever it is as has given you this shake-up, and then me and you and the son here, between us, we ought to be able to get to the bottom of it!'

'It began some nights ago,' said the shepherd. 'You know that cave up there—I never liked it, somehow, and the sheep never liked it neither, and when sheep don't like a thing there's generally some reason for it. Well, for some time past there's been faint noises coming from that cave—noises like heavy sighings, with grunts mixed up in them; and sometimes a snoring, far away down—*real* snoring, yet somehow not *honest* snoring, like you and me o'nights, you know!'

'*I* know,' remarked the Boy, quietly.

'Of course I was terrible frightened,' the shepherd went on; 'yet somehow I couldn't keep away. So this very evening, before I come down, I took a cast round by the cave, quietly. And there—O Lord! there I saw him at last, as plain as I see you!'

'Saw *who*?' said his wife, beginning to share in her husband's nervous terror.

'Why *him*, I'm a telling you!' said the shepherd. 'He was sticking half-way out of the cave, and seemed to be enjoying of the cool of the evening in a poetical sort of way. He was as big as four cart-horses, and all covered with shiny scales—deep-blue scales at the top of him, shading off to a tender sort o' green below. As he breathed, there was that sort of flicker over his nostrils that you see over our chalk roads on a baking windless day in summer. He had his chin on his paws, and I should say he was meditating about things. Oh, yes, a peaceable sort o' beast enough, and not ramping or carrying on or doing anything but what was quite right and proper. I admit all that. And yet, what am I to do? *Scales*, you know, and claws, and a tail for certain, though I didn't see that end of him—I ain't *used* to 'em, and I don't *hold* with 'em, and that's a fact!'

The Boy, who had apparently been absorbed in his book during his father's recital, now closed the volume, yawned, clasped his hands behind his head, and said sleepily:

'It's all right, father. Don't you worry. It's only a dragon.'

'Only a dragon?' cried his father. 'What do you mean, sitting there, you and your dragons? *Only* a dragon indeed! And what do *you* know about it?'

''Cos it *is*, and 'cos I *do* know,' replied the Boy, quietly. 'Look here, father, you know we've each of us got our line. *You* know about sheep, and weather, and things; *I* know about dragons. I always said, you know, that that cave up there was a dragon-cave. I always said it must have belonged to a dragon some time, and ought to belong to a dragon now, if rules count for anything. Well, now you tell me it *has* got a dragon, and so *that's* all right. I'm not half as much surprised as when you told me it *hadn't* got a dragon. Rules always come right if you wait quietly. Now, please, just leave this all to me. And I'll stroll up to-morrow morning—no, in the morning I can't, I've got a whole heap of things to do—well, perhaps in the evening, if I'm quite free, I'll go up and have a talk to him, and you'll find it'll be all right. Only, please, don't you go worrying round there without me. You don't understand 'em a bit, and they're very sensitive, you know!'

'He's quite right, father,' said the sensible mother. 'As he says, dragons is his line and not ours. He's wonderful knowing about book-beasts, as every one allows. And to tell the truth, I'm not half happy in my own mind, thinking of that poor animal lying alone up there, without a bit o' hot supper or anyone to change the news with; and maybe we'll be able to do something for him; and if he ain't quite respectable our Boy'll find it out quick enough. He's got a pleasant sort o' way with him that makes everybody tell him everything.'

Next day, after he'd had his tea, the Boy strolled up the chalky track that led to the summit of the Downs; and there, sure enough, he found the dragon, stretched lazily on the sward in front of his cave. The view from that point was a magnificent one. To the right and left, the bare and willowy leagues of Downs; in front, the vale, with its clustered homesteads, its threads of white roads running through orchards and well-tilled acreage, and, far away, a hint of grey old cities on the horizon. A cool breeze played over the surface of the grass and the silver shoulder of a large moon was showing above distant junipers. No wonder the dragon seemed in a peaceful and contented mood; indeed, as the Boy approached he could hear the beast purring with a happy regularity. 'Well, we live and learn!' he said to himself. 'None of my books ever told me that dragons purred!'

'Hullo, dragon!' said the Boy, quietly, when he had got up to him.

The dragon, on hearing the approaching footsteps, made the beginning of a courteous effort to rise. But when he saw it was a Boy, he set his eyebrows severely.

'Now don't you hit me,' he said; 'or bung stones, or squirt water, or anything. I won't have it, I tell you!'

'Not goin' to hit you,' said the Boy, wearily, dropping on the grass beside the beast: 'and don't, for goodness' sake, keep on saying "Don't"; I hear so much of it, and it's monotonous, and makes me tired. I've simply looked in to ask you how you were and all that sort of thing; but if I'm in the way I can easily clear out. I've lots of friends, and no one can say I'm in the habit of shoving myself in where I'm not wanted!'

'No, no, don't go off in a huff,' said the dragon, hastily; 'fact is,—I'm as happy up here as the day's long; never without an occupation, dear fellow, never without an occupation! And yet, between ourselves, it *is* a trifle dull at times.'

The Boy bit off a stalk of grass and chewed it. 'Going to make a long stay here?' he asked, politely.

'Can't hardly say at present,' replied the dragon. 'It seems a nice place enough—but I've only been here a short time, and one must look about and reflect and consider before settling down. It's rather a serious thing, settling down. Besides—now I'm going to tell you something! You'd never guess it if you tried ever so!—fact is, I'm such a confoundedly lazy beggar!'

'You surprise me,' said the Boy, civilly.

'It's the sad truth,' the dragon went on, settling down between his paws and evidently delighted to have found a listener at last: 'and I fancy that's really how I came to be here. You see all the other fellows were so active and *earnest* and all that sort of thing—always rampaging, and skirmishing, and scouring the desert sands, and pacing the margin of the sea, and chasing knights all over the place, and devouring damsels, and going on generally—whereas I liked to get my meals regular and then to prop my back against a bit of rock and snooze a bit, and wake up and think of things going on and how they kept going on just the same, you know! So when it happened I got fairly caught.'

'When *what* happened, please?' asked the Boy.

'That's just what I don't precisely know,' said the dragon. 'I suppose the earth sneezed, or shook itself, or the bottom dropped out of something. Anyhow there was a shake and a roar and a general stramash, and I found myself miles away underground and wedged in as tight as tight. Well, thank goodness, my wants are few, and at any rate I had peace and quietness and wasn't always being asked to come along and *do* something. And I've got such an active mind—always occupied, I assure you! But time went on, and there was a certain sameness about the life, and at last I began to think it would be fun to work my way upstairs and see what you other fellows were doing. So I scratched and burrowed, and worked this way and that way and at last I came out through this cave here. And I like the country, and the view, and the people—what I've seen of 'em—and on the whole I feel inclined to settle down here.'

'What's your mind always occupied about?' asked the Boy. 'That's what I want to know.'

The dragon coloured slightly and looked away. Presently he said bashfully:

'Did you ever—just for fun—try to make up poetry—verses, you know?'

''Course I have,' said the Boy. 'Heaps of it. And some of it's quite good, I feel sure, only there's no one here cares about it. Mother's very kind and all that, when I read it to her, and so's father for that matter. But somehow they don't seem to—'

'Exactly,' cried the dragon; 'my own case exactly. They don't seem to, and you can't argue with 'em about it. Now you've got culture, you have, I could tell it on you at once, and I should just like your candid opinion about some little things I threw off lightly, when I was down there. I'm awfully pleased to have met you, and I'm hoping the other neighbours will be equally agreeable. There was a very nice old gentleman up here only last night, but he didn't seem to want to intrude.'

'That was my father,' said the Boy, 'and he *is* a nice old gentleman, and I'll introduce you some day if you like.'

'Can't you two come up here and dine or something to-morrow?' asked the dragon, eagerly. 'Only, of course, if you've got nothing better to do,' he added politely.

'Thanks awfully,' said the Boy, 'but we don't go out anywhere without my mother, and, to tell you the truth, I'm afraid she mightn't quite approve of you. You see there's no getting over the hard fact that you're a dragon, is there? And when you talk of settling down, and the neighbours, and so on, I can't help feeling that you don't quite realise your position. You're an enemy of the human race, you see!'

'Haven't got an enemy in the world,' said the dragon, cheerfully. 'Too lazy to make 'em, to begin with. And if I *do* read other fellows my poetry, I'm always ready to listen to theirs!'

'Oh, dear!' cried the Boy, 'I wish you'd try and grasp the situation properly. When the other people find you out, they'll come after you with spears and swords and all sorts of things. You'll have to be exterminated, according to their way of looking at it! You're a scourge, and a pest, and a baneful monster!'

'Not a word of truth in it,' said the dragon, wagging his head solemnly. 'Character'll bear the strictest investigation. And now, there's a little sonnet-thing I was working on when you appeared on the scene—'

'Oh, if you *won't* be sensible,' cried the Boy, getting up, 'I'm going off home. No, I can't stop for sonnets; my mother's sitting up. I'll look you up to-morrow, sometime or other, and do for goodness' sake try and realise that you're a pestilential scourge, or you'll find yourself in a most awful fix. Good-night!'

The Boy found it an easy matter to set the mind of his parents at ease about his new friend. They had always left that branch to him, and they took his word without a murmur. The shepherd was formally introduced and many compliments and kind inquiries were exchanged. His wife, however, though expressing her willingness to do anything she could,—to mend things, or set the cave to rights, or cook a little something when the dragon had been poring over sonnets and forgotten his meals, as male things *will* do,—could not be brought to recognise him formally. The fact that he was a dragon and 'they didn't know who he was' seemed to count for everything with her. She made no objection, however, to her little son spending his evenings with the dragon quietly, so long as he was home by nine o'clock: and many a pleasant night they had, sitting on the sward, while the dragon told stories of old, old times, when dragons were quite plentiful and the world was a livelier place than it is now, and life was full of thrills and jumps and surprises.

What the Boy had feared, however, soon came to pass. The most modest and retiring dragon in the world, if he's as big as four cart-horses and covered with blue scales, cannot keep altogether out of the public view. And so in the village tavern of nights the fact that a real live dragon sat brooding in the cave on the Downs was naturally a subject for talk. Though the villagers were extremely frightened, they were rather proud as well. It was a distinction to have a dragon of your own, and it was felt to be a feather in the cap of the village. Still, all were agreed that this sort of thing couldn't be allowed to go on. The dreadful beast must be exterminated, the country-side must be freed from this pest, this terror, this destroying scourge. The fact that not even a hen-roost was the worse for the dragon's arrival wasn't allowed to have anything to do with it. He was a dragon, and he couldn't deny it, and if he didn't choose to behave as such that was his own lookout. But in spite of much valiant talk no hero was found willing to take sword and spear and free the suffering village and win deathless fame; and each night's heated discussion always ended in nothing. Meanwhile the dragon, a happy Bohemian, lolled on the turf, enjoyed the sunsets, told antediluvian anecdotes to the Boy, and polished his old verses while meditating on fresh ones.

One day the Boy, on walking in to the village, found everything wearing a festal appearance which was not to be accounted for in the calendar. Carpets and gay-coloured stuffs were hung out of the windows, the church-bells clamoured noisily, the little street was flower-strewn, and the whole population jostled each other along either side of it, chattering, shoving, and ordering each other to stand back. The Boy saw a friend of his own age in the crowd and hailed him.

'What's up?' he cried. 'Is it the players, or bears, or a circus, or what?'

'It's all right,' his friend hailed back. 'He's a-coming.'

'*Who's* a-coming?' demanded the Boy, thrusting into the throng.

'Why, St George, of course,' replied his friend. 'He's heard tell of our dragon, and he's comin' on purpose to slay the deadly beast, and free us from his horrid yoke. O my! won't there be a jolly fight!'

Here was news indeed! The Boy felt that he ought to make quite sure for himself, and he wriggled himself in between the legs of his good-natured elders, abusing them all the time for their unmannerly habit of shoving. Once in the front rank, he breathlessly awaited the arrival.

Presently from the far-away end of the line came the sound of cheering. Next, the measured tramp of a great war-horse made his heart beat quicker, and then he found himself cheering with the rest, as, amidst welcoming shouts, shrill cries of women, uplifting of babies and waving of handkerchiefs, St George paced slowly up the street. The Boy's heart stood still and he breathed with sobs, the beauty and the grace of the hero were so far beyond anything he had yet seen. His fluted armour was inlaid with gold, his plumed helmet hung at his saddle-bow, and his thick fair hair framed a face gracious and gentle beyond expression till you caught the sternness in his eyes. He drew rein in front of the little inn, and the villagers crowded round with greetings and thanks and voluble statements of their wrongs and grievances and oppressions. The Boy heard the grave gentle voice of the Saint, assuring them that all would be well now, and that he would stand by them and see them righted and free them from their foe; then he dismounted and passed through the doorway and the crowd poured in after him. But the Boy made off up the hill as fast as he could lay his legs to the ground.

'It's all up, dragon!' he shouted as soon as he was within sight of the beast. 'He's coming! He's here now! You'll have to pull yourself together and *do* something at last!'

The dragon was licking his scales and rubbing them with a bit of house-flannel the Boy's mother had lent him, till he shone like a great turquoise.

'Don't be *violent*, Boy,' he said without looking round. 'Sit down and get your breath, and try and remember that the noun governs the verb, and then perhaps you'll be good enough to tell me *who's* coming?'

'That's right, take it coolly,' said the Boy. 'Hope you'll be half as cool when I've got through with my news. It's only St George who's coming, that's all; he rode into the village half-an-hour ago. Of course you can lick him—a great big fellow like you! But I thought I'd warn you, 'cos he's sure to be round early, and he's got the longest, wickedest-looking spear you ever did see!' And the Boy got up and began to jump round in sheer delight at the prospect of the battle.

'O deary, deary me,' moaned the dragon; 'this is too awful. I won't see him, and that's flat. I don't want to know the fellow at all. I'm sure he's not nice. You must tell him to go away at once, please. Say he can write if he likes, but I can't give him an interview. I'm not seeing anybody at present.'

'Now dragon, dragon,' said the Boy, imploringly, 'don't be perverse and wrongheaded. You've *got* to fight him some time or other, you know, 'cos he's St George and you're the dragon. Better get it over, and then we can go on with the sonnets. And you ought to consider other people a little, too. If it's been dull up here for you, think how dull it's been for me!'

'My dear little man,' said the dragon, solemnly, 'just understand, once for all, that I can't fight and I won't fight. I've never fought in my life, and I'm not going to begin now, just to give you a Roman holiday. In old days I always let the other fellows—the *earnest* fellows—do all the fighting, and no doubt that's why I have the pleasure of being here now.'

'But if you don't fight he'll cut your head off!' gasped the Boy, miserable at the prospect of losing both his fight and his friend.

'Oh, I think not,' said the dragon in his lazy way. 'You'll be able to arrange something. I've every confidence in you, you're such a *manager*. Just run down, there's a dear chap, and make it all right. I leave it entirely to you.'

The Boy made his way back to the village in a state of great despondency. First of all, there wasn't going to be any fight; next, his dear and honoured friend the dragon hadn't shown up in quite such a heroic light as he would have liked; and lastly, whether the dragon was a hero at heart or not, it made no difference, for St George would most undoubtedly cut his head off. 'Arrange things indeed!' he said bitterly to himself. 'The dragon treats the whole affair as if it was an invitation to tea and croquet.'

The villagers were straggling homewards as he passed up the street, all of them in the highest spirits, and gleefully discussing the splendid fight that was in store. The Boy pursued his way to the inn, and passed into the principal chamber, where St George now sat alone, musing over the chances of the fight, and the sad stories of rapine and of wrong that had so lately been poured into his sympathetic ears.

'May I come in, St George?' said the Boy, politely, as he paused at the door. 'I want to talk to you about this little matter of the dragon, if you're not tired of it by this time.'

'Yes, come in, Boy,' said the Saint, kindly. 'Another tale of misery and wrong, I fear me. Is it a kind parent, then, of whom the tyrant has bereft you? Or some tender sister or brother? Well, it shall soon be avenged.'

'Nothing of the sort,' said the Boy. 'There's a misunderstanding somewhere, and I want to put it right. The fact is, this is a *good* dragon.'

'Exactly,' said St George, smiling pleasantly, 'I quite understand. A good *dragon*. Believe me, I do not in the least regret that he is an adversary worthy of my steel, and no feeble specimen of his noxious tribe.'

'But he's *not* a noxious tribe,' cried the Boy, distressedly. 'Oh dear, oh dear, how *stupid* men are when they get an idea into their heads! I tell you he's a *good* dragon, and a friend of mine, and tells me the most beautiful stories you ever heard, all about old times and when he was little. And he's been so kind to mother, and mother'd do anything for him. And father likes him too, though father doesn't hold with art and poetry much, and always falls asleep when the dragon starts talking about *style*. But the fact is, nobody can help liking him when once they know him. He's so engaging and so trustful, and as simple as a child!'

'Sit down, and draw your chair up,' said St George. 'I like a fellow who sticks up for his friends, and I'm sure the dragon has his good points, if he's got a friend like you. But that's not the question. All this evening I've been listening, with grief and anguish unspeakable, to tales of murder, theft, and wrong; rather too highly coloured, perhaps, not always quite convincing, but forming in the main a most serious roll of crime. History teaches us that the greatest rascals often possess all the domestic virtues; and I fear that your cultivated friend, in spite of the qualities which have won (and rightly) your regard, has got to be speedily exterminated.'

'Oh, you've been taking in all the yarns those fellows have been telling you,' said the Boy, impatiently. 'Why, our villagers are the biggest story-tellers in all the country round. It's a known fact. You're a stranger in these parts, or else you'd have heard it already. All they want is a *fight*. They're the most awful beggars for getting up fights—it's meat and drink to them. Dogs, bulls, dragons—anything so long as it's a fight. Why, they've got a poor innocent badger in the stable behind here, at this moment. They were going to have some fun with him to-day, but they're saving him up now till *your* little affair's over. And I've no doubt they've been telling you what a hero you were, and how you were bound to win, in the cause of right and justice, and so on; but let me tell you, I came down the street just now, and they were betting six to four on the dragon freely!'

'Six to four on the dragon!' murmured St George, sadly, resting his cheek on his hand. 'This is an evil world, and sometimes I begin to think that all the wickedness in it is not entirely bottled up inside the dragons. And yet—may not this wily beast have misled you as to his real character, in order that your good report of him may serve as a cloak for his evil deeds? Nay, may there not be, at this very moment, some hapless Princess immured within yonder gloomy cavern?'

The moment he had spoken, St George was sorry for what he had said, the Boy looked so genuinely distressed.

'I assure you, St George,' he said earnestly, 'there's nothing of the sort in the cave at all. The dragon's a real gentleman, every inch of him, and I may say that no one would be more shocked and grieved than he would, at hearing you talk in that—that *loose* way about matters on which he has very strong views!'

'Well, perhaps I've been over-credulous,' said St George. 'Perhaps I've misjudged the animal. But what are we to do? Here are the dragon and I, almost face to face, each supposed to be thirsting for each other's blood. I don't see any way out of it, exactly. What do you suggest? Can't you arrange things, somehow?'

'That's just what the dragon said,' replied the Boy, rather nettled. 'Really, the way you two seem to leave everything to me—I suppose you couldn't be persuaded to go away quietly, could you?'

'Impossible, I fear,' said the Saint. 'Quite against the rules. *You* know that as well as I do.'

'Well, then, look here,' said the Boy, 'it's early yet—would you mind strolling up with me and seeing the dragon and talking it over? It's not far, and any friend of mine will be most welcome.'

'Well, it's *irregular*,' said St George, rising, 'but really it seems about the most sensible thing to do. You're taking a lot of trouble on your friend's account,' he added, good-naturedly, as they passed out through the door together. 'But cheer up! Perhaps there won't have to be any fight after all.'

'Oh, but I hope there will, though!' replied the little fellow, wistfully.

'I've brought a friend to see you, dragon,' said the Boy, rather loud.

The dragon woke up with a start. 'I was just—er—thinking about things,' he said in his simple way. 'Very pleased to make your acquaintance, sir. Charming weather we're having!'

'This is St George,' said the Boy, shortly. 'St George, let me introduce you to the dragon. We've come up to talk things over quietly, dragon, and now for goodness' sake do let us have a little straight common-sense, and come to some practical business-like arrangement, for I'm sick of views and theories of life and personal tendencies, and all that sort of thing. I may perhaps add that my mother's sitting up.'

'So glad to meet you, St George,' began the dragon, rather nervously, 'because you've been a great traveller, I hear, and I've always been rather a stay-at-home. But I can show you many antiquities, many interesting features of our country-side, if you're stopping here any time—'

'I think,' said St George, in his frank, pleasant way, 'that we'd really better take the advice of our young friend here, and try to come to some understanding, on a business footing, about this little affair of ours. Now don't you think that after all the simplest plan would be just to fight it out, according to the rules, and let the best man win? They're betting on you, I may tell you, down in the village, but I don't mind that!'

'Oh, yes, *do*, dragon,' said the Boy, delightedly; 'it'll save such a lot of bother!'

'My young friend, you shut up,' said the dragon, severely. 'Believe me, St George,' he went on, 'there's nobody in the world I'd sooner oblige than you and this young gentleman here. But the whole thing's nonsense, and conventionality, and popular thick-headedness. There's absolutely nothing to fight about, from beginning to end. And anyhow I'm not going to, so that settles it!'

'But supposing I make you?' said St George, rather nettled.

'You can't,' said the dragon, triumphantly. 'I should only go into my cave and retire for a time down the hole I came up. You'd soon get heartily sick of sitting outside and waiting for me to come out and fight you. And as soon as you'd really gone away, why, I'd come up again gaily, for I tell you frankly, I like this place, and I'm going to stay here!'

St George gazed for a while on the fair landscape around them. 'But this would be a beautiful place for a fight,' he began again persuasively. 'These great bare rolling Downs for the arena,—and me in my golden armour showing up against your big blue scaly coils! Think what a picture it would make!'

'Now you're trying to get at me through my artistic sensibilities,' said the dragon. 'But it won't work. Not but what it would make a very pretty picture, as you say,' he added, wavering a little.

'We seem to be getting rather nearer to *business*,' put in the Boy. 'You must see, dragon, that there's got to be a fight of some sort, 'cos you can't want to have to go down that dirty old hole again and stop there till goodness knows when.'

'It might be arranged,' said St George, thoughtfully. 'I *must* spear you somewhere, of course, but I'm not bound to hurt you very much. There's such a lot of you that there must be a few *spare* places somewhere. Here, for instance, just behind your foreleg. It couldn't hurt you much, just here!'

'Now you're tickling, George,' said the dragon, coyly. 'No, that place won't do at all. Even if it didn't hurt,—and I'm sure it would, awfully,—it would make me laugh, and that would spoil everything.'

'Let's try somewhere else, then,' said St George, patiently. 'Under your neck, for instance,—all these folds of thick skin,—if I speared you here you'd never even know I'd done it!'

'Yes, but are you sure you can hit off the right place?' asked the dragon, anxiously.

'Of course I am,' said St George, with confidence. 'You leave that to me!'

'It's just because I've *got* to leave it to you that I'm asking,' replied the dragon, rather testily. 'No doubt you would deeply regret any error you might make in the hurry of the moment; but you wouldn't regret it half as much as I should! However, I suppose we've got to trust somebody, as we go through life, and your plan seems, on the whole, as good a one as any.'

'Look here, dragon,' interrupted the Boy, a little jealous on behalf of his friend, who seemed to be getting all the worst of the bargain: 'I don't quite see where *you* come in! There's to be a fight, apparently, and you're to be licked; and what I want to know is, what are *you* going to get out of it?'

'St George,' said the dragon, 'just tell him, please,—what will happen after I'm vanquished in the deadly combat?'

'Well, according to the rules I suppose I shall lead you in

triumph down to the market-place or whatever answers to it,' said St George.

'Precisely,' said the dragon. 'And then—'

'And then there'll be shoutings and speeches and things,' continued St George. 'And I shall explain that you're converted, and see the error of your ways, and so on.'

'Quite so,' said the dragon. 'And then—?'

'Oh, and then—' said St George, 'why, and then there will be the usual banquet, I suppose.'

'Exactly,' said the dragon; 'and that's where *I* come in. Look here,' he continued, addressing the Boy, 'I'm bored to death up here, and no one really appreciates me. I'm going into Society, I am, through the kindly aid of our friend here, who's taking such a lot of trouble on my account; and you'll find I've got all the qualities to endear me to people who entertain! So now that's all settled, and if you don't mind—I'm an old-fashioned fellow—don't want to turn you out, but—'

'Remember, you'll have to do your proper share of the fighting, dragon!' said St George, as he took the hint and rose to go; 'I mean ramping, and breathing fire, and so on!'

'I can *ramp* all right,' replied the dragon, confidently; 'as to breathing fire, it's surprising how easily one gets out of practice; but I'll do the best I can. Good-night!'

They had descended the hill and were almost back in the village again, when St George stopped short, '*Knew* I had forgotten something,' he said. 'There ought to be a Princess. Terror-stricken and chained to a rock, and all that sort of thing. Boy, can't you arrange a Princess?'

The Boy was in the middle of a tremendous yawn. 'I'm tired to death,' he wailed, 'and I *can't* arrange a Princess, or anything more, at this time of night. And my mother's sitting up, and *do* stop asking me to arrange more things till to-morrow!'

Next morning the people began streaming up to the Downs at quite an early hour, in their Sunday clothes and carrying baskets with bottle-necks sticking out of them, every one intent on securing good places for the combat. This was not exactly a simple matter, for of course it was quite possible that the dragon might win, and in that case even those who had put their money on him felt they could hardly expect him to deal with his backers on a different footing to the rest. Places were chosen, therefore, with circumspection and with a view to a speedy retreat in case of emergency; and the front rank was mostly composed of boys who had escaped from parental control and now sprawled and rolled about on the grass, regardless of the shrill threats and warnings discharged at them by their anxious mothers behind.

The Boy had secured a good front place, well up towards the cave, and was feeling as anxious as a stage-manager on a first night. Could the dragon be depended upon? He might change his mind and vote the whole performance rot; or else, seeing that the affair had been so hastily planned, without even a rehearsal, he might be too nervous to show up. The Boy looked narrowly at the cave, but it showed no sign of life or occupation. Could the dragon have made a moon-light flitting?

The higher portions of the ground were now black with sightseers, and presently a sound of cheering and a waving of handkerchiefs told that something was visible to them which the Boy, far up towards the dragon-end of the line as he was, could not yet see. A minute more and St George's red plumes topped the hill, as the Saint rode slowly forth on the great level space which stretched up to the grim mouth of the cave. Very gallant and beautiful he looked, on his tall war-horse, his golden armour glancing in the sun, his great spear held erect, the little white pennon, crimson-crossed, fluttering at its point. He drew rein and remained motionless. The lines of spectators began to give back a little, nervously; and even the boys in front stopped pulling hair and cuffing each other, and leaned forward expectant.

'Now then, dragon!' muttered the Boy, impatiently, fidgeting where he sat. He need not have distressed himself, had he only known. The dramatic possibilities of the thing had tickled the dragon immensely, and he had been up from an early hour, preparing for his first public appearance with as much heartiness as if the years had run backwards, and he had been again a little dragonlet, playing with his sisters on the floor of their mother's cave, at the game of saints-and-dragons, in which the dragon was bound to win.

A low muttering, mingled with snorts, now made itself heard; rising to a bellowing roar that seemed to fill the plain. Then a cloud of smoke obscured the mouth of the cave, and out of the midst of it the dragon himself, shining, sea-blue, magnificent, pranced splendidly forth; and everybody said, 'Oo-oo-oo!' as if he had been a mighty rocket! His scales were glittering, his long spiky tail lashed his sides, his claws tore up the turf and sent it flying high over his back, and smoke and fire incessantly jetted from his angry nostrils. 'Oh, well done, dragon!' cried the Boy, excitedly. 'Didn't think he had it in him!' he added to himself.

St George lowered his spear, bent his head, dug his heels into his horse's sides, and came thundering over the turf. The dragon charged with a roar and a squeal,—a great blue whirling combination of coils and snorts and clashing jaws and spikes and fire.

'Missed!' yelled the crowd. There was a moment's entanglement of golden armour and blue-green coils, and spiky tail, and then the great horse, tearing at his bit, carried the Saint, his spear swung high in the air, almost up to the mouth of the cave.

The dragon sat down and barked viciously, while St George with difficulty pulled his horse round into position.

'End of Round One!' thought the Boy. 'How well they managed it! But I hope the Saint won't get excited. I can trust the dragon all right. What a regular play-actor the fellow is!'

St George had at last prevailed on his horse to stand steady, and was looking round him as he wiped his brow. Catching sight of the Boy, he smiled and nodded, and held up three fingers for an instant.

'It seems to be all planned out,' said the Boy to himself. 'Round Three is to be the finishing one, evidently. Wish it could have lasted a bit longer. Whatever's that old fool of a dragon up to now?'

The dragon was employing the interval in giving a ramping-performance for the benefit of the crowd. Ramping, it should be explained, consists in running round and round in a wide circle, and sending waves and ripples of movement along the whole length of your spine, from your pointed ears right down to the spike at the end of your long tail. When you are covered with blue scales, the effect is particularly pleasing; and the Boy recollected the dragon's recently expressed wish to become a social success.

St George now gathered up his reins and began to move forward, dropping the point of his spear and settling himself firmly in the saddle.

'Time!' yelled everybody excitedly; and the dragon, leaving off his ramping, sat up on end, and began to leap from one side to the other with huge ungainly bounds, whooping like a Red Indian.

This naturally disconcerted the horse, who swerved violently, the Saint only just saving himself by the mane; and as they shot past the dragon delivered a vicious snap at the horse's tail which sent the poor beast careering madly far over the Downs, so that the language of the Saint, who had lost a stirrup, was fortunately inaudible to the general assemblage,

Round Two evoked audible evidence of friendly feeling towards the dragon. The spectators were not slow to appreciate a combatant who could hold his own so well and clearly wanted to show good sport; and many encouraging remarks reached the ears of our friend as he strutted to and fro, his chest thrust out and his tail in the air, hugely enjoying his new popularity.

St George had dismounted and was tightening his girths, and telling his horse, with quite an Oriental flow of imagery, exactly what he thought of him, and his relations, and his conduct on the present occasion; so the Boy made his way down to the Saint's end of the line, and held his spear for him.

'It's been a jolly fight, St George!' he said with a sigh. 'Can't you let it last a bit longer?'

'Well, I think I'd better not,' replied the Saint. 'The fact is, your simple-minded old friend's getting conceited, now they've begun cheering him, and he'll forget all about the arrangement and take to playing the fool, and there's no telling where he would stop. I'll just finish him off this round.'

He swung himself into the saddle and took his spear from the Boy. 'Now don't you be afraid,' he added kindly. 'I've marked my spot exactly, and *he's* sure to give me all the assistance in his power, because he knows it's his only chance of being asked to the banquet!'

St George now shortened his spear, bringing the butt well up under his arm; and, instead of galloping as before, trotted smartly towards the dragon, who crouched at his approach, flicking his tail till it cracked in the air like a great cart-whip. The Saint wheeled as he neared his opponent and circled warily round him, keeping his eye on the spare place; while the dragon, adopting similar tactics, paced with caution round the same circle, occasionally feinting with his head. So the two sparred for an opening, while the spectators maintained a breathless silence.

Though the round lasted for some minutes, the end was so swift that all the Boy saw was a lightning movement of the Saint's arm, and then a whirl and a confusion of spines, claws, tail, and flying bits of turf. The dust cleared away, the spectators whooped and ran in cheering, and the Boy made out that the dragon was down, pinned to the earth by the spear, while St George had dismounted, and stood astride of him.

It all seemed so genuine that the Boy ran in breathlessly, hoping the dear old dragon wasn't really hurt. As he approached, the dragon lifted one large eyelid, winked solemnly, and collapsed again. He was held fast to earth by the neck, but the Saint had hit him in the spare place agreed upon, and it didn't even seem to tickle.

'Bain't you goin' to cut 'is 'ed orf, master?' asked one of the applauding crowd. He had backed the dragon, and naturally felt a trifle sore.

'Well, not *to-day*, I think,' replied St George, pleasantly. 'You see, that can be done at *any* time. There's no hurry at all. I think we'll all go down to the village first, and have some refreshment, and then I'll give him a good talking-to, and you'll find he'll be a very different dragon!'

At that magic word *refreshment* the whole crowd formed up in procession and silently awaited the signal to start. The time for talking and cheering and betting was past, the hour for action had arrived. St George, hauling on his spear with both hands, released the dragon, who rose and shook himself and ran his eye over his spikes and scales and things, to see that they were all in order. Then the Saint mounted and led off the procession, the dragon following meekly in the company of the Boy, while the thirsty spectators kept at a respectful interval behind.

There were great doings when they got down to the village again, and had formed up in front of the inn. After refreshment St George made a speech, in which he informed his audience that he had removed their direful scourge, at a great deal of trouble and inconvenience to himself, and now they weren't to go about grumbling and fancying they'd got grievances, because they hadn't. And they shouldn't be so fond of fights, because next time they might have to do the fighting themselves, which would not be the same thing at all. And there was a certain badger in the inn stables which had got to be released at once, and he'd come and see it done himself. Then he told them that the dragon had been thinking over things, and saw that there were two sides to every question, and he wasn't going to do it any more, and if they were good perhaps he'd stay and settle down there. So they must make friends, and not be prejudiced, and go about fancying they knew everything there was to be known, because they didn't, not by a long way. And he warned them against the sin of romancing, and making up stories and fancying other people would believe them just because they were plausible and highly-coloured. Then he sat down, amidst much repentant cheering, and the dragon nudged the Boy in the ribs and whispered that he couldn't have done it better himself. Then every one went off to get ready for the banquet.

Banquets are always pleasant things, consisting mostly, as they do, of eating and drinking; but the specially nice thing about a banquet is, that it comes when something's over, and there's nothing more to worry about, and to-morrow seems a long way off. St George was happy because there had been a fight and he hadn't had to kill anybody; for he didn't really like killing, though he generally had to do it. The dragon was happy because there had been a fight, and so far from being hurt in it he had won popularity and a sure footing in society. The Boy was happy because there had been a fight, and in spite of it all his two friends were on the best of terms. And all the others were happy because there had been a fight, and—well, they didn't require any other reasons for their happiness. The dragon exerted himself to say the right thing to everybody, and proved the life and soul of the evening; while the Saint and the Boy, as they looked on, felt that they were only assisting at a feast of which the honour and the glory were entirely the dragon's. But they didn't mind that, being good fellows, and the dragon was not in the least proud or forgetful. On the contrary, every ten minutes or so he leant over towards the Boy and said impressively: 'Look here! you *will* see me home afterwards, won't you?' And the Boy always nodded, though he had promised his mother not to be out late.

At last the banquet was over, the guests had dropped away with many good-nights and congratulations and invitations, and the dragon, who had seen the last of them off the premises, emerged into the street followed by the Boy, wiped his brow, sighed, sat down in the road and gazed at the stars. 'Jolly night it's been!' he murmured. 'Jolly stars! Jolly little place this! Think I shall just stop here. Don't feel like climbing up any beastly hill. Boy's promised to see me home. Boy had better do it then! No responsibility on my part. Responsibility all Boy's!' And his chin sank on his broad chest and he slumbered peacefully.

'Oh, *get* up, dragon,' cried the Boy, piteously. 'You *know* my mother's sitting up, and I'm so tired, and you made me promise to see you home, and I never knew what it meant or I wouldn't have done it!' And the Boy sat down in the road by the side of the sleeping dragon, and cried.

The door behind them opened, a stream of light illumined the road, and St George, who had come out for a stroll in the cool night-air, caught sight of the two figures sitting there—the great motionless dragon and the tearful little Boy.

'What's the matter, Boy?' he inquired kindly, stepping to his side.

'Oh, it's this great lumbering *pig* of a dragon!' sobbed the Boy. 'First he makes me promise to see him home, and then he says I'd better do it, and goes to sleep! Might as well try to see a *haystack* home! And I'm so tired, and mother's—' here he broke down again.

'Now don't take on,' said St George. 'I'll stand by you, and we'll *both* see him home. Wake up, dragon!' he said sharply, shaking the beast by the elbow.

The dragon looked up sleepily. 'What a night, George!' he murmured; 'what a—'

'Now look here, dragon,' said the Saint, firmly. 'Here's this little fellow waiting to see you home, and you *know* he ought to have been in bed these two hours, and what his mother'll say *I* don't know, and anybody but a selfish pig would have *made* him go to bed long ago—'

'And he *shall* go to bed!' cried the dragon, starting up. 'Poor little chap, only fancy his being up at this hour! It's a shame, that's what it is, and I don't think, St George, you've been very considerate—but come along at once, and don't let us have any more arguing or shilly-shallying. You give me hold of your hand, Boy—thank you, George, an arm up the hill is just what I wanted!'

So they set off up the hill arm-in-arm, the Saint, the Dragon, and the Boy. The lights in the little village began to go out; but there were stars, and a late moon, as they climbed to the Downs together. And, as they turned the last corner and disappeared from view, snatches of an old song were borne back on the night-breeze. I can't be certain which of them was singing, but I *think* it was the Dragon!

1898

E. NESBIT

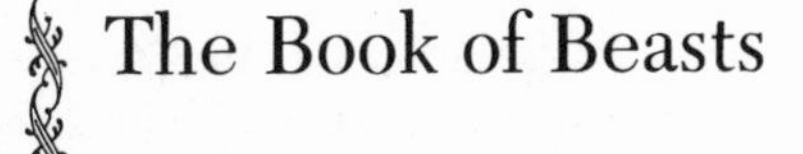

The Book of Beasts

He happened to be building a Palace when the news came, and he left all the bricks kicking about the floor for Nurse to clear up—but then the news was rather remarkable news. You see, there was a knock at the front door and voices talking downstairs, and Lionel thought it was the man come to see about the gas which had not been allowed to be lighted since the day when Lionel made a swing by tying his skipping-rope to the gas-bracket.

And then, quite suddenly, Nurse came in, and said, 'Master Lionel, dear, they've come to fetch you to go and be King.'

Then she made haste to change his smock and to wash his face and hands and brush his hair, and all the time she was doing it Lionel kept wriggling and fidgeting and saying, 'Oh, don't, Nurse,' and, 'I'm sure my ears are quite clean,' or, 'Never mind my hair, it's all right,' and 'That'll do.'

'You're going on as if you was going to be an eel instead of a King,' said Nurse.

The minute Nurse let go for a moment Lionel bolted off without waiting for his clean handkerchief, and in the drawing-room there were two very grave-looking gentlemen in red robes with fur, and gold coronets with velvet sticking up out of the middle like the cream in the very expensive jam tarts.

They bowed low to Lionel, and the gravest one said:

'Sire, your great-great-great-great-great-grandfather, the King of this country, is dead, and now you have got to come and be King.'

'Yes, please, sir,' said Lionel; 'when does it begin?'

'You will be crowned this afternoon,' said the grave gentleman who was not quite so grave-looking as the other.

'Would you like me to bring Nurse, or what time would you like me to be fetched, and hadn't I better put on my velvet suit with the lace collar?' said Lionel, who had often been out to tea.

'Your Nurse will be removed to the Palace later. No, never mind about changing your suit; the Royal robes will cover all that up.'

The grave gentlemen led the way to a coach with eight white horses, which was drawn up in front of the house where Lionel lived. It was No. 7, on the left-hand side of the street as you go up.

Lionel ran upstairs at the last minute, and he kissed Nurse and said:

'Thank you for washing me. I wish I'd let you do the other ear. No—there's no time now. Give me the hanky. Good-bye, Nurse.'

'Good-bye, ducky,' said Nurse; 'be a good little King now, and say "please" and "thank you," and remember to pass the cake to the little girls, and don't have more than two helps of anything.'

So off went Lionel to be made a King. He had never expected to be a King any more than you have, so it was all quite new to him—so new that he had never even thought of it. And as the coach went through the town he had to bite his tongue to be quite sure it was real, because if his tongue was real it showed he wasn't dreaming. Half an hour before he had been building with bricks in the nursery; and now—the streets were all fluttering with flags; every window was crowded with people waving handkerchiefs and scattering flowers; there were scarlet soldiers everywhere along the pavements, and all the bells of all the churches were ringing like mad, and like a great song to the music of their ringing he heard thousands of people shouting, 'Long live Lionel! Long live our little King!'

He was a little sorry at first that he had not put on his best clothes, but he soon forgot to think about that. If he had been a girl he would very likely have bothered about it the whole time.

As they went along, the grave gentlemen, who were the Chancellor and the Prime Minister, explained the things which Lionel did not understand.

'I thought we were a Republic,' said Lionel. 'I'm sure there hasn't been a King for some time.'

'Sire, your great-great-great-great-great-grandfather's death happened when my grandfather was a little boy,' said the Prime Minister, 'and since then your loyal people have been saving up to buy you a crown—so much a week, you know, according to people's means—sixpence a week from those who have first-rate pocket-money, down to a halfpenny a week from those who haven't so much. You know it's the rule that the crown must be paid for by the people.'

'But hadn't my great-great-however-much-it-is-grandfather a crown?'

'Yes, but he sent it to be tinned over, for fear of vanity, and he had had all the jewels taken out, and sold them to buy books. He was a strange man; a very good King he was, but he had his faults—he was fond of books. Almost with his latest breath he sent the crown to be tinned—and he never lived to pay the tinsmith's bill.'

Here the Prime Minister wiped away a tear, and just then the carriage stopped and Lionel was taken out of the carriage to be crowned. Being crowned is much more tiring work than you would suppose, and by the time it was over, and Lionel had worn the Royal robes for an hour or two and had had his hand kissed by everybody whose business it was to do it, he was quite worn out, and was very glad to get into the Palace nursery.

Nurse was there, and tea was ready: seedy cake and plummy cake, and jam and hot buttered toast, and the prettiest china with red and gold and blue flowers on it, and real tea, and as many cups of it as you liked. After tea Lionel said:

'I think I should like a book. Will you get me one, Nurse?'

'Bless the child,' said Nurse, 'you don't suppose you've lost the use of your legs with just being a King? Run along, do, and get your books yourself.'

So Lionel went down into the library. The Prime Minister and the Chancellor were there, and when Lionel came in they bowed very low, and were beginning to ask Lionel most politely what on earth he was coming bothering for now—when Lionel cried out:

'Oh, what a worldful of books! Are they yours?'

'They are yours, your Majesty,' answered the Chancellor. 'They were the property of the late King, your great-great——'

'Yes, I know,' Lionel interrupted. 'Well, I shall read them all. I love to read. I am so glad I learned to read.'

'If I might venture to advise your Majesty,' said the Prime Minister, 'I should *not* read these books. Your great——'

'Yes?' said Lionel, quickly.

'He was a very good King—oh, yes, really a very superior King in his way, but he was a little—well, strange.'

'Mad?' asked Lionel, cheerfully.

'No, no'—both the gentlemen were sincerely shocked. 'Not mad; but if I may express it so, he was—er—too clever by half. And I should not like a little King of mine to have anything to do with his books.'

Lionel looked puzzled.

'The fact is,' the Chancellor went on, twisting his red beard in an agitated way, 'your great——'

'Go on,' said Lionel.

'Was *called* a wizard.'

'But he wasn't?'

'Of course not—a most worthy King was your great——'

'I see.'

'But I wouldn't touch his books.'

'Just this one,' cried Lionel, laying his hands on the cover of a great brown book that lay on the study table. It had gold patterns on the brown leather, and gold clasps with turquoises and rubies in the twists of them, and gold corners, so that the leather should not wear out too quickly.

'I *must* look at this one,' Lionel said, for on the back in big letters he read: 'The Book of Beasts.'

The Chancellor said, 'Don't be a silly little King.'

But Lionel had got the gold clasps undone, and he opened the first page, and there was a beautiful Butterfly all red, and brown, and yellow, and blue, so beautifully painted that it looked as if it were alive.

'There,' said Lionel, 'isn't that lovely? Why——'

But as he spoke the beautiful Butterfly fluttered its many-coloured wings on the yellow old page of the book, and flew up and out of the window.

'Well!' said the Prime Minister, as soon as he could speak for the lump of wonder that had got into his throat and tried to choke him, 'that's magic, that is.'

But before he had spoken the King had turned the next page, and there was a shining bird complete and beautiful in every blue feather of him. Under him was written, 'Blue Bird of Paradise,' and while the King gazed enchanted at the charming picture the

Blue Bird fluttered his wings on the yellow page and spread them and flew out of the book.

Then the Prime Minister snatched the book away from the King and shut it up on the blank page where the bird had been, and put it on a very high shelf. And the Chancellor gave the King a good shaking, and said:

'You're a naughty, disobedient little King,' and was very angry indeed.

'I don't see that I've done any harm,' said Lionel. He hated being shaken, as all boys do; he would much rather have been slapped.

'No harm?' said the Chancellor. 'Ah—but what do you know about it? That's the question. How do you know what might have been on the next page—a snake or a worm, or a centipede or a revolutionist, or something like that.'

'Well, I'm sorry if I've vexed you,' said Lionel. 'Come, let's kiss and be friends.' So he kissed the Prime Minister, and they settled down for a nice quiet game of noughts and crosses, while the Chancellor went to add up his accounts.

But when Lionel was in bed he could not sleep for thinking of the book, and when the full moon was shining with all her might and light he got up and crept down to the library and climbed up and got the 'Book of Beasts.'

He took it outside on to the terrace, where the moonlight was as bright as day, and he opened the book, and saw the empty pages with 'Butterfly' and 'Blue Bird of Paradise' underneath, and then he turned the next page. There was some sort of red thing sitting under a palm tree, and under it was written 'Dragon.' The Dragon did not move, and the King shut up the book rather quickly and went back to bed.

But the next day he wanted another look, so he got the book out into the garden, and when he undid the clasps with the rubies and turquoises, the book opened all by itself at the picture with 'Dragon' underneath, and the sun shone full on the page. And then, quite suddenly, a great Red Dragon came out of the book, and spread vast scarlet wings and flew away across the garden to the far hills, and Lionel was left with the empty page before him, for the page was quite empty except for the green palm tree and the yellow desert, and the little streaks of red where the paint brush had gone outside the pencil outline of the Red Dragon.

And then Lionel felt that he had indeed done it. He had not been King twenty-four hours, and already he had let loose a Red Dragon to worry his faithful subjects' lives out. And they had been saving up so long to buy him a crown, and everything!

Lionel began to cry.

Then the Chancellor and the Prime Minister and the Nurse all came running to see what was the matter. And when they saw the book they understood, and the Chancellor said:

'You naughty little King! Put him to bed, Nurse, and let him think over what he's done.'

'Perhaps, my Lord,' said the Prime Minister, 'we'd better first find out just exactly what he *has* done.'

Then Lionel, in floods of tears, said:

'It's a Red Dragon, and it's gone flying away to the hills, and I *am* so sorry, and, oh, do forgive me!'

But the Prime Minister and the Chancellor had other things to think of than forgiving Lionel. They hurried off to consult the police and see what could be done. Everyone did what they could. They sat on committees and stood on guard, and lay in wait for the Dragon but he stayed up in the hills, and there was nothing more to *be* done. The faithful Nurse, meanwhile, did not neglect her duty. Perhaps she did more than anyone else, for she slapped the King and put him to bed without his tea, and when it got dark she would not give him a candle to read by.

'You are a naughty little King,' she said, 'and nobody will love you.'

Next day the Dragon was still quiet, though the more poetic of Lionel's subjects could see the redness of the Dragon shining through the green trees quite plainly. So Lionel put on his crown and sat on his throne and said he wanted to make some laws.

And I need hardly say that though the Prime Minister and the Chancellor and the Nurse might have the very poorest opinion of Lionel's private judgment, and might even slap him and send him to bed, the minute he got on his throne and set his crown on his head, he became infallible—which means that everything he said was right, and that he couldn't possibly make a mistake. So when he said:

'There is to be a law forbidding people to open books in schools or elsewhere'—he had the support of at least half of his subjects, and the other half—the grown-up half—pretended to think he was quite right.

Then he made a law that everyone should always have enough to eat. And this pleased everyone except the ones who had always had too much.

And when several other nice new laws were made and written down he went home and made mud-houses and was very happy. And he said to his Nurse:

'People will love me now I've made such a lot of pretty new laws for them.'

But Nurse said: 'Don't count your chickens, my dear. You haven't seen the last of that Dragon yet.'

Now the next day was Saturday. And in the afternoon the Dragon suddenly swooped down upon the common in all his hideous redness, and carried off the Football Players, umpires, goal-posts, football, and all.

Then the people were very angry indeed, and they said:

'We might as well be a Republic. After saving up all these years to get his crown, and everything!'

And wise people shook their heads and foretold a decline in the National Love of Sport. And, indeed, football was not at all popular for some time afterwards.

Lionel did his best to be a good King during the week, and the people were beginning to forgive him for letting the Dragon out of the book. 'After all,' they said, 'football is a dangerous game, and perhaps it is wise to discourage it.'

Popular opinion held that the Football Players, being tough and hard, had disagreed with the Dragon so much that he had gone away to some place where they only play cats' cradle and games that do not make you hard and tough.

All the same, Parliament met on the Saturday afternoon, a convenient time, when most of the Members would be free to attend, to consider the Dragon. But unfortunately the Dragon, who had only been asleep, woke up because it was Saturday, and he considered the Parliament, and afterwards there were not any Members left, so they tried to make a new Parliament, but being an M.P. had somehow grown as unpopular as football playing, and no one would consent to be elected, so they had to do without a Parliament. When the next Saturday came round everyone was a little nervous, but the Red Dragon was pretty quiet that day and only ate an Orphanage.

Lionel was very, very unhappy. He felt that it was his disobedience that had brought this trouble on the Parliament and the

Orphanage and the Football Players, and he felt that it was his duty to try and do something. The question was, what?

The Blue Bird that had come out of the book used to sing very nicely in the Palace rose-garden, and the Butterfly was very tame, and would perch on his shoulder when he walked among the tall lilies: so Lionel saw that *all* the creatures in the Book of Beasts could not be wicked, like the Dragon, and he thought:

'Suppose I could get another beast out who would fight the Dragon?'

So he took the Book of Beasts out into the rose-garden and opened the page next to the one where the Dragon had been just a tiny bit to see what the name was. He could only see 'cora,' but he felt the middle of the page swelling up thick with the creature that was trying to come out, and it was only by putting the book down and sitting on it suddenly, very hard, that he managed to get it shut. Then he fastened the clasps with the rubies and turquoises in them and sent for the Chancellor, who had been ill on Saturday week, and so had not been eaten with the rest of the Parliament, and he said:

'What animal ends in "cora"?'

The Chancellor answered:

'The Manticora, of course.'

'What is he like?' asked the King.

'He is the sworn foe of Dragons,' said the Chancellor. 'He drinks their blood. He is yellow, with the body of a lion and the face of a man. I wish we had a few Manticoras here now. But the last died hundreds of years ago—worse luck!'

Then the King ran and opened the book at the page that had 'cora' on it, and there was the picture-Manticora, all yellow, with a lion's body and a man's face, just as the Chancellor had said. And under the picture was written, 'Manticora.'

And in a few minutes the Manticora came sleepily out of the book, rubbing its eyes with its hands and mewing piteously. It seemed very stupid, and when Lionel gave it a push and said, 'Go along and fight the Dragon, do,' it put its tail between its legs and fairly ran away. It went and hid behind the Town Hall, and at night when the people were asleep it went round and ate all the pussy-cats in the town. And then it mewed more than ever. And on the Saturday morning, when people were a little timid about going out, because the Dragon had no regular hour for calling, the

Manticora went up and down the streets and drank all the milk that was left in the cans at the doors for people's teas, and it ate the cans as well.

And just when it had finished the very last little ha'porth, which was short measure, because the milkman's nerves were quite upset, the Red Dragon came down the street looking for the Manticora. It edged off when it saw him coming, for it was not at all the Dragon-fighting kind; and, seeing no other door open, the poor, hunted creature took refuge in the General Post Office, and there the Dragon found it, trying to conceal itself among the ten o'clock mail. The Dragon fell on the Manticora at once, and the mail was no defence. The mewings were heard all over the town. All the pussies and the milk the Manticora had had seemed to have strengthened its mew wonderfully. Then there was a sad silence, and presently the people whose windows looked that way saw the Dragon come walking down the steps of the General Post Office spitting fire and smoke, together with tufts of Manticora fur, and the fragments of the registered letters. Things were growing very serious. However popular the King might become during the week, the Dragon was sure to do something on Saturday to upset the people's loyalty.

The Dragon was a perfect nuisance for the whole of Saturday, except during the hour of noon, and then he had to rest under a tree or he would have caught fire from the heat of the sun. You see, he was very hot to begin with.

At last came a Saturday when the Dragon actually walked into the Royal nursery and carried off the King's own pet Rocking-Horse. Then the King cried for six days, and on the seventh he was so tired that he had to stop. Then he heard the Blue Bird singing among the roses and saw the Butterfly fluttering among the lilies, and he said:

'Nurse, wipe my face, please. I am not going to cry any more.'

Nurse washed his face, and told him not to be a silly little King. 'Crying,' said she, 'never did anyone any good yet.'

'I don't know,' said the little King. 'I seem to see better, and to hear better now that I've cried for a week. Now, Nurse, dear, I know I'm right, so kiss me in case I never come back. I *must* try if I can't save the people.'

'Well, if you must, you must,' said Nurse; 'but don't tear your clothes or get your feet wet.'

So off he went.

The Blue Bird sang more sweetly than ever, and the Butterfly shone more brightly, as Lionel once more carried the Book of Beasts out into the rose-garden, and opened it—very quickly, so that he might not be afraid and change his mind. The book fell open wide, almost in the middle, and there was written at the bottom of the page, 'The Hippogriff,' and before Lionel had time to see what the picture was, there was a fluttering of great wings and a stamping of hoofs, and a sweet, soft, friendly neighing; and there came out of the book a beautiful white horse with a long, long, white mane and a long, long, white tail, and he had great wings like swan's wings, and the softest, kindest eyes in the world, and he stood there among the roses.

The Hippogriff rubbed its silky-soft, milky-white nose against the little King's shoulder, and the little King thought: 'But for the wings you are very like my poor, dear, lost Rocking-Horse.' And the Blue Bird's song was very loud and sweet.

Then suddenly the King saw coming through the sky the great straggling, sprawling, wicked shape of the Red Dragon. And he knew at once what he must do. He caught up the Book of Beasts and jumped on the back of the gentle, beautiful Hippogriff, and leaning down he whispered in the sharp white ear:

'Fly, dear Hippogriff, fly your very fastest to the Pebbly Waste.'

And when the Dragon saw them start, he turned and flew after them, with his great wings flapping like clouds at sunset, and the Hippogriff's wide wings were snowy as clouds at the moon-rising.

When the people in the town saw the Dragon fly off after the Hippogriff and the King they all came out of their houses to look, and when they saw the two disappear they made up their minds to the worst, and began to think what would be worn for Court mourning.

But the Dragon could not catch the Hippogriff. The red wings were bigger than the white ones, but they were not so strong, and so the white-winged horse flew away and away and away, with the Dragon pursuing, till he reached the very middle of the Pebbly Waste.

Now, the Pebbly Waste is just like the parts of the seaside where there is no sand—all round, loose, shifting stones, and there is no grass there and no tree within a hundred miles of it.

Lionel jumped off the white horse's back in the very middle of the Pebbly Waste, and he hurriedly unclasped the Book of Beasts and laid it open on the pebbles. Then he clattered among the pebbles in his haste to get back on to his white horse, and had just jumped on when up came the Dragon. He was flying very feebly, and looking round everywhere for a tree, for it was just on the stroke of twelve, the sun was shining like a gold guinea in the blue sky, and there was not a tree for a hundred miles.

The white-winged horse flew round and round the Dragon as he writhed on the dry pebbles. He was getting very hot: indeed, parts of him even had begun to smoke. He knew that he must certainly catch fire in another minute unless he could get under a tree. He made a snatch with his red claws at the King and Hippogriff, but he was too feeble to reach them, and besides, he did not dare to over-exert himself for fear he should get any hotter.

It was then that he saw the Book of Beasts lying on the pebbles, open at the page with 'Dragon' written at the bottom. He looked and he hesitated, and he looked again, and then, with one last squirm of rage, the Dragon wriggled himself back into the picture, and sat down under the palm tree, and the page was a little singed as he went in.

As soon as Lionel saw that the Dragon had really been obliged to go and sit under his own palm tree because it was the only tree there, he jumped off his horse and shut the book with a bang.

'Oh, hurrah!' he cried. 'Now we really *have* done it.'

And he clasped the book very tight with the turquoise and ruby clasps.

'Oh, my precious Hippogriff,' he cried, 'you are the bravest, dearest, most beautiful——'

'Hush,' whispered the Hippogriff, modestly. 'Don't you see that we are not alone?'

And indeed there was quite a crowd round them on the Pebbly Waste: the Prime Minister and the Parliament and the Football Players and the Orphanage and the Manticora and the Rocking-Horse, and indeed everyone who had been eaten by the Dragon. You see, it was impossible for the Dragon to take them into the book with him—it was a tight fit even for one Dragon—so, of course, he had to leave them outside.

They all got home somehow, and all lived happy ever after.

When the King asked the Manticora where he would like to live he begged to be allowed to go back into the book. 'I do not care for public life,' he said.

Of course he knew his way on to his own page, so there was no danger of his opening the book at the wrong page and letting out a Dragon or anything. So he got back into his picture, and has never come out since: that is why you will never see a Manticora as long as you live, except in a picture-book. And of course he left the pussies outside, because there was no room for them in the book—and the milk-cans too.

Then the Rocking-Horse begged to be allowed to go and live on the Hippogriff's page of the book. 'I should like,' he said, 'to live somewhere where Dragons can't get at me.'

So the beautiful, white-winged Hippogriff showed him the way in, and there he stayed till the King had him taken out for his great-great-great-great-grandchildren to play with.

As for the Hippogriff, he accepted the position of the King's Own Rocking-Horse—a situation left vacant by the retirement of the wooden one. And the Blue Bird and the Butterfly sing and flutter among the lilies and roses of the Palace garden to this very day.

1900

L. F. BAUM

The Queen of Quok

A KING once died, as kings are apt to do, being as liable to shortness of breath as other mortals.

It was high time this king abandoned his earth life, for he had lived in a sadly extravagant manner, and his subjects could spare him without the slightest inconvenience.

His father had left him a full treasury, both money and jewels being in abundance. But the foolish king just deceased had squandered every penny in riotous living. He had then taxed his subjects until most of them became paupers, and this money vanished in more riotous living. Next he sold all the grand old furniture in the palace; all the silver and gold plate and bric-a-brac; all the rich carpets and furnishings and even his own kingly wardrobe, reserving only a soiled and moth-eaten ermine robe to fold over his threadbare raiment. And he spent the money in further riotous living.

Don't ask me to explain what riotous living is. I only know, from hearsay, that it is an excellent way to get rid of money. And so this spendthrift king found it.

He now picked all the magnificent jewels from his kingly crown and from the round ball on the top of his scepter, and sold them and spent the money. Riotous living, of course. But at last he was at the end of his resources. He couldn't sell the crown itself, because no one but the king had the right to wear it. Neither could he sell the royal palace, because only the king had the right to live there.

So, finally, he found himself reduced to a bare palace, containing only a big mahogany bedstead that he slept in, a small stool on which he sat to pull off his shoes and the moth-eaten ermine robe.

In this straight he was reduced to the necessity of borrowing an occasional dime from his chief counselor, with which to buy a ham sandwich. And the chief counselor hadn't many dimes. One who counseled his king so foolishly was likely to ruin his own prospects as well.

So the king, having nothing more to live for, died suddenly and left a ten-year-old son to inherit the dismantled kingdom, the moth-eaten robe and the jewel-stripped crown.

No one envied the child, who had scarcely been thought of until he became king himself. Then he was recognized as a personage of some importance, and the politicians and hangers-on, headed by the chief counselor of the kingdom, held a meeting to determine what could be done for him.

These folk had helped the old king to live riotously while his money lasted, and now they were poor and too proud to work. So they tried to think of a plan that would bring more money into the little king's treasury, where it would be handy for them to help themselves.

After the meeting was over the chief counselor came to the young king, who was playing peg-top in the courtyard, and said:

'Your majesty, we have thought of a way to restore your kingdom to its former power and magnificence.'

'All right,' replied his majesty, carelessly. 'How will you do it?'

'By marrying you to a lady of great wealth,' replied the counselor.

'Marrying me!' cried the king. 'Why, I am only ten years old!'

'I know; it is to be regretted. But your majesty will grow older, and the affairs of the kingdom demand that you marry a wife.'

'Can't I marry a mother, instead?' asked the poor little king, who had lost his mother when a baby.

'Certainly not,' declared the counselor. 'To marry a mother would be illegal; to marry a wife is right and proper.'

'Can't you marry her yourself?' inquired his majesty, aiming his peg-top at the chief counselor's toe, and laughing to see how he jumped to escape it.

'Let me explain,' said the other. 'You haven't a penny in the world, but you have a kingdom. There are many rich women who would be glad to give their wealth in exchange for a queen's coronet—even if the king is but a child. So we have decided to advertise that the one who bids the highest shall become the queen of Quok.'

'If I must marry at all,' said the king, after a moment's thought, 'I prefer to marry Nyana, the armorer's daughter.'

'She is too poor,' replied the counselor.

'Her teeth are pearls, her eyes are amethysts, and her hair is gold,' declared the little king.

'True, your majesty. But consider that your wife's wealth must be used. How would Nyana look after you have pulled her teeth of pearls, plucked out her amethyst eyes and shaved her golden head?'

The boy shuddered.

'Have your own way,' he said, despairingly. 'Only let the lady be as dainty as possible and a good playfellow.'

'We shall do our best,' returned the chief counselor, and went away to advertise throughout the neighboring kingdoms for a wife for the boy king of Quok.

There were so many applicants for the privilege of marrying the little king that it was decided to put him up at auction, in order that the largest possible sum of money should be brought into the kingdom. So, on the day appointed, the ladies gathered at the palace from all the surrounding kingdoms—from Bilkon, Mulgravia, Junkum and even as far away as the republic of Macvelt.

The chief counselor came to the palace early in the morning and had the king's face washed and his hair combed; and then he padded the inside of the crown with old newspapers to make it small enough to fit his majesty's head. It was a sorry looking crown, having many big and little holes in it where the jewels had once been; and it had been neglected and knocked around until it was quite battered and tarnished. Yet, as the counselor said, it was the king's crown, and it was quite proper he should wear it on the solemn occasion of his auction.

Like all boys, be they kings or paupers, his majesty had torn and soiled his one suit of clothes, so that they were hardly presentable; and there was no money to buy new ones. Therefore the counselor wound the old ermine robe around the king and sat him upon the stool in the middle of the otherwise empty audience chamber.

And around him stood all the courtiers and politicians and hangers-on of the kingdom, consisting of such people as were too proud or lazy to work for a living. There was a great number of them, you may be sure, and they made an imposing appearance.

Then the doors of the audience chamber were thrown open, and the wealthy ladies who aspired to being queen of Quok came trooping in. The king looked them over with much anxiety, and decided they were each and all old enough to be his grandmother, and ugly enough to scare away the crows from the royal cornfields. After which he lost interest in them.

But the rich ladies never looked at the poor little king squatting upon his stool. They gathered at once about the chief counselor, who acted as auctioneer.

'How much am I offered for the coronet of the queen of Quok?' asked the counselor, in a loud voice.

'Where is the coronet?' inquired a fussy old lady who had just buried her ninth husband and was worth several millions.

'There isn't any coronet at present,' explained the chief counselor, 'but whoever bids highest will have the right to wear one, and she can then buy it.'

'Oh,' said the fussy old lady, 'I see.' Then she added: 'I'll bid fourteen dollars.'

'Fourteen thousand dollars!' cried a sour-looking woman who was thin and tall and had wrinkles all over her skin—'like a frosted apple,' the king thought.

The bidding now became fast and furious, and the poverty-stricken courtiers brightened up as the sum began to mount into the millions.

'He'll bring us a very pretty fortune, after all,' whispered one to his comrade, 'and then we shall have the pleasure of helping him spend it.'

The king began to be anxious. All the women who looked at all kind-hearted or pleasant had stopped bidding for lack of money, and the slender old dame with the wrinkles seemed determined to get the coronet at any price, and with it the boy husband. This ancient creature finally became so excited that her wig got crosswise of her head and her false teeth kept slipping out, which horrified the little king greatly; but she would not give up.

At last the chief counselor ended the auction by crying out:

'Sold to Mary Ann Brodjinsky de la Porkus for three million, nine hundred thousand, six hundred and twenty-four dollars and sixteen cents!' And the sour-looking old woman paid the money in cash and on the spot, which proves this is a fairy story.

The king was so disturbed at the thought that he must marry this hideous creature that he began to wail and weep; whereupon the woman boxed his ears soundly. But the counselor reproved her for punishing her future husband in public, saying:

'You are not married yet. Wait until to-morrow, after the

wedding takes place. Then you can abuse him as much as you wish. But at present we prefer to have people think this is a love match.'

The poor king slept but little that night, so filled was he with terror of his future wife. Nor could he get the idea out of his head that he preferred to marry the armorer's daughter, who was about his own age. He tossed and tumbled around upon his hard bed until the moonlight came in at the window and lay like a great white sheet upon the bare floor. Finally, in turning over for the hundredth time, his hand struck against a secret spring in the headboard of the big mahogany bedstead, and at once, with a sharp click, a panel flew open.

The noise caused the king to look up, and, seeing the open panel, he stood upon tiptoe, and, reaching within, drew out a folded paper. It had several leaves fastened together like a book, and upon the first page was written:

When the king is in trouble
This leaf he must double
And set it on fire
To obtain his desire.

This was not very good poetry, but when the king had spelled it out in the moonlight he was filled with joy.

'There's no doubt about my being in trouble,' he exclaimed; 'so I'll burn it at once, and see what happens.'

He tore off the leaf and put the rest of the book in its secret hiding place. Then, folding the paper double, he placed it on the top of his stool, lighted a match and set fire to it.

It made a horrid smudge for so small a paper, and the king sat on the edge of the bed and watched it eagerly.

When the smoke cleared away he was surprised to see, sitting upon the stool, a round little man, who, with folded arms and crossed legs, sat calmly facing the king and smoking a black briarwood pipe.

'Well, here I am,' said he.

'So I see,' replied the little king. 'But how did you get here?'

'Didn't you burn the paper?' demanded the round man, by way of answer.

'Yes, I did,' acknowledged the king.

'Then you are in trouble, and I've come to help you out of it. I'm the Slave of the Royal Bedstead.'

'Oh!' said the king. 'I didn't know there was one.'

'Neither did your father, or he would not have been so foolish as to sell everything he had for money. By the way, it's lucky for you he did not sell this bedstead. Now, then, what do you want?'

'I'm not sure what I want,' replied the king; 'but I know what I don't want, and that is the old woman who is going to marry me.'

'That's easy enough,' said the Slave of the Royal Bedstead. 'All you need do is to return her the money she paid the chief counselor and declare the match off. Don't be afraid. You are the king, and your word is law.'

'To be sure,' said his majesty. 'But I am in great need of money. How am I going to live if the chief counselor returns to Mary Ann Brodjinski her millions?'

'Phoo! that's easy enough,' again answered the man, and, putting his hand in his pocket, he drew out and tossed to the king an old-fashioned leather purse. 'Keep that with you,' said he, 'and you will always be rich, for you can take out of the purse as many twenty-five-cent silver pieces as you wish, one at a time. No matter how often you take one out, another will instantly appear in its place within the purse.'

'Thank you,' said the king, gratefully. 'You have rendered me a rare favor; for now I shall have money for all my needs and will not be obliged to marry anyone. Thank you a thousand times!'

'Don't mention it,' answered the other, puffing his pipe slowly and watching the smoke curl into the moonlight. 'Such things are easy to me. Is that all you want?'

'All I can think of just now,' returned the king.

'Then, please close that secret panel in the bedstead,' said the man; 'the other leaves of the book may be of use to you some time.'

The boy stood upon the bed as before and, reaching up, closed the opening so that no one else could discover it. Then he turned to face his visitor, but the Slave of the Royal Bedstead had disappeared.

'I expected that,' said his majesty; 'yet I am sorry he did not wait to say good-by.'

With a lightened heart and a sense of great relief the boy king placed the leathern purse underneath his pillow, and climbing into bed again slept soundly until morning.

When the sun rose his majesty rose also, refreshed and comforted, and the first thing he did was to send for the chief counselor.

That mighty personage arrived looking glum and unhappy, but the boy was too full of his own good fortune to notice it. Said he:

'I have decided not to marry anyone, for I have just come into a fortune of my own. Therefore I command you to return to that old woman the money she has paid you for the right to wear the coronet of the queen of Quok. And make public declaration that the wedding will not take place.'

Hearing this the counselor began to tremble, for he saw the young king had decided to reign in earnest; and he looked so guilty that his majesty inquired:

'Well! what is the matter now?'

'Sire.' replied the wretch, in a shaking voice, 'I cannot return the woman her money, for I have lost it!'

'Lost it!' cried the king, in mingled astonishment and anger.

'Even so, your majesty. On my way home from the auction last night I stopped at the drug store to get some potash lozenges for my throat, which was dry and hoarse with so much loud talking; and your majesty will admit it was through my efforts the woman was induced to pay so great a price. Well, going into the drug store I carelessly left the package of money lying on the seat of my carriage, and when I came out again it was gone. Nor was the thief anywhere to be seen.'

'Did you call the police?' asked the king.

'Yes, I called; but they were all on the next block, and although they have promised to search for the robber I have little hope they will ever find him.'

The king sighed.

'What shall we do now?' he asked.

'I fear you must marry Mary Ann Brodjinski,' answered the chief counselor; 'unless, indeed, you order the executioner to cut her head off.'

'That would be wrong,' declared the king. 'The woman must not be harmed. And it is just that we return her money, for I will not marry her under any circumstances.'

'Is that private fortune you mentioned large enough to repay her?' asked the counselor.

'Why, yes,' said the king, thoughtfully, 'but it will take some time to do it, and that shall be your task. Call the woman here.'

The counselor went in search of Mary Ann, who, when she

heard she was not to become a queen, but would receive her money back, flew into a violent passion and boxed the chief counselor's ears so viciously that they stung for nearly an hour. But she followed him into the king's audience chamber, where she demanded her money in a loud voice, claiming as well the interest due upon it over night.

'The counselor has lost your money,' said the boy king, 'but he shall pay you every penny out of my own private purse. I fear, however, you will be obliged to take it in small change.'

'That will not matter,' she said, scowling upon the counselor as if she longed to reach his ears again; 'I don't care how small the change is so long as I get every penny that belongs to me, and the interest. Where is it?'

'Here,' answered the king, handing the counselor the leathern purse. 'It is all in silver quarters, and they must be taken from the purse one at a time; but there will be plenty, to pay your demands, and to spare.'

So, there being no chairs, the counselor sat down upon the floor in one corner and began counting out silver twenty-five-cent pieces from the purse, one by one. And the old woman sat upon the floor opposite him and took each piece of money from his hand.

It was a large sum: three million nine hundred thousand six hundred and twenty-four dollars and sixteen cents. And it takes four times as many twenty-five-cent pieces as it would dollars to make up the amount.

The king left them sitting there and went to school, and often thereafter he came to the counselor and interrupted him long enough to get from the purse what money he needed to reign in a proper and dignified manner. This somewhat delayed the counting, but as it was a long job, anyway, that did not matter much.

The king grew to manhood and married the pretty daughter of the armorer, and they now have two lovely children of their own. Once in awhile they go into the big audience chamber of the palace and let the little ones watch the aged, hoary-headed counselor count out silver twenty-five-cent pieces to a withered old woman, who watches his every movement to see that he does not cheat her.

It is a big sum, three million, nine hundred thousand, six

hundred and twenty-four dollars and sixteen cents in twenty-five-cent pieces.

But this is how the counselor was punished for being so careless with the woman's money. And this is how Mary Ann Brodjinski de la Porkus was also punished for wishing to marry a ten-year-old king in order that she might wear the coronet of the queen of Quok.

1901

H. G. WELLS

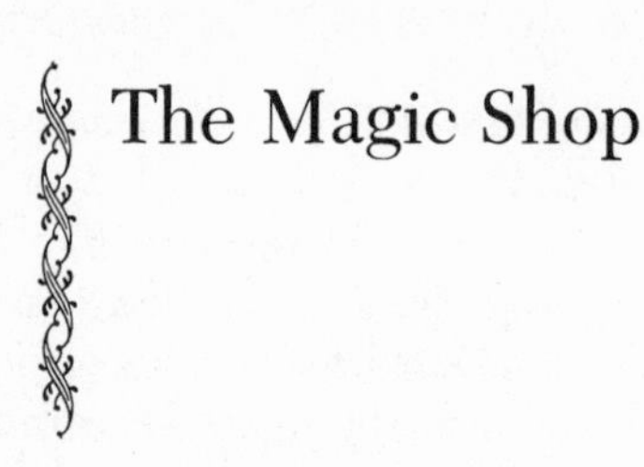

The Magic Shop

I HAD seen the Magic Shop from afar several times; I had passed it once or twice, a shop window of alluring little objects, magic balls, magic hens, wonderful cones, ventriloquist dolls, the material of the basket trick, packs of cards that *looked* all right, and all that sort of thing, but never had I thought of going in until one day, almost without warning, Gip hauled me by my finger right up to the window, and so conducted himself that there was nothing for it but to take him in. I had not thought the place was there, to tell the truth—a modest-sized frontage in Regent Street, between the picture shop and the place where the chicks run about just out of patent incubators—but there it was sure enough. I had fancied it was down nearer the Circus, or round the corner in Oxford Street, or even in Holborn; always over the way and a little inaccessible it had been, with something of the mirage in its position; but here it was now quite indisputably, and the fat end of Gip's pointing finger made a noise upon the glass.

'If I was rich,' said Gip, dabbing a finger at the Disappearing Egg, 'I'd buy myself that. And that'—which was The Crying Baby, Very Human—'and that,' which was a mystery, and called, so a neat card asserted, 'Buy One and Astonish Your Friends.'

'Anything,' said Gip, 'will disappear under one of those cones. I have read about it in a book.

'And there, dadda, is the Vanishing Halfpenny—only they've put it this way up so's we can't see how it's done.'

Gip, dear boy, inherits his mother's breeding, and he did not propose to enter the shop or worry in any way; only, you know, quite unconsciously he lugged my finger doorward, and he made his interest clear.

'That,' he said, and pointed to the Magic Bottle.

'If you had that?' I said; at which promising inquiry he looked up with a sudden radiance.

'I could show it to Jessie,' he said, thoughtful as ever of others.

'It's less than a hundred days to your birthday, Gibbles,' I said, and laid my hand on the door-handle.

Gip made no answer, but his grip tightened on my finger, and so we came into the shop.

It was no common shop this; it was a magic shop, and all the prancing precedence Gip would have taken in the matter of mere toys was wanting. He left the burthen of the conversation to me.

It was a little, narrow shop, not very well lit, and the door-bell pinged again with a plaintive note as we closed it behind us. For a moment or so we were alone and could glance about us. There was a tiger in *papier-mâché* on the glass case that covered the low counter—a grave, kind-eyed tiger that waggled his head in a methodical manner; there were several crystal spheres, a china hand holding magic cards, a stock of magic fish-bowls in various sizes, and an immodest magic hat that shamelessly displayed its springs. On the floor were magic mirrors; one to draw you out long and thin, one to swell your head and vanish your legs, and one to make you short and fat like a draught; and while we were laughing at these the shopman, as I suppose, came in.

At any rate, there he was behind the counter—a curious, sallow, dark man, with one ear larger than the other and a chin like the toe-cap of a boot.

'What can we have the pleasure?' he said, spreading his long, magic fingers on the glass case; and so with a start we were aware of him.

'I want,' I said, 'to buy my little boy a few simple tricks.'

'Legerdemain?' he asked. 'Mechanical? Domestic?'

'Anything amusing?' said I.

'Um!' said the shopman, and scratched his head for a moment as if thinking. Then, quite distinctly, he drew from his head a glass ball. 'Something in this way?' he said, and held it out.

The action was unexpected. I had seen the trick done at entertainments endless times before—it's part of the common stock of conjurers—but I had not expected it here. 'That's good,' I said, with a laugh.

'Isn't it?' said the shopman.

Gip stretched out his disengaged hand to take this object and found merely a blank palm.

'It's in your pocket,' said the shopman, and there it was!

'How much will that be?' I asked.

'We make no charge for glass balls,' said the shopman, politely. 'We get them'—he picked one out of his elbow as he spoke—'free.' He produced another from the back of his neck, and laid it beside its predecessor on the counter. Gip regarded his glass ball sagely, then directed a look of inquiry at the two on the counter, and finally brought his round-eyed scrutiny to the shopman, who smiled. 'You may have those too,' said the shopman, 'and if you *don't* mind, one from my mouth—*So!*'

Gip counselled me mutely for a moment, and then in a profound silence put away the four balls, resumed my reassuring finger, and nerved himself for the next event.

'We get all our smaller tricks in that way,' the shopman remarked.

I laughed in the manner of one who subscribes to a jest. 'Instead of going to the wholesale shop,' I said. 'Of course, it's cheaper.'

'In a way,' the shopman said. 'Though we pay in the end. But not so heavily—as people suppose. . . . Our larger tricks, and our daily provisions and all the other things we want, we get out of that hat. . . . And you know, sir, if you'll excuse my saying it, there *isn't* a wholesale shop, not for Genuine Magic goods, sir. I don't know if you noticed our inscription—the Genuine Magic shop.' He drew a business-card from his cheek and handed it to me. 'Genuine,' he said, with his finger on the word, and added, 'There is absolutely no deception, sir.'

He seemed to be carrying out the joke pretty thoroughly, I thought.

He turned to Gip with a smile of remarkable affability. 'You, you know, are the Right Sort of Boy.'

I was surprised at his knowing that, because, in the interests of discipline, we keep it rather a secret even at home; but Gip received it in unflinching silence, keeping a steadfast eye on him.

'It's only the Right Sort of Boy gets through that doorway.'

And as if by way of illustration, there came a rattling at the door, and a squeaking little voice could be faintly heard. 'Nyar! I *warn* 'a go in there dadda, I WARN 'a go in there. Nya-a-a-ah!' and then the accents of a down-trodden parent urging consolations and propitiations. 'It's locked, Edward,' he said.

'But it isn't,' said I.

'It is, sir,' said the shopman, 'always—for that sort of child,' and as he spoke we had a glimpse of the other youngster, a small, white face, pallid from sweet-eating and over-sapid food, and distorted by evil passions, a ruthless little egotist, pawing at the enchanted pane. 'It's no good, sir,' said the shopman, as I moved, with my natural helpfulness, doorward, and presently the spoilt child was carried off howling.

'How do you manage that?' I said, breathing more freely.

'Magic!' said the shopman, with a careless wave of the hand, and behold! sparks of coloured fire flew out of his fingers and vanished into the shadows of the shop.

'You were saying,' he said, addressing himself to Gip, 'before you came in, that you would like one of our "Buy One and Astonish your Friends" boxes?'

Gip, after a gallant effort, said 'Yes.'

'It's in your pocket.'

And leaning over the counter—he really had an extraordinarily long body—this amazing person produced the article in the customary conjurer's manner. 'Paper,' he said, and took a sheet out of the empty hat with the springs; 'string,' and behold his mouth was a string-box, from which he drew an unending thread, which when he had tied his parcel he bit off—and, it seemed to me, swallowed the ball of string. And then he lit a candle at the nose of one of the ventriloquist's dummies, stuck one of his fingers (which had become sealing-wax red) into the flame, and so sealed the parcel. 'Then there was the Disappearing Egg,' he remarked, and produced one from within my coat-breast and packed it, and also The Crying Baby, Very Human. I handed each parcel to Gip as it was ready, and he clasped them to his chest.

He said very little, but his eyes were eloquent; the clutch of his arms was eloquent. He was the playground of unspeakable emotions. These, you know, were *real* Magics.

Then, with a start, I discovered something moving about in my hat—something soft and jumpy. I whipped it off, and a ruffled pigeon—no doubt a confederate—dropped out and ran on the counter, and went, I fancy, into a cardboard box behind the *papier-mâché* tiger.

'Tut, tut!' said the shopman, dexterously relieving me of my headdress; 'careless bird, and—as I live—nesting!'

He shook my hat, and shook out into his extended hand two or three eggs, a large marble, a watch, about half-a-dozen of the inevitable glass balls, and the crumpled, crinkled paper, more and more and more, talking all the time of the way in which people neglect to brush their hats *inside* as well as out, politely, of course, but with a certain personal application. 'All sorts of things accumulate, sir. . . . Not *you*, of course, in particular. . . . Nearly every customer. . . . Astonishing what they carry about with them. . . .' The crumpled paper rose and billowed on the counter more and more and more, until he was nearly hidden from us, until he was altogether hidden, and still his voice went on and on. 'We none of us know what the fair semblance of a human being may conceal, sir. Are we all then no better than brushed exteriors, whited sepulchres—'

His voice stopped—exactly like when you hit a neighbour's gramophone with a well-aimed brick, the same instant silence, and the rustle of the paper stopped, and everything was still. . . .

'Have you done with my hat?' I said, after an interval.

There was no answer.

I stared at Gip, and Gip stared at me; and there were our distortions in the magic mirrors, looking very rum, and grave, and quiet. . . .

'I think we'll go now,' I said. 'Will you tell me how much all this comes to? . . .

'I say,' I said, on a rather louder note, 'I want the bill; and my hat, please.'

It might have been a sniff from behind the paper pile. . . .

'Let's look behind the counter, Gip,' I said. 'He's making fun of us.'

I led Gip round the head-wagging tiger, and what do you think there was behind the counter? No one at all! Only my hat on the floor, and a common conjurer's lop-eared white rabbit lost in meditation, and looking as stupid and crumpled as only a conjurer's rabbit can do. I resumed my hat, and the rabbit lolloped a lollop or so out of my way.

'Dadda!' said Gip, in a guilty whisper.

'What is it, Gip?' said I.

'I *do* like this shop, dadda.'

'So should I,' I said to myself, 'if the counter wouldn't suddenly extend itself to shut one off from the door.' But I didn't call Gip's attention to that. 'Pussy!' he said, with a hand out to the rabbit as it came lolloping past us; 'Pussy, do Gip a magic!' and his eyes followed it as it squeezed through a door I had certainly not remarked a moment before. Then this door opened wider, and the man with one ear larger than the other appeared again. He was smiling still, but his eye met mine with something between amusement and defiance. 'You'd like to see our showroom, sir,' he said, with an innocent suavity. Gip tugged my finger forward. I glanced at the counter and met the shopman's eye again. I was beginning to think the magic just a little too genuine. 'We haven't *very* much time,' I said. But somehow we were inside the show-room before I could finish that.

'All goods of the same quality,' said the shopman, rubbing his flexible hands together, 'and that is the Best. Nothing in the place that isn't genuine Magic, and warranted thoroughly rum. Excuse me, sir!'

I felt him pull at something that clung to my coat-sleeve, and then I saw he held a little, wriggling red demon by the tail—the little creature bit and fought and tried to get at his hand—and in a moment he tossed it carelessly behind a counter. No doubt the thing was only an image of twisted indiarubber, but for the moment—! And his gesture was exactly that of a man who handles some petty biting bit of vermin. I glanced at Gip, but Gip was looking at a magic rocking-horse. I was glad he hadn't seen the thing. 'I say,' I said, in an undertone, and indicating Gip and the red demon with my eyes, 'you haven't many things like *that* about, have you?'

'None of ours! Probably brought it with you,' said the shopman—also in an undertone, and with a more dazzling smile than ever. 'Astonishing what people *will* carry about with them unawares!' And then to Gip, 'Do you see anything you fancy here?'

There were many things that Gip fancied there.

He turned to this astonishing tradesman with mingled confidence and respect. 'Is that a Magic Sword?' he said.

'A Magic Toy Sword. It neither bends, breaks, nor cuts the fingers. It renders the bearer invincible in battle against anyone under eighteen. Half-a-crown to seven and sixpence, according to size. These panoplies on cards are for juvenile knights-errant and very useful—shield of safety, sandals of swiftness, helmet of invisibility.'

'Oh, dadda!' gasped Gip.

I tried to find out what they cost, but the shopman did not heed me. He had got Gip now; he had got him away from my finger; he had embarked upon the exposition of all his confounded stock, and nothing was going to stop him. Presently I saw with a qualm of distrust and something very like jealousy that Gip had hold of this person's finger as usually he has hold of mine. No doubt the fellow was interesting, I thought, and had an interestingly faked lot of stuff, really *good* faked stuff, still—

I wandered after them, saying very little, but keeping an eye on this prestidigital fellow. After all, Gip was enjoying it. And no doubt when the time came to go we should be able to go quite easily.

It was a long, rambling place, that showroom, a gallery broken up by stands and stalls and pillars, with archways leading off to other departments, in which the queerest-looking assistants loafed and stared at one, and with perplexing mirrors and curtains. So perplexing, indeed, were these that I was presently unable to make out the door by which we had come.

The shopman showed Gip magic trains that ran without steam or clockwork, just as you set the signals, and then some very, very valuable boxes of soldiers that all came alive directly you took off the lid and said—I myself haven't a very quick ear and it was a tongue-twisting sound, but Gip—he has his mother's ear—got it in no time. 'Bravo!' said the shopman, putting the men back into the box unceremoniously and handing it to Gip. 'Now,' said the shopman, and in a moment Gip had made them all alive again.

'You'll take that box?' asked the shopman.

'We'll take that box,' said I, 'unless you charge its full value. In which case it would need a Trust Magnate—'

'Dear heart! *No!*' and the shopman swept the little men back again, shut the lid, waved the box in the air, and there it was, in brown paper, tied up and—*with Gip's full name and address on the paper!*

The shopman laughed at my amazement.

'This is the genuine magic,' he said. 'The real thing.'

'It's almost too genuine for my taste,' I said again.

After that he fell to showing Gip tricks, odd tricks, and still odder the way they were done. He explained them, he turned them inside out, and there was the dear little chap nodding his busy bit of a head in the sagest manner.

I did not attend as well as I might. 'Hey, presto!' said the Magic Shopman, and then would come the clear, small 'Hey, presto!' of the boy. But I was distracted by other things. It was being borne in upon me just how tremendously rum this place was; it was, so to speak, inundated by a sense of rumness. There was something vaguely rum about the fixtures even, about the ceiling, about the floor, about the casually distributed chairs. I had a queer feeling that whenever I wasn't looking at them straight they went askew, and moved about, and played a noiseless puss-in-the-corner behind my back. And the cornice had a serpentine design with masks—masks altogether too expressive for proper plaster.

Then abruptly my attention was caught by one of the odd-looking assistants. He was some way off and evidently unaware of my presence—I saw a sort of three-quarter length of him over a pile of toys and through an arch—and, you know, he was leaning against a pillar in an idle sort of way doing the most horrid things with his features! The particularly horrid thing he did was with his nose. He did it just as though he was idle and wanted to amuse himself. First of all it was a short, blobby nose, and then suddenly he shot it out like a telescope, and then out it flew and became thinner and thinner until it was like a long, red, flexible whip. Like a thing in a nightmare it was! He flourished it about and flung it forth as a fly-fisher flings his line.

My instant thought was that Gip mustn't see him. I turned about, and there was Gip quite preoccupied with the shopman, and thinking no evil. They were whispering together and looking at me. Gip was standing on a stool, and the shopman was holding a sort of big drum in his hand.

'Hide and seek, dadda!' cried Gip. 'You're He!'

And before I could do anything to prevent it, the shopman had clapped the big drum over him.

I saw what was up directly. 'Take that off,' I cried, 'this instant! You'll frighten the boy. Take it off!'

The shopman with the unequal ears did so without a word, and held the big cylinder towards me to show its emptiness. And the stool was vacant! In that instant my boy had utterly disappeared! . . .

You know, perhaps, that sinister something that comes like a hand out of the unseen and grips your heart about. You know it takes your common self away and leaves you tense and deliberate, neither slow nor hasty, neither angry nor afraid. So it was with me.

I came up to this grinning shopman and kicked his stool aside.

'Stop this folly!' I said. 'Where is my boy?'

'You see,' he said, still displaying the drum's interior, 'there is no deception—'

I put out my hand to grip him, and he eluded me by a dexterous movement. I snatched again, and he turned from me and pushed open a door to escape. 'Stop!' I said, and he laughed, receding. I leapt after him—into utter darkness.

Thud!

'Lor' bless my 'eart! I didn't see you coming, sir!'

I was in Regent Street, and I had collided with a decent-looking working man; and a yard away, perhaps, and looking extremely perplexed with himself, was Gip. There was some sort of apology, and then Gip had turned and come to me with a bright little smile, as though for a moment he had missed me.

And he was carrying four parcels in his arm!

He secured immediate possession of my finger.

For the second I was rather at a loss. I stared round to see the door of the magic shop, and, behold, it was not there! There was no door, no shop, nothing, only the common pilaster between the shop where they sell pictures and the window with the chicks! . . .

I did the only thing possible in that mental tumult; I walked straight to the kerbstone and held up my umbrella for a cab.

''Ansoms,' said Gip, in a note of culminating exultation.

I helped him in, recalled my address with an effort, and got in also. Something unusual proclaimed itself in my tail-coat pocket, and I felt and discovered a glass ball. With a petulant expression I flung it into the street.

Gip said nothing.

For a space neither of us spoke.

'Dadda!' said Gip, at last, 'that *was* a proper shop!'

I came round with that to the problem of just how the whole thing had seemed to him. He looked completely undamaged—so far, good; he was neither scared nor unhinged, he was simply tremendously satisfied with the afternoon's entertainment, and there in his arms were the four parcels.

Confound it! what could be in them?

'Um!' I said. 'Little boys can't go to shops like that every day.'

He received this with his usual stoicism, and for a moment I was sorry I was his father and not his mother, and so couldn't suddenly there, *coram publico*, in our hansom, kiss him. After all, I thought, the thing wasn't so very bad.

But it was only when we opened the parcels that I really began to be reassured. Three of them contained boxes of soldiers, quite ordinary lead soldiers, but of so good a quality as to make Gip altogether forget that originally these parcels had been Magic Tricks of the only genuine sort, and the fourth contained a kitten, a little living white kitten, in excellent health and appetite and temper.

I saw this unpacking with a sort of provisional relief. I hung about in the nursery for quite an unconscionable time. . . .

That happened six months ago. And now I am beginning to believe it is all right. The kitten had only the magic natural to all kittens, and the soldiers seem as steady a company as any colonel could desire. And Gip—?

The intelligent parent will understand that I have to go cautiously with Gip.

But I went so far as this one day. I said, 'How would you like your soldiers to come alive, Gip, and march about by themselves?'

'Mine do,' said Gip. 'I just have to say a word I know before I open the lid.'

'Then they march about alone?'

'Oh, *quite*, dadda. I shouldn't like them if they didn't do that.'

I displayed no unbecoming surprise, and since then I have taken occasion to drop in upon him once or twice, unannounced, when the soldiers were about, but so far I have never discovered them performing in anything like a magical manner. . . .

It's so difficult to tell.

There's also a question of finance. I have an incurable habit of paying bills. I have been up and down Regent Street several times, looking for that shop. I am inclined to think, indeed, that in that matter honour is satisfied, and that, since Gip's name and address are known to them, I may very well leave it to these people, whoever they may be, to send in their bill in their own time.

1903

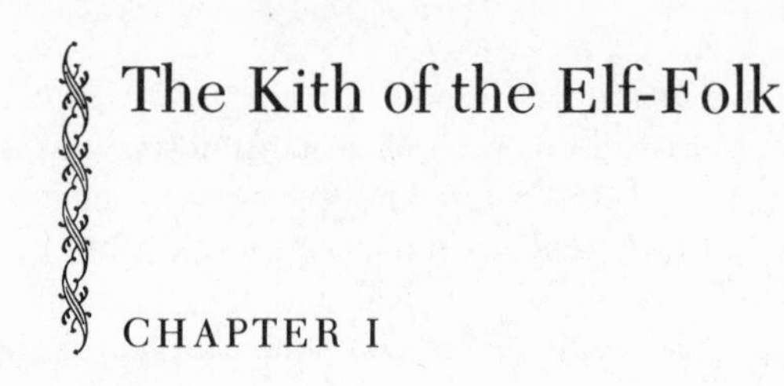
LORD DUNSANY

The Kith of the Elf-Folk

CHAPTER I

THE north wind was blowing, and red and golden the last days of Autumn were streaming hence. Solemn and cold over the marshes arose the evening.

It became very still.

Then the last pigeon went home to the trees on the dry land in the distance, whose shapes already had taken upon themselves a mystery in the haze.

Then all was still again.

As the light faded and the haze deepened, mystery crept nearer from every side.

Then the green plover came in crying, and all alighted.

And again it became still, save when one of the plover arose and flew a little way uttering the cry of the waste. And hushed and silent became the earth, expecting the first star. Then the duck came in, and the widgeon, company by company: and all the light of day faded out of the sky saving one red band of light. Across the light appeared, black and huge, the wings of a flock of geese beating up wind to the marshes. These, too, went down among the rushes.

Then the stars appeared and shone in the stillness, and there was silence in the great spaces of the night.

Suddenly the bells of the cathedral in the marshes broke out, calling to evensong.

Eight centuries ago on the edge of the marsh men had built the huge cathedral, or it may have been seven centuries ago, or perhaps nine—it was all one to the Wild Things.

So evensong was held, and candles lighted, and the lights through the windows shone red and green in the water, and the sound of the organ went roaring over the marshes. But from the deep and perilous places, edged with bright mosses, the Wild

Things came leaping up to dance on the reflection of the stars, and over their heads as they danced the marsh-lights rose and fell.

The Wild Things are somewhat human in appearance, only all brown of skin and barely two feet high. Their ears are pointed like the squirrel's, only far larger, and they leap to prodigious heights. They live all day under deep pools in the loneliest marshes, but at night they come up and dance. Each Wild Thing has over its head a marsh-light, which moves as the Wild Thing moves; they have no souls, and cannot die, and are of the kith of the Elf-folk.

All night they dance over the marshes treading upon the reflection of the stars (for the bare surface of the water will not hold them by itself); but when the stars begin to pale, they sink down one by one into the pools of their home. Or if they tarry longer, sitting upon the rushes, their bodies fade from view as the marsh-fires pale in the light, and by daylight none may see the Wild Things of the kith of the Elf-folk. Neither may any see them even at night unless they were born, as I was, in the hour of dusk, just at the moment when the first star appears.

Now, on the night that I tell of, a little Wild Thing had gone drifting over the waste, till it came right up to the walls of the cathedral and danced upon the images of the coloured saints as they lay in the water among the reflection of the stars. And as it leaped in its fantastic dance, it saw through the painted windows to where the people prayed, and heard the organ roaring over the marshes. The sound of the organ roared over the marshes, but the song and prayers of the people streamed up from the cathedral's highest tower like thin gold chains, and reached to Paradise, and up and down them went the angels from Paradise to the people, and from the people to Paradise again.

Then something akin to discontent troubled the Wild Thing for the first time since the making of the marshes; and the soft grey ooze and the chill of the deep water seemed to be not enough, nor the first arrival from northwards of the tumultuous geese, nor the wild rejoicing of the wings of the wildfowl when every feather sings, nor the wonder of the calm ice that comes when the snipe depart and beards the rushes with frost and clothes the hushed waste with a mysterious haze where the sun goes red and low, nor even the dance of the Wild Things in the marvellous night; and the little Wild Thing longed to have a soul, and to go and worship God.

And when evensong was over and the lights were out, it went back crying to its kith.

But on the next night, as soon as the images of the stars appeared in the water, it went leaping away from star to star to the farthest edge of the marshlands, where a great wood grew where dwelt the Oldest of the Wild Things.

And it found the Oldest of Wild Things sitting under a tree, sheltering itself from the moon.

And the little Wild Thing said: 'I want to have a soul to worship God, and to know the meaning of music, and to see the inner beauty of the marshlands and to imagine Paradise.'

And the Oldest of the Wild Things said to it: 'What have we to do with God? We are only Wild Things, and of the kith of the Elf-folk.'

But it only answered, 'I want to have a soul.'

Then the Oldest of the Wild Things said: 'I have no soul to give you; but if you got a soul, one day you would have to die, and if you knew the meaning of music you would learn the meaning of sorrow, and it is better to be a Wild Thing and not to die.'

So it went weeping away.

But they that were kin to the Elf-folk were sorry for the little Wild Thing; and though the Wild Things cannot sorrow long, having no souls to sorrow with, yet they felt for awhile a soreness where their souls should be when they saw the grief of their comrade.

So the kith of the Elf-folk went abroad by night to make a soul for the little Wild Thing. And they went over the marshes till they came to the high fields among the flowers and grasses. And there they gathered a large piece of gossamer that the spider had laid by twilight; and the dew was on it.

Into this dew had shone all the lights of the long banks of the ribbed sky, as all the colours changed in the restful spaces of evening. And over it the marvellous night had gleamed with all its stars.

Then the Wild Things went with their dew-bespangled gossamer down to the edge of their home. And there they gathered a piece of the grey mist that lies by night over the marshlands. And into it they put the melody of the waste that is borne up and down the marshes in the evening on the wings of the golden plover. And they put into it, too, the mournful songs that the reeds are compelled to sing before the presence of the arrogant North Wind. Then each of the Wild Things gave some treasured memory of the old marshes, 'For we can spare it,' they said. And to all this they added a few images of the stars that they gathered out of the water. Still the soul that the kith of the Elf-folk were making had no life.

Then they put into it the low voices of two lovers that went walking in the night, wandering late alone. And after that they waited for the dawn. And the queenly dawn appeared, and the marsh-lights of the Wild Things paled in the glare, and their bodies faded from view; and still they waited by the marsh's edge. And to them waiting came over field and marsh, from the ground and out of the sky, the myriad song of the birds.

This, too, the Wild Things put into the piece of haze that they had gathered in the marshlands, and wrapped it all up in their dew-bespangled gossamer. Then the soul lived.

And there it lay in the hands of the Wild Things no larger than a hedgehog; and wonderful lights were in it, green and blue; and they changed ceaselessly, going round and round, and in the grey midst of it was a purple flare.

And the next night they came to the little Wild Thing and showed her the gleaming soul. And they said to her: 'If you must have a soul and go and worship God, and become a mortal and die, place this to your left breast a little above the heart, and it will enter and you will become a human. But if you take it you can never be rid of it to become immortal again unless you pluck it out and give it to another; and *we* will not take it, and most of the humans have a soul already. And if you cannot find a human without a soul you will one day die, and your soul cannot go to Paradise because it was only made in the marshes.'

Far away the little Wild Thing saw the cathedral windows alight for evensong, and the song of the people mounting up to Paradise, and all the angels going up and down. So it bid farewell with tears and thanks to the Wild Things of the kith of Elf-folk, and went leaping away towards the green dry land, holding the soul in its hands.

And the Wild Things were sorry that it had gone, but could not be sorry long because they had no souls.

At the marsh's edge the little Wild Thing gazed for some moments over the water to where the marsh-fires were leaping up and down, and then pressed the soul against its left breast a little above the heart.

Instantly it became a young and beautiful woman, who was cold and frightened. She clad herself somehow with bundles of reeds, and went towards the lights of a house that stood close by. And she pushed open the door and entered, and found a farmer and a farmer's wife sitting over their supper.

And the farmer's wife took the little Wild Thing with the soul of the marshes up to her room, and clothed her and braided her hair, and brought her down again, and gave her the first food that she had ever eaten. Then the farmer's wife asked many questions.

'Where have you come from?' she said.

'Over the marshes.'

'From what direction?' said the farmer's wife.

'South,' said the little Wild Thing with the new soul.

'But none can come over the marshes from the south,' said the farmer's wife.

'No, they can't do that,' said the farmer.

'I lived in the marshes.'

'Who are you?' asked the farmer's wife.

'I am a Wild Thing, and have found a soul in the marshes, and we are kin to the Elf-folk.'

Talking it over afterwards, the farmer and his wife agreed that she must be a gipsy who had been lost, and that she was queer with hunger and exposure.

So that night the little Wild Thing slept in the farmer's house, but her new soul stayed awake the whole night long dreaming of the beauty of the marshes.

As soon as dawn came over the waste and shone on the farmer's house, she looked from the window towards the glittering waters, and saw the inner beauty of the marsh. For the Wild Things only love the marsh and know its haunts, but now she perceived the mystery of its distances and the glamour of its perilous pools, with their fair and deadly mosses, and felt the marvel of the North Wind who comes dominant out of unknown icy lands, and the wonder of that ebb and flow of life when the wildfowl whirl in at evening to the marshlands and at dawn pass out to sea. And she knew that over her head above the farmer's house stretched wide Paradise, where perhaps God was now imagining a sunrise while angels played low on lutes, and the sun came rising up on the world below to gladden fields and marshes.

And all that heaven thought, the marsh thought too; for the blue of the marsh was as the blue of heaven, and the great cloud shapes in heaven became the shapes in the marsh, and through each ran momentary rivers of purple, errant between banks of gold. And the stalwart army of reeds appeared out of the gloom with all their pennons waving as far as the eye could see. And from another window she saw the vast cathedral gathering its ponderous strength together, and lifting it up in towers out of the marshlands.

She said, 'I will never, never leave the marsh.'

An hour later she dressed with great difficulty and went down to eat the second meal of her life. The farmer and his wife were kindly folk, and taught her how to eat.

'I suppose the gipsies don't have knives and forks,' one said to the other afterwards.

After breakfast the farmer went and saw the Dean, who lived near his cathedral, and presently returned and brought back to the Dean's house the little Wild Thing with the new soul.

'This is the lady,' said the farmer. 'This is Dean Murnith.' Then he went away.

'Ah,' said the Dean, 'I understand you were lost the other night in the marshes. It was a terrible night to be lost in the marshes.'

'I love the marshes,' said the little Wild Thing with the new soul.

'Indeed! How old are you?' said the Dean.

'I don't know,' she answered.

'You must know about how old you are,' he said.

'Oh, about ninety,' she said, 'or more.'

'Ninety years!' exclaimed the Dean.

'No, ninety centuries,' she said; 'I am as old as the marshes.'

Then she told her story—how she had longed to be a human and go and worship God, and have a soul and see the beauty of the world, and how all the Wild Things had made her a soul of gossamer and mist and music and strange memories.

'But if this is true,' said Dean Murnith, 'this is very wrong. God cannot have intended you to have a soul. What is your name?'

'I have no name,' she answered.

'We must find a Christian name and a surname for you. What would you like to be called?'

'Song of the Rushes,' she said.

'That won't do at all,' said the Dean.

'Then I would like to be called Terrible North Wind, or Star in the Waters,' she said.

'No, no, no,' said Dean Murnith; 'that is quite impossible. We could call you Miss Rush if you like. How would Mary Rush do? Perhaps you had better have another name—say Mary Jane Rush.'

So the little Wild Thing with the soul of the marshes took the names that were offered her, and became Mary Jane Rush.

'And we must find something for you to do,' said Dean Murnith. 'Meanwhile we can give you a room here.'

'I don't want to do anything,' replied Mary Jane; 'I want to worship God in the cathedral and live beside the marshes.'

Then Mrs Murnith came in, and for the rest of that day Mary Jane stayed at the house of the Dean.

And there with her new soul she perceived the beauty of the world; for it came grey and level out of misty distances, and widened into grassy fields and ploughlands right up to the edge of an old gabled town; and solitary in the fields far off an ancient windmill stood, and his honest hand-made sails went round and round in the free East Anglian winds. Close by, the gabled houses leaned out over the streets, planted fair upon sturdy timbers that grew in the olden time, all glorying among themselves upon their beauty. And out of them, buttress by buttress, growing and going upwards, aspiring tower by tower, rose the cathedral.

And she saw the people moving in the streets all leisurely and slow, and unseen among them, whispering to each other, unheard by living men and concerned only with bygone things, drifted the ghosts of very long ago. And wherever the streets ran eastwards, wherever were gaps in the houses, always there broke into view the sight of the great marshes, like to some bar of music weird and strange that haunts a melody, arising again and again, played on the violin by one musician only, who plays no other bar, and he is swart and lank about the hair and bearded about the lips, and his moustache droops long and low, and no one knows the land from which he comes.

All these were good things for a new soul to see.

Then the sun set over green fields and ploughlands and the night came up. One by one the merry lights of cheery lamp-lit windows took their stations in the solemn night.

Then the bells rang, far up in a cathedral tower, and their

melody fell on the roofs of the old houses and poured over their eaves until the streets were full, and then flooded away over green fields and ploughlands till it came to the sturdy mill and brought the miller trudging to evensong, and far away eastwards and seawards the sound rang out over the remoter marshes. And it was all as yesterday to the old ghosts in the streets.

Then the Dean's wife took Mary Jane to evening service, and she saw three hundred candles filling all the aisle with light. But sturdy pillars stood there in unlit vastnesses; great colonnades going away into the gloom where evening and morning, year in year out, they did their work in the dark, holding the cathedral roof aloft. And it was stiller than the marshes are still when the ice has come and the wind that brought it has fallen.

Suddenly into this stillness rushed the sound of the organ, roaring, and presently the people prayed and sang.

No longer could Mary Jane see their prayers ascending like thin gold chains, for that was but an elfin fancy, but she imagined clear in her new soul the seraphs passing in the ways of Paradise, and the angels changing guard to watch the World by night.

When the Dean had finished service, a young curate, Mr Millings, went up into the pulpit.

He spoke of Abana and Pharpar, rivers of Damascus: and Mary Jane was glad that there were rivers having such names, and heard with wonder of Nineveh, that great city, and many things strange and new.

And the light of the candles shone on the curate's fair hair, and his voice went ringing down the aisle, and Mary Jane rejoiced that he was there.

But when his voice stopped she felt a sudden loneliness, such as she had not felt since the making of the marshes; for the Wild Things never are lonely and never unhappy, but dance all night on the reflection of the stars, and having no souls desire nothing more.

After the collection was made, before any one moved to go, Mary Jane walked up the aisle to Mr Millings.

'I love you,' she said.

CHAPTER II

Nobody sympathised with Mary Jane. 'So unfortunate for Mr Millings,' every one said; 'such a promising young man.'

Mary Jane was sent away to a great manufacturing city of the Midlands, where work had been found for her in a cloth factory. And there was nothing in that town that was good for a soul to see. For it did not know that beauty was to be desired; so it made many things by machinery, and became hurried in all its ways, and boasted its superiority over other cities and became richer and richer, and there was none to pity it.

In this city Mary Jane had had lodgings found for her near the factory.

At six o'clock on those November mornings, about the time that, far away from the city, the wildfowl rose up out of the calm marshes and passed to the troubled spaces of the sea, at six o'clock the factory uttered a prolonged howl and gathered the workers together, and there they worked, saving two hours for food, the whole of the daylit hours and into the dark till the bells tolled six again.

There Mary Jane worked with other girls in a long dreary room, where giants sat pounding wool into a long thread-like strip with iron, rasping hands. And all day long they roared as they sat at their soulless work. But the work of Mary Jane was not with these, only their roar was ever in her ears as their clattering iron limbs went to and fro.

Her work was to tend a creature smaller, but infinitely more cunning.

It took the strip of wool that the giants had threshed, and whirled it round and round until it had twisted it into hard thin thread. Then it would make a clutch with fingers of steel at the thread that it had gathered, and waddle away about five yards and come back with more.

It had mastered all the subtlety of skilled workers, and had gradually displaced them; one thing only it could not do, it was unable to pick up the ends if a piece of the thread broke, in order to tie them together again. For this a human soul was required, and it was Mary Jane's business to pick up broken ends; and the moment she placed them together the busy soulless creature tied them for itself.

All here was ugly; even the green wool as it whirled round and round was neither the green of the grass nor yet the green of the rushes, but a sorry muddy green that befitted a sullen city under a murky sky.

When she looked out over the roofs of the town, there too was ugliness; and well the houses knew it, for with hideous stucco they aped in grotesque mimicry the pillars and temples of old Greece, pretending to one another to be that which they were not. And emerging from these houses and going in, and seeing the pretence of paint and stucco year after year until it all peeled away, the souls of the poor owners of those houses sought to be other souls until they grew weary of it.

At evening Mary Jane went back to her lodgings. Only then, after the dark had fallen, could the soul of Mary Jane perceive any beauty in that city, when the lamps were lit and here and there a star shone through the smoke. Then she would have gone abroad and beheld the night, but this the old woman to whom she was confided would not let her do. And the days multiplied themselves by seven and became weeks, and the weeks passed by, and all days were the same. And all the while the soul of Mary Jane was crying for beautiful things, and found not one, saving on Sundays, when she went to church, and left it to find the city greyer than before.

One day she decided that it was better to be a Wild Thing in the lonely marshes than to have a soul that cried for beautiful things and found not one. From that day she determined to be rid of her soul, so she told her story to one of the factory girls, and said to her:

'The other girls are poorly clad and they do soulless work; surely some of them have no souls and would take mine.'

But the factory girl said to her: 'All the poor have souls. It is all they have.'

Then Mary Jane watched the rich whenever she saw them, and vainly sought for some one without a soul.

One day at the hour when the machines rested and the human beings that tended them rested too, the wind being at that time from the direction of the marshlands, the soul of Mary Jane lamented bitterly. Then, as she stood outside the factory gates, the soul irresistibly compelled her to sing, and a wild song came from her lips hymning the marshlands. And into her song came crying her yearning for home and for the sound of the shout of the North Wind, masterful and proud, with his lovely lady the snow; and she sang of tales that the rushes murmured to one another, tales that the teal knew and the watchful heron. And over the crowded streets her song went crying away, the song of waste places and of wild free lands, full of wonder and magic, for she had in her elf-made soul the song of the birds and the roar of the organ in the marshes.

At this moment Signor Thompsoni, the well-known English tenor, happened to go by with a friend. They stopped and listened; every one stopped and listened.

'There has been nothing like this in Europe in my time,' said Signor Thompsoni.

So a change came into the life of Mary Jane.

People were written to, and finally it was arranged that she should take a leading part in the Covent Garden Opera in a few weeks.

So she went to London to learn.

London and singing lessons were better than the City of the Midlands and those terrible machines. Yet still Mary Jane was not free to go and live as she liked by the edge of the marshlands, and she was still determined to be rid of her soul, but could find no one that had not a soul of their own.

One day she was told that the English people would not listen to her as Miss Rush, and was asked what more suitable name she would like to be called by.

'I would like to be called Terrible North Wind,' said Mary Jane, 'or Song of the Rushes.'

When she was told that this was impossible and Signorina Maria Russiano was suggested, she acquiesced at once, as she had acquiesced when they took her away from her curate; she knew nothing of the ways of humans.

At last the day of the Opera came round, and it was a cold day of the winter.

And Signorina Russiano appeared on the stage before a crowded house.

And Signorina Russiano sang.

And into the song went all the longing of her soul, the soul that could not go to Paradise, but could only worship God and know the meaning of music, and the longing pervaded that Italian song as the infinite mystery of the hills is borne along the sound of distant sheep-bells. Then in the souls that were in that crowded house arose little memories of a great while since that were quite, quite dead, and lived awhile again during that marvellous song.

And a strange chill went into the blood of all that listened, as though they stood on the border of bleak marshes and the North Wind blew.

And some it moved to sorrow and some to regret, and some to an unearthly joy,—then suddenly the song went wailing away like the winds of the winter from the marshlands when Spring appears from the South.

So it ended. And a great silence fell fog-like over all that house, breaking in upon the end of a chatty conversation that Celia, Countess of Birmingham, was enjoying with a friend.

In the dead hush Signorina Russiano rushed from the stage; she appeared again running among the audience, and dashed up to Lady Birmingham.

'Take my soul,' she said; 'it is a beautiful soul. It can worship God, and knows the meaning of music and can imagine Paradise. And if you go to the marshlands with it you will see beautiful things; there is an old town there built of lovely timbers, with ghosts in its streets.'

Lady Birmingham stared. Every one was standing up. 'See,' said Signorina Russiano, 'it is a beautiful soul.'

And she clutched at her left breast a little above the heart, and there was the soul shining in her hand, with the green and blue lights going round and round and the purple flare in the midst.

'Take it,' she said, 'and you will love all that is beautiful, and know the four winds, each one by his name, and the songs of the birds at dawn. I do not want it, because I am not free. Put it to your left breast a little above the heart.'

Still everybody was standing up, and Lady Birmingham felt uncomfortable.

'Please offer it to some one else,' she said.

'But they all have souls already,' said Signorina Russiano.

And everybody went on standing up. And Lady Birmingham took the soul in her hand.

'Perhaps it is lucky,' she said.

She felt that she wanted to pray.

She half-closed her eyes, and said 'Unberufen.' Then she put the soul to her left breast a little above the heart, and hoped that the people would sit down and the singer go away.

Instantly a heap of clothes collapsed before her. For a moment; in the shadow among the seats, those who were born in the dusk hour might have seen a little brown thing leaping free from the clothes, then it sprang into the bright light of the hall, and became invisible to any human eye.

It dashed about for a little, then found the door, and presently was in the lamplit streets.

To those that were born in the dusk hour it might have been seen leaping rapidly wherever the streets ran northwards and eastwards, disappearing from human sight as it passed under the lamps and appearing again beyond them with a marsh-light over its head.

Once a dog perceived it and gave chase, and was left far behind.

The cats of London, who are all born in the dusk hour, howled fearfully as it went by.

Presently it came to the meaner streets, where the houses are smaller. Then it went due north-eastwards, leaping from roof to roof. And so in a few minutes it came to more open spaces, and then to the desolate lands, where market gardens grow, which are neither town nor country. Till at last the good black trees came into view, with their demonaic shapes in the night, and the grass was cold and wet, and the night-mist floated over it. And a great white owl came by, going up and down in the dark. And at all these things the little Wild Thing rejoiced elvishly.

And it left London far behind it, reddening the sky, and could distinguish no longer its unlovely roar, but heard again the noises of the night.

And now it would come through a hamlet glowing and comfortable in the night; and now to the dark, wet, open fields again; and many an owl it overtook as they drifted through the night, a people friendly to the Elf-folk. Sometimes it crossed wide rivers, leaping from star to star; and, choosing its way as it went, to avoid the hard rough roads, came before midnight to the East Anglian lands.

And it heard there the shout of the North Wind, who was dominant and angry, as he drove southwards his adventurous geese; while the rushes bent before him chaunting plaintively and low, like enslaved rowers of some fabulous trireme, bending and swinging under blows of the lash, and singing all the while a doleful song.

And it felt the good dank air that clothes by night the broad East Anglian lands, and came again to some old perilous pool where the soft green mosses grew, and there plunged downward and downward into the dear dark water till it felt the homely ooze once more coming up between its toes. Thence, out of the lovely chill that is in the heart of the ooze, it arose renewed and rejoicing to dance upon the image of the stars.

I chanced to stand that night by the marsh's edge, forgetting in my mind the affairs of men; and I saw the marsh-fires come leaping up from all the perilous places. And they came up by flocks the whole night long to the number of a great multitude. And danced away together over the marshes.

And I believe that there was a great rejoicing all that night among the kith of the Elf-folk.

1910

CARL SANDBURG

The Story of Blixie Bimber and the Power of the Gold Buckskin Whincher

BLIXIE Bimber grew up looking for luck. If she found a horseshoe she took it home and put it on the wall of her room with a ribbon tied to it. She would look at the moon through her fingers, under her arms, over her right shoulder but never—never over her *left* shoulder. She listened and picked up everything anybody said about the ground hog and whether the ground hog saw his shadow when he came out the second of February.

If she dreamed of onions she knew the next day she would find a silver spoon. If she dreamed of fishes she knew the next day she would meet a strange man who would call her by her first name. She grew up looking for luck.

She was sixteen years old and quite a girl, with her skirts down to her shoe tops, when something happened. She was going to the postoffice to see if there was a letter for her from Peter Potato Blossom Wishes, her best chum, or a letter from Jimmy the Flea, her best friend she kept steady company with.

Jimmy the Flea was a climber. He climbed skyscrapers and flagpoles and smokestacks and was a famous steeplejack. Blixie Bimber liked him because he was a steeplejack, a little, but more because he was a whistler.

Every time Blixie said to Jimmy, 'I got the blues—whistle the blues out of me,' Jimmy would just naturally whistle till the blues just naturally went away from Blixie.

On the way to the postoffice, Blixie found a gold buckskin *whincher*. There it lay in the middle of the sidewalk. How and why it came to be there she never knew and nobody ever told her. 'It's luck,' she said to herself as she picked it up quick.

And so—she took it home and fixed it on a little chain and wore it around her neck.

She did not know and nobody ever told her a gold buckskin whincher is different from just a plain common whincher. It has a *power*. And if a thing has a power over you then you just naturally can't help yourself.

So—around her neck fixed on a little chain Blixie Bimber wore the gold buckskin whincher and never knew it had a power and all the time the power was working.

'The first man you meet with an X in his name you must fall head over heels in love with him,' said the silent power in the gold buckskin whincher.

And that was why Blixie Bimber stopped at the postoffice and went back again asking the clerk at the postoffice window if he was sure there wasn't a letter for her. The name of the clerk was Silas Baxby. For six weeks he kept steady company with Blixie Bimber. They went to dances, hayrack rides, picnics and high jinks together.

All the time the power in the gold buckskin whincher was working. It was hanging by a little chain around her neck and always working. It was saying, 'The next man you meet with two X's in his name you must leave all and fall head over heels in love with him.'

She met the high school principal. His name was Fritz Axenbax. Blixie dropped her eyes before him and threw smiles at him. And for six weeks he kept steady company with Blixie Bimber. They went to dances, hayrack rides, picnics and high jinks together.

'Why do you go with him for steady company?' her relatives asked.

'It's a power he's got,' Blixie answered, 'I just can't help it—it's a power.'

'One of his feet is bigger than the other—how can you keep steady company with him?' they asked again.

All she would answer was, 'It's a power.'

All the time, of course, the gold buckskin whincher on the little chain around her neck was working. It was saying, 'If she meets a man with three X's in his name she must fall head over heels in love with him.'

At a band concert in the public square one night she met James Sixbixdix. There was no helping it. She dropped her eyes and threw her smiles at him. And for six weeks they kept steady company going to band concerts, dances, hayrack rides, picnics and high jinks together.

'Why do you keep steady company with him? He's a musical soup eater,' her relatives said to her. And she answered, 'It's a power—I can't help myself.'

Leaning down with her head in a rain water cistern one day, listening to the echoes against the strange wooden walls of the cistern, the gold buckskin whincher on the little chain around her neck slipped off and fell down into the rain water.

'My luck is gone,' said Blixie. Then she went into the house and made two telephone calls. One was to James Sixbixdix telling him she couldn't keep the date with him that night. The other was to Jimmy the Flea, the climber, the steeplejack.

'Come on over—I got the blues and I want you to whistle 'em away,' was what she telephoned Jimmy the Flea.

And so—if you ever come across a gold buckskin whincher, be careful. It's got a power. It'll make you fall head over heels in love with the next man you meet with an X in his name. Or it will do other strange things because different whinchers have different powers.

1922

WALTER
DE LA MARE

The Lovely Myfanwy

In an old castle under the forested mountains of the Welsh Marches there lived long ago Owen ap Gwythock, Lord of Eggleyseg. He was a short, burly, stooping man with thick black hair on head and face, large ears, and small restless eyes. And he lived in his great castle alone, except for one only daughter, the lovely Myfanwy.

Lovely indeed was she. Her hair, red as red gold, hung in plaits to her knees. When she laughed, it was like bells in a far-away steeple. When she sang, Echo forgot to reply. And her spirit would sit gently looking out of her blue eyes like cushats out of their nest in an ivy bush.

Myfanwy was happy, too—in most things. All that her father could give her for her ease and pleasure was hers—everything indeed but her freedom. She might sing, dance, think and say; eat, drink, and delight in whatsoever she wished or willed. Indeed her father loved her so dearly that he would sit for hours together merely watching her—as you may watch wind over wheat, reflections in water, or clouds in the heavens. So long as she was safely and solely his all was well.

But ever since Myfanwy had been a child, a miserable foreboding had haunted his mind. Supposing she should some day leave him? Supposing she were lost or decoyed away? Supposing she fell ill and died? What then? The dread of this haunted his mind day and night. His dark brows loured at the very thought of it. It made him morose and sullen; it tied up the tongue in his head.

For this sole reason he had expressly forbidden Myfanwy even to stray but a few paces beyond the precincts of his castle; with its battlemented towers, its galleries and corridors and multitudinous apartments, its high garden and courtyard, its alleys, fountains, fish-pools and orchards. He could trust nobody. He couldn't bear her out of his sight. He spied, he watched, he walked in his sleep, he listened and peeped; and all for fear of losing Myfanwy.

So although she might have for company the doves and swans and peacocks, the bees and butterflies, the swallows and swifts and jackdaws and the multitude of birds of every song and flight and feather that haunted the castle; humans, except her father, she had none. The birds and butterflies could fly away at will wherever their wings could carry them. Even the fishes in the fish-pools and in the fountains had their narrow alleys of marble and alabaster through which on nimble fin they could win back to the great river at last. Not so Myfanwy.

She was her father's unransomable prisoner; she was a bird in a cage. She might feast her longing eyes on the distant horizon beyond whose forests lay the sea, but knew she could not journey thither. While as for the neighbouring township, with its busy streets and market-place—not more than seven country miles away—she had only dreamed of its marvels and dreamed in vain. A curious darkness at such times came into her eyes, and her spirit would look out of them not like a dove but as might a dumb nightingale out of its nest—a nightingale that has had its tongue cut out for a delicacy to feed some greedy prince.

How criss-cross a thing is the heart of man. Solely because this lord loved his daughter so dearly, if ever she so much as sighed for change or adventure, like some stubborn beast of burden he would set his feet together and refuse to budge an inch. Beneath his heavy brows he would gaze at the brightness of her unringleted hair as if mere looking could keep that gold secure; as if earth were innocent of moth and rust and change and chance, and had never had course to dread and tremble at sound of the unrelenting footfall of Time.

All he could think of that would keep her his own was hers without the asking: delicate raiment and meats and strange fruits and far-fetched toys and devices and pastimes, and as many books as would serve a happy scholar a long life through. He never tired of telling her how much he loved and treasured her. But there is a hunger of the heart no *thing* in the world can ever satisfy. And Myfanwy listened, and sighed.

Besides which, Myfanwy grew up and grew older as a green-tressed willow grows from a sapling; and now that she had come to her eighteenth spring she was lovelier than words could tell. This only added yet another and sharper dread and foreboding to her father's mind. It sat like a skeleton at his table whenever he broke bread or sipped wine. Even the twittering of a happy swallow from distant Africa reminded him of it like a knell. It was this: that some day a lover, a suitor, would come and carry her off.

Why, merely to *see* her, even with her back turned—to catch a glimpse of her slim shoulders, of her head stooping over a rose-bush would be enough. Let her but laugh—two notes—and you listened! Nobody—prince nor peasant, knight nor squire—brave, foolish, young or weary, would be able to resist her. Owen ap Gwythock knew it in his bones. But one look, and instantly the looker's heart would be stolen out of his body. He would fall in love with her—fall as deep and irrevocably as the dark sparkling foaming water crashing over into the gorge of Modwr-Eggleyseg, scarcely an arrow's flight beyond his walls.

And supposing any such suitor should *tell* Myfanwy that he loved her, might she not—forgetting all his own care and loving-kindness—be persuaded to flee away and leave him to his solitude? Solitude—now that old age was close upon him! At thought of this, for fear of it, he would sigh and groan within: and he would bid the locksmiths double their locks and bolts and bars; and he would sit for hours watching the highroad that swept up past his walls, and scowling at sight of every stranger who passed that way.

He even at last forbade Myfanwy to walk in the garden except with an immense round mushroom hat on her head, a hat so wide in the brim that it concealed from any trespasser who might be spying over the wall even the glinting of her hair—everything of her indeed except her two velvet shoes beneath the hem of her dress as they stepped in turn—and softly as moles—one after the other from blossoming alley to alley and from lawn to lawn.

And because Myfanwy loved her father almost as dearly as he loved her, she tried her utmost to be gay and happy and not to fret or complain or grow pale and thin and pine. But as a caged bird with a kind mistress may hop and sing and flutter behind its bars as if it were felicity itself, and yet be sickening at heart for the wild wood and its green haunts, so it was with Myfanwy.

If only she might but just once venture into the town, she would think to herself; but just to see the people in the streets, and the pedlars in the market-place, and the cakes and sweetmeats and honey-jars in the shops, and strangers passing to and fro, and the sunshine in the high gables, and the talking and the laughing and the bargaining and the dancing—the horses, the travellers, the bells, the starshine.

Above all, it made her heart ache to think her father should have so little faith in her duty and love for him that he would not consent to let her wander even a snail's journey out of his sight. When, supper over, she leaned over his great chair as he sat there in his crimson—his black hair dangling on his shoulders, his beard hunched up on his chest—to kiss him good night, this thought would be in her eyes even if not on the tip of her tongue. And at such times he himself—as if he knew in his heart what he would never dare to confess—invariably shut down his eyelids or looked the other way.

Now servants usually have long tongues, and gossip flits from place to place like seeds of thistledown. Simply because Myfanwy was never seen abroad, the fame of her beauty had long since spread through all the countryside. Minstrels sang of it, and had even carried their ballads to countries and kingdoms and principalities far beyond Wales.

Indeed, however secret and silent men may be concerning rare beauty and goodness, somehow news of it sows itself over the wide world. A saint may sit in his cave or his cell, scarcely ever seen by mortal eye, quiet as sunshine in a dingle of the woods or seabirds in the hollows of the Atlantic, doing his deeds of pity and loving-kindness, and praying his silent prayers. And he may live to be a withered-up, hollow-cheeked old man with a long white beard, and die, and his body be shut up in a tomb. But nevertheless, little by little, the fame of his charity, and of the miracles of his compassion will spread abroad, and at last you may even chance on his image in a shrine thousands of leagues distant from the hermitage where he lived and died, and centuries after he has gone on his way.

Like this it was with the loveliness and gentleness of Myfanwy. That is why, when the Lord of Eggleyseg himself rode through the streets of the neighbouring town, he perceived out of the corner of his eye strangers in outlandish disguise whom he suspected at once must be princes and noblemen from foreign climes come thither even if merely to set eyes on his daughter. That is why the streets were so full of music and singing that of a summer evening you could scarcely hear the roar of its cataracts. That is why its townsfolk were entertained with tumblers and acrobats and fortune-tellers and soothsayers and tale-tellers almost the whole year long. Ever and again, indeed, grandees visited it *without* disguise. They lived for weeks there, with their retinues of servants, their hawks and hounds and tasselled horses in some one of its high ancient houses. And their one sole hope and desire was to catch but a glimpse of the far-famed Myfanwy.

But as they came, so they went away. However they might plot and scheme to gain a footing in the castle—it was in vain. The portcullis was always down; there were watchmen perpetually on the look-out in its turrets; and the gates of the garden were festooned with heavy chains. There was not in its frowning ancient walls a single window less than twenty feet above the ground that was not thickly, rustily, and securely barred.

None the less, Myfanwy occasionally found herself in the garden alone. Occasionally she stole out if but for one breath of freedom, sweeter by far to those who pine for it than that of pink, or mint, or jasmine, or honeysuckle. And one such early evening in May, when her father—having nodded off to sleep, wearied out after so much watching and listening and prying and peering—was snoring in an arbour or summerhouse, she came to its western gates, and having for a moment lifted the brim of her immense hat to look at the sunset, she gazed wistfully a while through its bars out into the green woods beyond.

The leafy boughs in the rosy light hung still as pictures in deep water. The skies resembled a tent of silk, blue as the sea. Deer were browsing over the dark turf; and a wonderful charm and carolling of birds was rising out of the glades and coverts of the woods.

But what Myfanwy had now fixed her dark eyes on was none of these, but the figure of a young man leaning there, erect but fast asleep, against the bole of a gigantic beech tree, not twenty paces distant from the gate at which she stood. He must, she fancied, have been keeping watch there for some little time. His eyelids were dark with watching; his face pale. Slim and gentle does were treading close beside him; the birds had clean forgotten his presence; and a squirrel was cracking the nut it held between its clawed forepaws not a yard above his head.

Myfanwy had never before set eyes on human stranger in this

valley beyond the gates. Her father's serving men were ancients who had been in his service in the castle years before she was born. This young man looked, she imagined, like a woodman, or a forester, or a swine-herd. She had read of them in a hand-written book of fantastic tales which she had chanced on among her mother's belongings.

And as Myfanwy, finger on brim of her hat, stood intently gazing, a voice in her heart told her that whoever and whatever this stranger might be, he was someone she had been waiting for, and even dreaming about, ever since she was a child. All else vanished out of her mind and her memory. It was as if her eyes were intent on some such old story itself, and one well known to her. This unconscious stranger was that story. Yet he himself—stiff as a baulk of wood against the beech-trunk, as if indeed he had been nailed to its bark—slumbered on.

So he might have continued to do, now so blessedly asleep, until she had vanished as she had come. But at that moment the squirrel there, tail for parasol immediately above his head, having suddenly espied Myfanwy beyond the bars of the gate, in sheer astonishment let fall its nut, and the young man—as if at a tiny knock on the door of his mind—opened his eyes.

For Myfanwy it was like the opening of a door into a strange and wonderful house. Her heart all but ceased to beat. She went cold to her fingertips. And the stranger too continued to gaze at Myfanwy—as if out of a dream.

And if everything could be expressed in words that this one quiet look between them told Myfanwy of things strange that yet seemed more familiar to her than the pebbles on the path and the thorns on the rose-bushes and the notes of the birds in the air and the first few drops of dew that were falling in the evening air, then it would take a book ten times as long as this in which to print it.

But even as she gazed Myfanwy suddenly remembered her father. She sighed; her fingers let fall the wide brim of her hat; she turned away. And oddly enough, by reason of this immense ridiculous hat, her father who but a few moments before had awakened in his arbour and was now hastening along the path of the rosery in pursuit of her, caught not a single glimpse of the stranger under the beech-tree. Indeed, before the squirrel could scamper off into hiding, the young man had himself vanished round the trunk of the tree and out of sight like a serpent into the grass.

In nothing except in this, however, did he resemble a serpent. For that very evening at supper her father told Myfanwy that yet another letter had been delivered at the castle, from some accursed Nick Nobody, asking permission to lay before him his suit for her hand. His rage was beyond words. He spilt his wine and crumbled his bread—his face a storm of darkness; his eyes like smouldering coals.

Myfanwy sat pale and trembling. Hitherto, such epistles, though even from princes of renowned estate and of realms even of the Orient, had carried much less meaning to her heart than the cuckooing of a cuckoo, or the whispering of the wind. Indeed, the cuckoo of those Welsh mountains and the wind from over their seas were voices of a language which, though secret, was not one past the heart's understanding. Not so these pompous declarations. Myfanwy would laugh at them—as though at the clumsy gambollings of a bear. She would touch her father's hand, and smile into his face, to assure him they had no meaning, that she was still as safe as safe could be.

But *this* letter—not for a single moment had the face of the young stranger been out of her mind. Her one sole longing and despair was the wonder whether she would ever in this world look upon him again. She sat like stone.

'Ay, ay, my dear,' said her father at last, laying his thick, square hand on hers as she sat beside him in her high-backed velvet chair—'ay, ay, my gentle one. It shows us yet again how full the world is of insolence and adventurers. This is a *cave*, a warning, an *alarum*, my dear—maledictions on his bones! We must be ten times more cautious; we must be wary; we must be lynx and fox and Argus—all eyes! And remember, my all, my precious one, remember this, that while I, your father, am alive, no harm, no ill can approach or touch you. Believe only in my love, beloved, and all is well with us.'

Her cold lips refused to speak. Myfanwy could find no words with which to answer him. With face averted she sat in a woeful daydream, clutching her father's thumb, and only vaguely listening to his transports of fury and affection, revenge and adoration. For her mind and heart now welled over with such a medley of thoughts and hopes and fears and sorrows that she could find no other way but this dumb clutch of expressing that she loved her father too.

At length, his rage not one whit abated, he rose from his chair, and having torn the insolent letter into thirty-two tiny pieces he flung them into the huge log fire burning in the stone chimney. 'Let me but lay a finger on the shameless popinjay,' he muttered to himself; 'I'll—I'll cut his tongue out!'

Now the first thing Myfanwy did when the chance offered was to hasten off towards the Western Gate if only to warn the stranger of her father's rage and menaces, and bid him go hide himself away and never never, never come back again.

But when once more she approached its bars the deer were still grazing in the forest, the squirrel was nibbling another nut, the beech had unfolded yet a few more of its needle-pointed leaves into the calm evening light; but of the stranger—not a sign. Where he had stood was now only the assurance that he was indeed gone for ever. And Myfanwy turned from the quiet scene, from the forest, its sunlight faded, all its beauty made forlorn. Try as she might in the days that followed to keep her mind and her thoughts fixed on her needle and her silks, her lute and her psalter, she could see nothing else but that long look of his.

And now indeed she began to pine and languish in body, haunted by the constant fear that her stranger might have met with some disaster. And simply because her father loved her so jealously, he knew at once what worm was in her mind, and he never ceased to watch and spy upon her, and to follow her every movement.

Now Myfanwy's bed-chamber was in the southern tower of this lord's castle, beneath which a road from the town to the eastward wound round towards the forests and distant mountains. And it being set so high above the ground beneath, there was no need for bars to its windows. While then, from these window-slits Myfanwy could see little more than the tops of the wayfarers' heads on the turf below, they were wide and lofty enough to let the setting sun in its due hour pour in its beams upon her walls and pictures and curtained Arabian bed. But the stone walls being so thick, in order to see out of her chamber at all, she must needs lie along a little on the cold inward sill, and peer out over the wide verdant country-side as if through the port-hole of a ship.

And one evening, as Myfanwy sat sewing a seam—and singing the while a soft tune to herself, if only to keep her thoughts from pining—she heard the murmur of many voices. And, though at first she knew not why, her heart for an instant or two stopped beating. Laying her slip of linen down, she rose, stole over the mats on the flagstones, and gently pushing her narrow shoulders onwards, peeped out and down at last through the window to look at the world below. And this was what she saw. In an old velvet cloak, his black hair dangling low upon his shoulders, there in the evening light beneath her window was a juggler standing, and in a circle round and about him was gathered a throng of gaping country-folk and idlers and children, some of whom must even have followed him out of the town. And one and all they were lost in wonder at his grace and skill.

Myfanwy herself indeed could not have imagined such things could be, and so engrossed did she become in watching him that she did not catch the whisper of a long-drawn secret sigh at her keyhole; nor did she hear her father as he turned away on tip-toe to descend the staircase again into the room below.

Indeed one swift glance from Myfanwy's no longer sorrowful eyes had pierced the disguise—wig, cloak, hat, and hose—of the juggler. And as she watched him she all but laughed aloud. Who would have imagined that the young stranger, whom she had seen for the first time leaning dumb, blind, and fast asleep against the trunk of a beech-tree could be possessed of such courage and craft and cunning as this!

His head was at the moment surrounded by a halo of glittering steel—so fast the daggers with which he was juggling whisked on from hand to hand. And suddenly the throng around him broke into a roar, for in glancing up and aside he had missed a dagger. It was falling—falling: but no, in a flash he had twisted back the sole of his shoe, and the point had stuck quivering in his heel, while he continued to whirl its companions into the golden air.

In that instant, however, his upward glance had detected the one thing in the world he had come out in hope to see—Myfanwy. He flung his daggers aside and fetched out of his travelling box a netful of coloured balls. Holloing out a string of outlandish gibberish to the people, he straightaway began to juggle with these. Higher and higher the seven of them soared into the mellow air, but one of the colour of gold soared on ever higher and higher than any. So high, indeed, that at last the people could watch it no longer because of the dazzle of the setting sun in their eyes. Presently, indeed, it swooped so loftily into the air that Myfanwy need but thrust out her hand to catch it as it paused for a breath of an instant before falling, and hung within reach of her stone window-sill.

And even as she watched, enthralled, a whispering voice within her cried, 'Take it!' She breathed a deep breath, shut her eyes, paused, and the next instant she had stretched out her hand into the air. The ball was hers.

Once more she peeped down and over, and once more the juggler was at his tricks. This time with what appeared to be a medley of all kinds of varieties of fruits; pomegranates, quinces, citrons, lemons, oranges and nectarines, and soaring high above them, nothing more unusual than an English apple. Once again the whisperer in Myfanwy's mind cried, 'Take it!' And she put out her hand and took the apple too.

Yet again she peeped and peered over, and this time it seemed that the juggler was flinging serpents into the air, for they writhed and looped and coiled around him as they whirled whiffling on from hand to hand. There was a hissing, too, and the people drew back a little, and a few of the timider children ran off to the other side of the highroad. And now, yet again, one of the serpents was soaring higher and higher above the rest. And Myfanwy could see from her coign of vantage that it was no live serpent but a strand of silken rope. And yet again and for the third time the whisperer whispered, 'Take it!' And Myfanwy put out her hand and took that too.

And, it happening that a little cloud was straying across the sun at this moment, the throng below had actually seen the highest-most of the serpents thus mysteriously disappear and they cried out as if with one voice, 'Gone!' 'Vanished!' 'Vanished!' 'Gone!' 'Magician, magician!' And the coins that came dancing into the juggler's tambourine in the moments that followed were enough to make him for that one minute the richest man in the world.

And now the juggler was solemnly doffing his hat to the people. He gathered his cloak around him more closely, put away his daggers, his balls, his fruits, his serpents, and all that was his, into a long green narrow box. Then he hoisted its strap over his shoulder, and doffing his cap once more, he clasped his tambourine under his elbow and seizing his staff, turned straight from the castle tower towards the hazy sun-bathed mountains. And, it beginning to be towards nightfall, the throng of people soon dispersed and melted away; the maids and scullions, wooed out by this spectacle from the castle, returned to their work; and the children ran off home to tell their mothers of these marvels and to mimic the juggler's tricks as they gobbled up their supper-crusts and were packed off to bed.

In the stillness that followed after the juggler's departure, Myfanwy found herself kneeling in her chamber in the tranquil golden twilight beside a wooden chair, her hands folded in her lap and her dark eyes fixed in wonderment and anxiety on the ball, and the apple and the rope; while in another such narrow stone chamber only ten or twelve stone steps beneath, her father was crouching at his window shaken with fury, and seeing in his imagination these strange gifts from the air almost as clearly as Myfanwy could see them with her naked eye.

For though the sun had been as much a dazzle to himself as to the common people in the highway, he had kept them fastened on the juggler's trickeries none the less, and had counted every coloured ball and every fruit and every serpent as they rose and fell in their rhythmical maze-like network of circlings in the air. And when each marvellous piece of juggling in turn was over, he knew that in the first place a golden ball was missing, and that in the second place a fruit like an English apple was missing, and that in the third place a silken cord with a buckle-hook to it like the head of a serpent had been flung into the air but had never come down to earth again. And at the cries and the laughter and the applause of the roaring common people and children beneath his walls, tears of rage and despair had burst from his eyes. Myfanwy was deceiving him. His dreaded hour was come.

But there again he was wrong. The truth is, his eyes were so green with jealousy and his heart so black with rage that his wits had become almost useless. Not only his wits either, but his courtesy and his spirit; for the next moment he was actually creeping up again like a thief from stair to stair, and presently had fallen once more on to his knees outside his beloved Myfanwy's chamber door and had fixed on her one of those green dark eyes of his at its little gaping cut-out pin-hole. And there he saw a strange sight indeed.

The evening being now well advanced, and the light of the afterglow too feeble to make more than a glimmer through her narrow stone window-slits, Myfanwy had lit with her tinder box (for of all things she loved light) no less than seven wax candles on a seven-branched candlestick. This she had stood on a table beside a high narrow mirror. And at the moment when the Baron fixed his eye to the pin-hole, she was standing, a little astoop, the apple in her hand, looking first at it, and then into the glass at the bright-lit reflected picture of herself holding the apple in her hand.

So now there were two Myfanwys to be seen—herself and her image in the glass. And which was the lovelier not even the juggler could have declared. Crouching there at the door-crack, her father could all but catch the words she was softly repeating to herself as she gazed at the reflected apple: 'Shall I, shan't I? Shall I, shan't I?' And then suddenly—and he dared not stir or cry out—she had raised the fruit to her lips and had nibbled its rind.

What happened then he could not tell, for the secret and sovereign part of that was deep in Myfanwy herself. The sharp juice of the fruit seemed to dart about in her veins like flashing fishes in her father's crystal fountains and water-conduits. It was as if happiness had begun gently to fall out of the skies around her, like dazzling flakes of snow. They rested on her hair, on her shoulders, on her hands, all over her. And yet not snow, for there was no coldness, but a scent as it were of shadowed woods at noonday, or of a garden when a shower has fallen. Even her bright eyes grew brighter; a radiance lit her cheek; her lips parted in a smile.

And it is quite certain if Myfanwy had been the Princess of Anywhere-in-the-World-at-All, she would then and there—like Narcissus stooping over his lilied water-pool—have fallen head over ears in love with herself! 'Wonder of wonders!' cried she in the quiet; 'but if this is what a mere nibble of my brave juggler's apple can do, then it were wiser indeed to nibble no more.' So she laid the apple down.

The Baron gloated on through the pin-hole—watching her as she stood transfixed like some lovely flower growing in the inmost silent solitude of a forest and blossoming before his very eyes.

And then, as if at a sudden thought, Myfanwy turned and took up the golden ball, which—as she had suspected and now discovered—was no ball, but a small orb-shaped box of rare inlaid woods, covered with golden thread. At touch of the tiny spring that showed itself in the midst, its lid at once sprang open, and Myfanwy put in finger and thumb and drew out into the crystal light a silken veil—but of a gossamer silk so finely spun that when its exquisite meshes had wreathed themselves downward to the floor the veil looked to be nothing more than a silvery grey mist in the candlelight.

It filmed down from her fingers to the flagstones beneath, almost as light as the air in which it floated. Marvellous that what would easily cover her, head to heel, could have been packed into so close a room as that two-inch ball! She gazed in admiration of this exquisite handiwork. Then, with a flick of her thumb, she had cast its cloudlike folds over her shoulders.

And lo!—as the jealous lord gloated on—of a sudden there was nothing to be seen where Myfanwy had stood but seven candles burning in their stick, and seven more in the mirror. She had vanished.

She was not gone very far, however. For presently he heard—as if out of nowhere—a low chuckling childlike peal of laughter which willy-nilly had broken from her lips at seeing that this Veil of Invisibility had blanked her very glass. She gazed steadily on into its clear vacancy, lost in wonder. Nothing at all of her whatsoever was now reflected there!—not the tip of her nose, not a thumb, not so much as a button or a silver tag. Myfanwy had vanished; and yet, as she well knew, here she truly was in her own body and no other, though tented in beneath the folds of the veil, as happy as flocks on April hills, or mermaids in the deep blue sea. It was a magic thing indeed, to be there and yet not there; to hear herself and yet remain transparent as water.

Motionless though she stood, her thoughts were at the same time flitting about like quick and nimble birds in her mind. This veil, too, was the gift of the juggler; her young sleeping stranger of the beech-tree in a strange disguise. And she could guess in her heart what use he intended her to make of it, even though at thought of it that heart misgave her. A moment after and as swiftly as she had gone, she had come back again—the veil in her fingers. Laughing softly to herself she folded and refolded it and replaced it in its narrow box. Then turning, she took up from the chair the silken cord, and as if in idle fancy twined it twice about her slender neck. And it seemed the cord took life into itself, for lo, showing there in the mirror, calm now as a statue of coloured ivory, stood Myfanwy; and couched over her left temple the swaying head of the Serpent of Wisdom, whispering in her ear.

Owen ap Gwythock could watch no more. Groping his way with

trembling fingers through the thick gloom of the staircase he crept down to the Banqueting Hall where already his Chief Steward awaited his coming to announce that supper was prepared.

To think that his Lovely One, his pearl of price, his gentle innocent, *his* Myfanwy—the one thing on earth he treasured most, and renowned for her gentleness and beauty in all countries of the world, had even for an instant forgotten their loves, forgotten her service and duty, was in danger of leaving and forsaking him for ever! In his jealousy and despair tears rolled down his furrowed cheeks as he ground his teeth together, thinking of the crafty enemy that was decoying her away.

Worse still; he knew in his mind's mind that in certain things in this world even the most powerful are powerless. He knew that against true love all resistance, all craft, all cunning at last prove of no avail. But in this grief and despair the bitterest of all the thoughts that were now busy in his brain was the thought that Myfanwy should be cheating and deceiving him, wantonly beguiling him; keeping things secret that should at once be told.

A dark and dismal mind was his indeed. To distrust one so lovely!—*that* might be forgiven him. But to creep about in pursuit of her like a weasel; to spy on her like a spy; to believe her guilty before she could prove her innocence! Could *that* be forgiven? And even at this very moment the avenger was at his heels.

For here was Myfanwy herself. Lovely as a convolvulus wreathing a withered stake, she was looking in at him from the doorpost, searching his face. For an instant she shut her eyes as if to breathe a prayer, then she advanced into the room, and, with her own hand, lay before him on the oak table beside his silver platter, first the nibbled apple, next the golden ball, and last the silken cord. And looking at him with all her usual love in her eyes and in her voice, she told him how these things had chanced into her hands, and whence they had come.

Her father listened; but durst not raise his eyes from his plate. The scowl on his low forehead grew blacker and blacker; even his beard seemed to bristle. But he heard her in silence to the end.

'So you see, dear father,' she was saying, 'how can I but be grateful and with all my heart to one who takes so much thought for me? And if you had seen the kindness and courtesy of his looks, even you yourself could not be angry. There never was, as you well know, anybody else in the whole wide world whom I wished to speak to but to you. And now there is none other than you except this stranger. I know nothing but that. Can you suppose indeed he meant these marvellous gifts for me? And why for me and no other, father dear? And what would you counsel me to do with them?'

Owen ap Gwythock stooped his head lower. Even the sight of his eyes had dimmed. The torches faintly crackled in their sconces, the candles on the table burned unfalteringly on.

He turned his cheek aside at last like a snarling dog. 'My dear,' he said, 'I have lived long enough in this world to know the perils that beset the young and fair. I grant you that this low mountebank must be a creature of infinite cunning. I grant you that his tricks, if harmless, would be worth a charitable groat. If, that is, he were only what he seems to be. But that is not so. For this most deadly stranger is a Deceiver and a Cheat. His lair, as I guess well, is in the cruel and mysterious East, and his one desire and stratagem is to snare you into his company. Once within reach of his claws, his infamous slaves will seize on you and bear you away to some evil felucca moored in the river. It seems, beloved, that your gentle charms are being whispered of in this wicked world. Even the beauty of the gentlest of flowers may be sullied by idle tongues. But once securely in the hands of this nefarious mountebank, he will put off to Barbary, perchance, or to the horrid regions of the Turk, perchance, there to set you up in the scorching market-place and to sell you for a slave. My child, the danger, the peril is gross and imminent. Dismiss at once this evil wretch from your mind and let his vile and dangerous devices be flung into the fire. The apple is pure delusion; the veil which you describe is a mere toy; and the cord is a device of the devil.'

Myfanwy looked at her father, stooping there, with sorrow in her eyes, in spite of the gladness sparkling and dancing in her heart. Why, if all that he was saying he thought true—why could he not lift his eyes and meet her face to face?

'Well then, that being so, dear father,' she said softly at last, 'and you knowing ten thousand times more of God's world than I have ever had opportunity of knowing, whatever my desire, I must ask you but this one small thing. Will you promise me not to have these pretty baubles destroyed at once, before, I mean, you have thought once more of *me*? If I had deceived you, then indeed I should be grieved beyond endurance. But try as I may to darken my thoughts of him, the light slips in, and I see in my very heart that this stranger cannot by any possibility of nature or heaven be all that you tell me of him. I have a voice at times that whispers me yes or no: and I obey. And of him it has said only yes. But I am young, and the walls of this great house are narrow, and you, dear father, as you have told me so often are wise. Do but then invite this young man into your presence! Question him, test him, gaze on him, hearken to him. And that being done, you will believe in him as I do. As I know I am happy, I know he is honest. It would afflict me beyond all telling to swerve by a hair's-breadth from my dear obedience to you. But, alas, if I never see him again, I shall wither up and die. And that—would it not—' she added smilingly—'that would be a worse disobedience yet? If you love me, then, as from my first hour in the world I *know* you have loved me, and I have loved you, I pray you think of me with grace and kindness—and in compassion too.'

And with that, not attempting to brush away the tears that had sprung into her eyes, and leaving the juggler's three gifts amid the flowers and fruit of the long table before him, Myfanwy hastened out of the room and returned to her chamber, leaving her father alone.

For a while her words lay like a cold refreshing dew on the dark weeds in his mind. For a while he pondered them, even; while his own gross fables appeared in all their ugly falseness.

But alas for himself and his pride and stubbornness, these gentler ruminations soon passed away. At thought once more of the juggler—of whom his spies had long since brought him far other tidings than he had expressed—rage, hatred and envy again boiled up in him and drowned everything else. He forgot his courtesy, his love for Myfanwy, his desire even to keep her love for him. Instead, on and on he sipped and sipped, and sat fuming and plotting and scheming with but one notion in his head—by hook or by crook to defeat this juggler and so murder the love of his innocent Myfanwy.

'Lo, now,' broke out at last a small shrill voice inside him. 'Lo, now, if thou taste of the magic apple, may it not be that it will give thee courage and skill to contend against him, and so bring all his hopes to ruin? Remember what a marvel but one merest nibble of the outer rind of it wrought in thy Myfanwy!'

And the foolish creature listened heedfully to this crafty voice, not realizing that the sole virtue of the apple was that of making any human who tasted it more like himself than ever. He sat there—his fist over his mouth—staring intently at the harmless-looking fruit. Then he tiptoed like a humpback across the room and listened at the entry. Then having poured out, and drained at a draught, yet another cup of wine, he cautiously picked up the apple by its stalk between finger and ringed thumb and once more squinted close and steadily at its red and green, and at the very spot where Myfanwy's small teeth had rasped away the skin.

It is in a *moment* that cities fall in earthquake, stars collide in the wastes of space, and men choose between good and evil. For suddenly—his mind made up, his face all turned a reddish purple—this foolish lord lifted the apple to his mouth and, stalk to dried blossom, bit it clean in half. And he munched and he munched and he munched.

He had chawed for but a few moments, however, when a dreadful and continuous change and transformation began to appear upon him. It seemed to him that his whole body and frame was being kneaded and twisted and wrung in much the same fashion as dough being made into bread, or clay in a modeller's fingers. Not knowing what these aches and stabbings and wrenchings meant, he had dropped as if by instinct upon his hands and knees, and thus stood munching, while gazing blankly and blindly, lost in some inward horror, into the great fire on the hearth.

And meanwhile, though he knew it not in full, there had been sprouting upon him grey coarse hairs—a full thick coat and hide of them—in abundance. There had come a tail to him with a sleek, dangling tassel; long hairy ears had jutted out upon his temples; the purple face turned grey, lengthening as it did so until it was at least full eighteen inches long, with a great jawful of large teeth. Hoofs for his hands, hoofs where his feet used to be, and behold!—standing there in his own banqueting hall—this poor deluded Owen ap Gwythock, Lord of Eggleyseg, transmogrified into an ass!

For minutes together the dazed creature stood in utter dismay—the self within unable to realize the change that had come over its outer shape. But, happening to stretch his shaggy and unfamiliar neck a little outward, he perceived his own image in a scoured and polished suit of armour that stood on one side of the great chimney. He shook his head, the ass's head replied. He shook himself, the long ears flapped together like a wood-pigeon's wings. He lifted his hand—a hoof clawed at nowhere!

At this the poor creature's very flesh seemed to creep upon his bones as he turned in horror and dismay in search of an escape from the fate that had overtaken him. That ass *he*? he *himself*? His poor wits in vain endeavoured to remain calm and cool. A panic of fear all but swept him away. And at this moment his full, lustrous, long-lashed, asinine eyes fell by chance upon the golden ball lying ajar on the table beside his wine-cup—the Veil of Invisibility glinting like money-spider's web from within.

Now no ass is quite such a donkey as he looks. And this Owen ap Gwythock, though now completely shut up in this uncouth hairy body was in his *mind* no more (though as much) of a donkey than he had ever been. His one thought, then, was to conceal his dreadful condition from any servant that might at any moment come that way, while he himself could seek out a quiet secluded corner in the dark wherein to consider how to rid himself of his ass's frame and to regain his own usual shape. And there lay the veil! What thing sweeter could there be than to defeat the juggler with his own devices.

Seizing the veil with his huge front teeth, he jerked it out of the ball and flung it as far as he could over his shaggy shoulders. But alas, his donkey's muzzle was far from being as deft as Myfanwy's delicate fingers. The veil but half concealed him. Tail, rump and back legs were now vanished from view; head, neck, shoulders and fore-legs remained in sight. In vain he tugged; in vain he wriggled and wrenched; his hard hoofs thumping on the hollow flagstones beneath. One half of him stubbornly remained in sight; the rest had vanished. For the time being he was no more even than half an ass.

At last, breathless and wearied out with these exertions, trembling and shuddering, and with not a vestige of sense left in his poor donkey's noddle, he wheeled himself about once more and caught up with his teeth the silken cord. It was his last hope.

But this having been woven of wisdom—it being indeed itself the Serpent of Wisdom in disguise—at touch of his teeth it at once converted itself into a strong hempen halter, and, before he could so much as rear out of the way to escape its noose or even bray for help, it had tethered him to a large steel hook in his own chimneypiece.

Bray he did, none the less: 'Hee-haw! Hee-haw!! Hee-ee-ee-ee Haw-aw-aw!!!' His prolonged, see-saw, dismal lamentations shattered the silence so harshly and so hoarsely that the sound rose up through the echoing stone walls and even pierced into Myfanwy's own bedchamber, where she sat in the darkness at her window, looking out half in sorrow, half in unspeakable happiness, at the stars.

Filled with alarm at this dreadful summons, in an instant or two she had descended the winding stone steps; and a strange scene met her eyes.

There, before her, in the full red light of the flaming brands in the hearth and the torches on the walls, stood the fore-legs, the neck, head, and ears of a fine, full-grown ass, and a yard or so behind them just nothing at all. Only vacancy!

Poor Myfanwy—she could but wring her hands in grief and despair; for there could be no doubt in her mind of who it was in truth now stood before her—her own dear father. And on his face such a look of rage, entreaty, shame and stupefaction as never man has seen on ass's countenance before. At sight of her the creature tugged even more furiously at his halter, and shook his shaggy shoulders; but still in vain. His mouth opened and a voice beyond words to describe, brayed out upon the silence these words: 'Oh, Myfanwy, see into what a pass your sorceries and deceits have reduced me!'

'Oh, my dear father,' she cried in horror, 'speak no more, I beseech you—not one syllable—or we shall be discovered. Or, if you utter a sound, let it be but in a whisper.'

She was at the creature's side in an instant, had flung her arms about his neck, and was whispering into his long hairy ear all the comfort and endearments and assurances that loving and tender heart could conceive. 'Listen, listen, dear father,' she was entreating him, 'I see indeed that you have been meddling with the apple, and the ball, and the cord. And I do assure you, with all my heart and soul, that I am thinking of nothing else but how to help you in this calamity that has overtaken us. Have patience. Struggle no more. All will be well. But oh, beloved, was it quite just to me to speak of my deceits?'

Her bright eyes melted with compassion as she looked upon one whom she had loved ever since she could remember, so dismally transmogrified.

'How can you hesitate, ungrateful creature?' the see-saw voice once more broke out. 'Relieve me of this awful shape, or I shall be strangled on my own hearthstone in this pestilent halter.'

But now, alas, footsteps were sounding outside the door. Without an instant's hesitation Myfanwy drew the delicate veil completely over the trembling creature's head, neck and fore-quarters and thus altogether concealed him from view. So—though it was not an instant too soon—when the Lord of Eggleyseg's Chief Steward appeared in the doorway, nothing whatever was changed within, except that his master no longer sat in his customary chair, Myfanwy stood solitary at the table, and a mysterious cord was stretched out between her hand and the hook in the chimney-piece.

'My father,' said Myfanwy, 'has withdrawn for a while. He is indisposed, and bids me tell you that not even a whisper must disturb his rest. Have a hot posset prepared at once, and see that the room beneath is left vacant.'

The moment the Steward had gone to do her bidding Myfanwy turned at once to her father, and lifting the veil, whispered into the long hairy ear again that he must be of good cheer. 'For you see, dear father, the only thing now to be done is that we set out together at once in search of the juggler who, meaning no unkindness, presented me with these strange gifts. He alone can and will, I am assured, restore you to your own dear natural shape. So I pray you to be utterly silent—not a word, not a murmur—while I lead you gently forth into the forest. Once there I have no doubt I shall be able to find our way to where he is. Indeed he may be already expectant of my coming.'

Stubborn and foolish though the Baron might be, he realized, even in his present shape, that this was his only wisdom. Whereupon, withdrawing the end of the bridle from the hook to which it was tethered, Myfanwy softly led the now invisible creature to the door, and so, gently onward down the winding stone staircase, on the stones of which his shambling hoofs sounded like the hollow beating of a drum.

The vast room beneath was already deserted by its usual occupants, and without more ado the two of them, father and daughter, were soon abroad in the faint moonlight that now by good fortune bathed the narrow bridle-path that led into the forest.

Never before in all her years on earth had Myfanwy strayed beyond the Castle walls; never before had she stood lost in wonder beneath the dark emptiness of the starry skies. She breathed in the sweet fresh night air, her heart blossoming within her like an evening primrose, refusing to be afraid. For she knew well that the safety of them both—this poor quaking animal's and her own—depended now solely on her own courage and resource, and that to be afraid would almost certainly lead them only from one disaster into another.

Simply, however, because a mere ownerless ass wandering by itself in the moonlit gloom of the forest would be a spectacle less strange than that of a solitary damsel like herself, she once more drew down her father's ear to her lips and whispered into it, explaining to him that it was she who must now be veiled, and that if he would forgive her such boldness—for after all, he had frequently carried her pickaback when she was a child—she would mount upon his back and in this way they would together make better progress on their journey.

Her father dared not take offence at her words, whatever his secret feelings might be. 'So long as you hasten, my child,' he gruffed out in the hush, striving in vain to keep his tones no louder than a human whisper, 'I will forgive you all.' In a moment then there might be seen jogging along the bridle-path, now in moonlight, now in shadow, a sleek and handsome ass, a halter over its nose, making no stay to browse the dewy grass at the wayside, but apparently obeying its own whim as it wandered steadily onward.

Now it chanced that night there was a wild band of mountain robbers encamped within the forest. And when of a sudden this strange and pompous animal unwittingly turned out of a thicket into the light of their camp fire, and raised its eyes like glowing balls of emerald to gaze in horror at its flames, they lifted their voices together in an uproarious peal of laughter. And one of them at once started up from where he lay in the bracken, to seize the creature's halter and so make it his prize.

Their merriment, however, was quickly changed into dismay when the robbers saw the strange creature being guided, as was evident, by an invisible and mysterious hand. He turned this way, he turned that, with an intelligence that was clearly not his own and not natural even to his kind, and so eluded every effort made by his enemy to get a hold on his halter, his teeth and eyeballs gleaming in the firelight.

At this, awe and astonishment fell upon these outlaws. Assuredly sorcery alone could account for such ungainly and unasslike antics and manœuvres. Assuredly some divine being must have the beast in keeping, and to meddle with it further might only prove their own undoing.

Fortunate indeed was it that Myfanwy's right foot, which by mischance remained uncovered by the veil, happened to be on the side of the animal away from the beams of the camp fire. For certainly had these malefactors seen the precious stones blazing in its buckle, their superstitions would have melted away like morning mist, their fears have given place to cupidity, and they would speedily have made the ass their own and held its rider to an incalculable ransom.

Before, however, the moon had glided more than a soundless pace or two on her night journey, Myfanwy and her incomparable ass were safely out of sight: and the robbers had returned to their carousals. What impulse bade her turn first this way, then that, in the wandering and labyrinthine glades and tracks of the forest, she could not tell. But even though her father—not daring to raise his voice in the deep silence—ever and again stubbornly tugged upon his halter in the belief that the travellers had taken a wrong turning and were irrevocably lost, Myfanwy kept steadily on her way.

With a touch of her heel or a gentle persuasive pat of her hand on his hairy neck she did her best to reassure and to soothe him. 'Only trust in me, dear father: I am sure all will be well.'

Yet she was haunted with misgivings. So that when at last a twinkling light, sprinkling its beams between the boughs, showed in the forest, it refreshed her heart beyond words to tell. She was reaching her journey's end. It was as if that familiar voice in the secrecy of her heart had murmured, 'Hst! He draws near!'

There and then she dismounted from off her father's hairy back and once more communed with him through that long twitching ear. 'Remain here in patience a while, dear father,' she besought him, 'without straying by a hair's-breadth from where you are; for everything tells me our Stranger is not far distant now, and no human being on earth, no living creature, even, must see you in this sad and unseemly disguise. I will hasten on to assure myself that the light which I perceive beaming through the thicket yonder is his, and no other's. Meanwhile—and this veil shall go with me in case of misadventure—meanwhile do you remain quietly beneath this spreading beech-tree, nor even stir unless you are over-wearied after our long night journey and you should feel inclined to rest a while on the softer turf in the shadow there under that bush of fragrant roses, or to refresh yourself at the brook whose brawling I hear welling up from that dingle in the hollow. In that case, return here, I pray you; contain yourself in patience, and be your tongue as dumb as a stone. For though you may *design* to speak softly, dearest father, that long sleek throat and those great handsome teeth will not admit of it.'

And her father, as if not even the thick hairy hide he wore could endure his troubles longer, opened his mouth as if to groan aloud. But restraining himself, he only sighed, while an owl out of the quiet breathed its mellow night-call as if in response. For having passed the last hour in a profound and afflicted reverie, this poor ass had now regained in part his natural human sense and sagacity. But pitiful was the eye, however asinine the grin, which he now bestowed as if in promise on Myfanwy who, with veil held delicately in her fingers stood there, radiant as snow, beside him in the moonlight.

And whether it was because of her grief for his own condition or because of the expectancy in her face at the thought of her meeting with the Stranger, or because maybe the ass feared in his despair and dejection that he might never see her again, he could not tell; but true it was that she had never appeared in a guise so brave and gay and passionate and tender. It might indeed be a youthful divinity gently treading the green sward beside this uncouth beast in the chequered light and shadow of that unearthly moonshine.

Having thus assured herself that all would be well until her return, Myfanwy kissed her father on his flat hairy brow, and veil in hand withdrew softly in the direction of the twinkling light.

Alas, though the Baron thirsted indeed for the chill dark waters whose song rose in the air from the hollow beneath, he could not contain himself in her absence, but unmindful of his mute promise followed after his daughter at a distance as she made her way to the light, his hoofs scarce sounding in the turf. Having come near, by peering through the dense bushes that encircled the juggler's nocturnal retreat in the forest, he could see and hear all that passed.

As soon as Myfanwy had made sure that this stranger sitting by his glowing watch-fire was indeed the juggler and no man else—and one strange leap of her heart assured her of this even before her eyes could carry their message—she veiled herself once more, and so, all her loveliness made thus invisible, she drew stealthily near and a little behind him, as he crouched over the embers. Then pausing, she called gently and in a still low voice, 'I beseech you, Stranger, to take pity on one in great distress.'

The juggler lifted his dreaming face, ruddied and shadowed in the light of his fire, and peered cautiously but in happy astonishment all around him.

'I beseech you, Stranger,' cried again the voice from the unseen, 'to take pity on one in great distress.'

And at this it seemed to the juggler that now ice was running through his veins and now fire. For he knew well that this was the voice of one compared with whom all else in the world to him was nought. He knew also that she must be standing near, though made utterly invisible to him by the veil of his own enchantments.

'Draw near, traveller. Have no fear,' he cried out softly into the darkness. 'All will be well. Tell me only how I may help you.'

But Myfanwy drew not a hair's-breadth nearer. Far from it. Instead, she flitted a little across the air of the glade, and now her voice came to him from up the wind towards the south, and fainter in the distance.

'There is one with me,' she replied, 'who by an evil stratagem has been transformed into the shape of a beast, and that beast a poor patient ass. Tell me this, sorcerer—how I may restore him to his natural shape, and mine shall be an everlasting gratitude. For it is my own father of whom I speak.'

Her voice paused and faltered on the word. She longed almost beyond bearing to reveal herself to this unknown one, trusting without the least doubt or misgiving that he would serve her faithfully in all she asked of him.

'But *that*, gentle lady,' replied the juggler, 'is not within my power, unless he of whom you speak draws near to show himself. Nor—though the voice with which you speak to me is sweeter than the music of harp-strings twangling on the air—nor is it within my power to make promises to a bodiless sound only. For how am I to be assured that the shape who utters the words I hear is not some dangerous demon of the darkness who is bent on mocking and deluding me, and who will bring sorcery on myself?'

There was silence for a while in the glade, and then 'No, no!' cried the juggler. 'Loveliest and bravest of all that is, I need not see thy shape to know thee. Thou art most assuredly the lovely Myfanwy, and all that I am, have ever been, and ever shall be is at thy service. Tell me, then, where is this poor ass that was once thy noble father?'

And at this, and at one and the same moment, Myfanwy, withdrawing the veil from her head and shoulders, disclosed her fair self standing there in the faint rosy glow of the slumbering fire, and there broke also from the neighbouring thicket so dreadful and hideous a noise of rage and anguish—through the hoarse and unpractised throat of the eavesdropper near by—that it might be supposed the clamour was not of one but of a chorus of demons—though it was merely our poor ass complaining of his fate.

'Oh, sir,' sighed Myfanwy, 'my dear father, I fear, in his grief and anxiety has been listening to what has passed between us. See, here he comes.'

Galloping hoofs were indeed now audible as the Lord of Eggleyseg in ass's skin and shape drew near to wreak his vengeance on the young magician. But being at this moment in his stubborn rage and folly more ass than human, the glaring of the watch-fire dismayed his heavy wits, and he could do no else but paw with his fore-legs, lifting his smooth nose with its gleaming teeth into the night air, snuffing his rage and defiance some twenty paces distant from the fire.

The young magician, being of a nature as courteous as he was bold, did not so much as turn his head to scan the angry shivering creature, but once more addressed Myfanwy. She stood bowed down a little, tears in her eyes; in part for grief at her father's broken promise and the humiliation he had brought upon himself, in part for joy that their troubles would soon be over and that she was now in the very company of the stranger who unwittingly had been the cause of them all.

'Have no fear,' he said, 'the magic that has changed the noble Baron your father into a creature more blest in its docility, patience, and humbleness than any other in the wide world, can as swiftly restore him to his natural shape.'

'Ah then, sir,' replied the maid, 'it is very certain that my father will wish to bear witness to your kindness with any small gift that is in our power. For, as he well knows, it was not by any design but his own that he ate of the little green apple of enchantment. I pray you, sir, moreover, to forgive me for first stealing that apple, and also the marvellous golden ball, *and* the silken cord from out of the air.'

The juggler turned and gazed strangely at Myfanwy. 'There is only one thing I desire in all this starry universe,' he answered. 'But I ask it not of *him*—for it is not of his giving. It is for your own forgiveness, lady.'

'*I* forgive you!' she cried. 'Alas, my poor father!'

But even as she spoke a faint smile was on her face, and her eyes wandered to the animal standing a few paces beyond the margin of the glow cast by the watch-fire, sniffing the night air the while, and twitching dismally the coarse grey mane behind his ears. For now that her father was so near his deliverance her young heart grew entirely happy again, and the future seemed as sweet with promise as wild flowers in May.

Without further word the juggler drew from out of his pouch, as if he always carried about with him a little privy store of vegetables, a fine, tapering, ripe, red carrot.

'This, lady,' said he, 'is my only wizardry. I make no bargain. My love for you will never languish, even if I never more again refresh my sleepless eyes with the vision of your presence in this solitary glade. Let your noble father the Lord of Eggleyseg draw near without distrust. There is but little difference, it might be imagined, between a wild apple and a carrot. But then, when all is said, there is little difference in the long sum between any living thing and another in this strange world. There are creatures in the world whose destiny it is in spite of their gentleness and humility and lowly duty and obedience to go upon four legs and to be in service of masters who deserve far less than *they* deserve, while there are men in high places of whom the reverse might truly be said. It is a mystery beyond my unravelling. But now all I ask is that you bid the ass who you tell me is hearkening at this moment to all that passes between us to nibble of this humble but useful and wholesome root. It will instantly restore him to his proper shape. Meanwhile, if you bid, I will myself be gone.'

Without further speech between them, Myfanwy accepted the magic carrot, and returned once more to the ass.

'Dear father,' she cried softly, 'here is a root that seems to be only a carrot; yet nibble of it and you will be at once restored, and will forget you were ever an—as you are. For many days to come, I fear, you will not wish to look upon the daughter that has been the unwilling cause of this night's woeful experience. There lives, as I have been told, in a little green arbour of the forest yonder, a hermit. This young magician will, I am truly certain, place me in his care a while until all griefs are forgotten between us. You will of your kindness consent, dear father, will you not?' she pleaded.

A long prodigious bray resounded dolefully in the hollows of the far-spread forest's dells and thickets. The Lord of Eggleyseg had spoken.

'Indeed, father,' smiled Myfanwy, 'I have never before heard you say "Yes" so heartily. What further speech is needed?'

Whereupon the ass, with more dispatch than gratitude, munched up the carrot, and in a few hours Owen ap Gwythock, once more restored to his former, though hardly his more appropriate shape, returned in safety to his Castle. There for many a day he mourned his woeful solitude, but learned, too, not only how true and faithful a daughter he had used so ill, but the folly of a love that is fenced about with mistrust and suspicion and is poisoned with jealousy.

And when May was come again, a prince, no longer in the disguise of a wandering juggler, drew near with his adored Myfanwy to the Lord of Eggleyseg's ancient castle. And Owen ap Gwythock, a little older but a far wiser man, greeted them with such rejoicings and entertainment, with such feastings and dancing and minstrelsy and jubilations as had never been heard of before. Indeed he would have been ass unadulterated if he had done else.

1925

T. H. WHITE

The Troll

My father used to say that an experience like the one I am about to relate was apt to shake one's interest in mundane matters. Naturally he did not expect to be believed, and he did not mind whether he was or not. He did not himself believe in the supernatural, but the thing happened, and he proposed to tell it as simply as possible. It was stupid of him to say that it shook his faith in mundane matters, for it was just as mundane as anything else. Indeed, the really frightening part about it was the horribly tangible atmosphere in which it took place. None of the outlines wavered in the least. The creature would have been less remarkable if it had been less natural. It seemed to overcome the usual laws without being immune to them.

My father was a keen fisherman, and used to go to all sorts of places for his fish. On one occasion he made Abisko his Lapland base, a comfortable railway hotel, one hundred and fifty miles within the Arctic circle. He travelled the prodigious length of Sweden (I believe it is as far from the South of Sweden to the North, as it is from the South of Sweden to the South of Italy) in the electric railway, and arrived tired out. He went to bed early, sleeping almost immediately, although it was bright daylight outside; as it is in those parts throughout the night at that time of the year. Not the least shaking part of his experience was that it should all have happened under the sun.

He went to bed early, and slept, and dreamt. I may as well make it clear at once, as clear as the outlines of that creature in the northern sun, that his story did not turn out to be a dream in the last paragraph. The division between sleeping and waking was abrupt, although the feeling of both was the same. They were both in the same sphere of horrible absurdity, though in the former he was asleep and in the latter almost terribly awake. He tried to be asleep several times.

My father always used to tell one of his dreams, because it somehow seemed of a piece with what was to follow. He believed that it was a consequence of the thing's presence in the next room. My father dreamed of blood.

It was the vividness of the dreams that was impressive, their minute detail and horrible reality. The blood came through the keyhole of a locked door which communicated with the next room. I suppose the two rooms had originally been designed *en suite*. It ran down the door panel with a viscous ripple, like the artificial one created in the conduit of Trumpingdon Street. But it was heavy, and smelt. The slow welling of it sopped the carpet and reached the bed. It was warm and sticky. My father woke up with the impression that it was all over his hands. He was rubbing his first two fingers together, trying to rid them of the greasy adhesion where the fingers joined.

My father knew what he had got to do. Let me make it clear that he was now perfectly wide awake, but he knew what he had got to do. He got out of bed, under this irresistible knowledge, and looked through the keyhole into the next room.

I suppose the best way to tell the story is simply to narrate it, without an effort to carry belief. The thing did not require belief. It was not a feeling of horror in one's bones, or a misty outline, or anything that needed to be given actuality by an act of faith. It was as solid as a wardrobe. You don't have to believe in wardrobes. They are there, with corners.

What my father saw through the keyhole in the next room was a Troll. It was eminently solid, about eight feet high, and dressed in brightly ornamented skins. It had a blue face, with yellow eyes, and on its head there was a woolly sort of nightcap with a red bobble on top. The features were Mongolian. Its body was long and sturdy, like the trunk of a tree. Its legs were short and thick, like the elephant's feet that used to be cut off for umbrella stands, and its arms were wasted: little rudimentary members like the forelegs of a kangaroo. Its head and neck were very thick and massive. On the whole, it looked like a grotesque doll.

That was the horror of it. Imagine a perfectly normal golliwog (but without the association of a Christie minstrel) standing in the corner of a room, eight feet high. The creature was as ordinary as that, as tangible, as stuffed, and as ungainly at the joints: but it could move itself about.

The Troll was eating a lady. Poor girl, she was tightly clutched to its breast by those rudimentary arms, with her head on a level with its mouth. She was dressed in a nightdress which had crumpled up under her armpits, so that she was a pitiful naked offering, like a classical picture of Andromeda. Mercifully, she appeared to have fainted.

Just as my father applied his eye to the keyhole, the Troll opened its mouth and bit off her head. Then, holding the neck between the bright blue lips, he sucked the bare meat dry. She shrivelled, like a squeezed orange, and her heels kicked. The creature had a look of thoughtful ecstasy. When the girl seemed to have lost succulence as an orange she was lifted into the air. She vanished in two bites. The Troll remained leaning against the wall, munching patiently and casting its eyes about it with a vague benevolence. Then it leant forward from the low hips, like a jacknife folding in half, and opened its mouth to lick the blood up from the carpet. The mouth was incandescent inside, like a gas fire, and the blood evaporated before its tongue, like dust before a vacuum cleaner. It straightened itself, the arms dangling before it in patient uselessness, and fixed its eyes upon the keyhole.

My father crawled back to bed, like a hunted fox after fifteen miles. At first it was because he was afraid that the creature had seen him through the hole, but afterwards it was because of his reason. A man can attribute many night-time appearances to the imagination, and can ultimately persuade himself that creatures of the dark did not exist. But this was an appearance in a sunlit room, with all the solidity of a wardrobe and unfortunately almost none of its possibility. He spent the first ten minutes making sure that he was awake, and the rest of the night trying to hope that he was asleep. It was either that, or else he was mad.

It is not pleasant to doubt one's sanity. There are no satisfactory tests. One can pinch oneself to see if one is asleep, but there are no means of determining the other problem. He spent some time opening and shutting his eyes, but the room seemed normal and remained unaltered. He also soused his head in a basin of cold water, without result. Then he lay on his back, for hours, watching the mosquitoes on the ceiling.

He was tired when he was called. A bright Scandinavian maid

admitted the full sunlight for him and told him that it was a fine day. He spoke to her several times, and watched her carefully, but she seemed to have no doubts about his behaviour. Evidently, then, he was not badly mad: and by now he had been thinking about the matter for so many hours that it had begun to get obscure. The outlines were blurring again, and he determined that the whole thing must have been a dream or a temporary delusion, something temporary, anyway, and finished with; so that there was no good in thinking about it longer. He got up, dressed himself fairly cheerfully, and went down to breakfast.

These hotels used to be run extraordinarily well. There was a hostess always handy in a little office off the hall, who was delighted to answer any questions, spoke every conceivable language, and generally made it her business to make the guests feel at home. The particular hostess at Abisko was a lovely creature into the bargain. My father used to speak to her a good deal. He had an idea that when you had a bath in Sweden one of the maids was sent to wash you. As a matter of fact this sometimes used to be the case, but it was always an old maid and highly trusted. You had to keep yourself underwater and this was supposed to confer a cloak of invisibility. If you popped your knee out she was shocked. My father had a dim sort of hope that the hostess would be sent to bath him one day: and I dare say he would have shocked her a good deal. However, this is beside the point. As he passed through the hall something prompted him to ask about the room next to his. Had anybody, he enquired, taken number 23?

'But, yes,' said the lady manager with a bright smile, '23 is taken by a doctor professor from Upsala and his wife, such a charming couple!'

My father wondered what the charming couple had been doing, whilst the Troll was eating the lady in the nightdress. However, he decided to think no more about it. He pulled himself together, and went in to breakfast. The professor was sitting in an opposite corner (the manageress had kindly pointed him out), looking mild and short-sighted, by himself. My father thought he would go out for a long climb on the mountains, since exercise was evidently what his constitution needed.

He had a lovely day. Lake Torne blazed a deep blue below him, for all its thirty miles, and the melting snow made a lacework of filigree round the tops of the surrounding mountain basin. He got away from the stunted birch trees, and the mossy bogs with the reindeer in them, and the mosquitoes, too. He forded something that might have been a temporary tributary of the Abiskojokk, having to take off his trousers to do so and tucking his shirt up round his neck. He wanted to shout, bracing himself against the glorious tug of the snow water, with his legs crossing each other involuntarily as they passed, and the boulders turning under his feet. His body made a bow wave in the water, which climbed and feathered on his stomach, on the upstream side. When he was under the opposite bank a stone turned in earnest, and he went in. He came up, shouting with laughter, and made out loud a remark which has since become a classic in my family, 'Thank God,' he said, 'I rolled up my sleeves.' He wrung out everything as best he could, and dressed again in the wet clothes, and set off up the shoulder of Niakatjavelk. He was dry and warm again in half a mile. Less than a thousand feet took him over the snow line, and there, crawling on hands and knees, he came face to face with what seemed to be the summit of ambition. He met an ermine. They were both on all fours, so that there was a sort of equality about the encounter, especially as the ermine was higher up than he was. They looked at each other for a fifth of a second, without saying anything, and then the ermine vanished. He searched for it everywhere in vain, for the snow was only patchy. My father sat down on a dry rock, to eat his well soaked luncheon of chocolate and rye bread.

Life is such unutterable hell, solely because it is sometimes beautiful. If we could only be miserable all the time, if there could be no such things as love or beauty or faith or hope, if I could be absolutely certain that my love would never be returned: how much more simple life would be. One could plod through the Siberian salt mines of existence without being bothered about happiness. Unfortunately the happiness is there. There is always the chance (about eight hundred and fifty to one) that another heart will come to mine. I can't help hoping, and keeping faith, and loving beauty. Quite frequently I am not so miserable as it would be wise to be. And there, for my poor father sitting on his boulder above the snow, was stark happiness beating at the gates.

The boulder on which he was sitting had probably never been sat upon before. It was a hundred and fifty miles within the Arctic circle, on a mountain five thousand feet high, looking down on a blue lake. The lake was so long that he could have sworn it sloped away at the ends, proving to the naked eye that the sweet earth was round. The railway line and the half-dozen houses of Abisko were hidden in the trees. The sun was warm on the boulder, blue on the snow, and his body tingled smooth from the spate water. His mouth watered for the chocolate, just behind the tip of his tongue.

And yet, when he had eaten the chocolate—perhaps it was heavy on his stomach—there was the memory of the Troll. My father fell suddenly into a black mood, and began to think about the supernatural. Lapland was beautiful in the summer, with the sun sweeping round the horizon day and night, and the small tree leaves twinkling. It was not the sort of place for wicked things. But what about the winter? A picture of the Arctic night came before him, with the silence and the snow. Then the legendary wolves and bears snuffled at the far encampments, and the nameless winter spirits moved on their darkling courses. Lapland had always been associated with sorcery, even by Shakespeare. It was at the outskirts of the world that the Old Things accumulated, like drift wood round the edges of the sea. If one wanted to find a wise woman, one went to the rims of the Hebrides; on the coast of Brittany one sought the mass of St Secaire. And what an outskirt Lapland was! It was an outskirt not only of Europe, but of civilisation. It had no boundaries. The Lapps went with the reindeer, and where the reindeer were, was Lapland. Curiously indefinite region, suitable to the indefinite things. The Lapps were not Christians. What a fund of power they must have had behind them, to resist the march of mind. All through the missionary centuries they had held to something: something had stood behind them, a power against Christ. My father realised with a shock that he was living in the age of the reindeer, a period contiguous to the mammoth and the fossil.

Well, this was not what he had come out to do. He dismissed the nightmares with an effort, got up from his boulder, and began the scramble back to his hotel. It was impossible that a Professor from Abisko could become a troll.

As my father was going in to dinner that evening the manageress stopped him in the hall.

'We have had a day so sad,' she said. 'The poor Dr Professor has disappeared his wife. She has been missing since last night. The Dr Professor is inconsolable.'

My father then knew for certain that he had lost his reason.

He went blindly to dinner, without making any answer, and began to eat a thick sour-cream soup that was taken cold with pepper and sugar. The Professor was still sitting in his corner, a sandy-headed man with thick spectacles and a desolate expression. He was looking at my father, and my father, with the soup spoon half-way to his mouth, looked at him. You know that eye-to-eye recognition, when two people look deeply into each other's pupils, and burrow to the soul? It usually comes before love. I mean the clear, deep, milk-eyed recognition expressed by the poet Donne. Their eyebeams twisted and did thread their eyes upon a double string. My father recognised that the Professor was a troll, and the Professor recognised my father's recognition. Both of them knew that the Professor had eaten his wife.

My father put down his soup spoon, and the Professor began to grow. The top of his head lifted and expanded, like a great loaf rising in an oven; his face went red and purple, and finally blue; the whole ungainly upperworks began to sway and topple towards the ceiling. My father looked about him. The other diners were eating unconcernedly. Nobody else could see it, and he was definitely mad at last. When he looked at the Troll again, the creature bowed. The enormous superstructure inclined itself towards him from the hips, and grinned seductively.

My father got up from his table experimentally, and advanced towards the Troll, arranging his feet on the carpet with excessive care. He did not find it easy to walk, or to approach the monster, but it was a question of his reason. If he was mad, he was mad; and it was essential that he should come to grips with the thing, in order to make certain.

He stood before it like a small boy, and held out his hand, saying, 'Good-evening.'

'Ho! Ho!' said the Troll, 'little mannikin. And what shall I have for my supper to-night?'

Then it held out its wizened furry paw and took my father by the hand.

My father went straight out of the dining-room, walking on air. He found the manageress in the passage and held out his hand to her.

'I am afraid I have burnt my hand,' he said. 'Do you think you could tie it up?'

The manageress said, 'But it is a very bad burn. There are blisters all over the back. Of course, I will bind it up at once.'

He explained that he had burnt it on one of the spirit lamps at the sideboard. He could scarcely conceal his delight. One cannot burn oneself by being insane.

'I saw you talking to the Dr Professor,' said the manageress, as she was putting on the bandage. 'He is a sympathetic gentleman, is he not?'

The relief about his sanity soon gave place to other troubles. The Troll had eaten its wife and given him a blister, but it had also made an unpleasant remark about its supper that evening. It proposed to eat my father. Now very few people can have been in a position to decide what to do when a troll earmarks them for its next meal. To begin with, although it was a tangible troll in two ways, it had been invisible to the other diners. This put my father in a difficult position. He could not, for instance, ask for protection. He could scarcely go to the Manageress and say, 'Professor Skål is an odd kind of werewolf, ate his wife last night, and proposes to eat me this evening.' He would have found himself in a looney-bin at once. Besides, he was too proud to do this, and still too confused. Whatever the proofs and blisters, he did not find it easy to believe in professors that turned into trolls. He had lived in the normal world all his life, and, at his age, it was difficult to start learning afresh. It would have been quite easy for a baby, who was still co-ordinating the world, to cope with the troll situation: for my father, not. He kept trying to fit it in somewhere, without disturbing the universe. He kept telling himself that it was nonsense: one did not get eaten by professors. It was like having a fever, and telling oneself that it was all right, really, only a delirium, only something that would pass.

There was that feeling on the one side, the desperate assertion of all the truths that he had learned so far, the tussle to keep the world from drifting, the brave but intimidated refusal to give in or to make a fool of himself.

On the other side there was stark terror. However much one struggled to be merely deluded, or hitched up momentarily in an odd pocket of space-time, there was panic. There was the urge to go away as quickly as possible, to flee the dreadful Troll. Unfortunately the last train had left Abisko, and there was nowhere else to go.

My father was not able to distinguish these trends of thought. For him they were at the time intricately muddled together. He was in a whirl. A proud man, and an agnostic, he stuck to his muddled guns alone. He was terribly afraid of the Troll, but he could not afford to admit its existence. All his mental processes remained hung up, whilst he talked on the terrace, in a state of suspended animation, with an American tourist who had come to Abisko to photograph the midnight sun.

The American told my father that the Abisko railway was the northernmost electric railway in the world, that twelve trains passed through it every day travelling between Upsala and Narvik, that the population of Abo was 12,000 in 1862, and that Gustavus Adolphus ascended the throne of Sweden in 1611. He also gave some facts about Greta Garbo.

My father told the American that a dead baby was required for the mass of St Secaire, that an elemental was a kind of mouth in space that sucked at you and tried to gulp you down, that homoeopathic magic was practised by the aborigines of Australia, and that a Lapland woman was careful at her confinement to have no knots or loops about her person, lest these should make the delivery difficult.

The American, who had been looking at my father in a strange way for some time, took offence at this and walked away; so that there was nothing for it but to go to bed.

Sir Hastings Utterwood walked upstairs on will power alone. His faculties seemed to have shrunk and confused themselves. He had to help himself with the bannister. He seemed to be navigating himself by wireless, from a spot about a foot above his forehead. The issues that were involved had ceased to have any meaning, but he went on doggedly up the stairs, moved forward by pride and contrariety. It was physical fear that alienated him from his body, the same fear that he had felt as a boy, walking down long corridors to be beaten. He walked firmly up the stairs.

Oddly enough, he went to sleep at once. He had climbed all day and been awake all night and suffered emotional extremes. Like a condemned man, who was to be hanged in the morning, my father gave the whole business up and went to sleep.

He was woken at midnight exactly. He heard the American on

the terrace below his window, explaining excitedly that there had been a cloud on the last two nights at 11.58, thus making it impossible to photograph the Midnight Sun. He heard the camera click.

There seemed to be a sudden storm of hail and wind. It roared at his window-sill, and the window curtains lifted themselves taut, pointing horizontally into the room. The shriek and rattle of the tempest framed the window in a crescendo of growing sound, an increasing blizzard directed towards himself. A blue paw came over the sill.

My father turned over and hid his head in the pillow. He could feel the domed head dawning at the window and the eyes fixing themselves upon the small of his back. He could feel the places physically, about four inches apart. They itched. Or else the rest of his body itched, except those places. He could feel the creature growing into the room, glowing like ice, and giving off a storm. His mosquito curtains rose in its afflatus, uncovering him, leaving him defenceless. He was in such an ecstasy of terror that he almost enjoyed it. He was like a bather plunging for the first time into freezing water and unable to articulate. He was trying to yell, but all he could do was to throw a series of hooting noises from his paralysed lungs. He became a part of the blizzard. The bed clothes were gone. He felt the Troll put out its hands.

My father was an agnostic, but, like most idle men, he was not above having a bee in his bonnet. His favourite bee was the psychology of the Catholic Church. He was ready to talk for hours about psycho-analysis and the confession. His greatest discovery had been the rosary.

The rosary, my father used to say, was intended solely as a factual occupation which calmed the lower centres of the mind. The automatic telling of the beads liberated the higher centres to meditate upon the mysteries. They were a sedative, like knitting or counting sheep. There was no better cure for insomnia than a rosary. For several years he had given up deep breathing or regular counting. When he was sleepless he lay on his back and told his beads, and there was a small rosary in the pocket of his pyjama coat.

The Troll put out its hands, to take him round the waist. He became completely paralysed, as if he had been winded. The Troll put its hand upon the beads.

They met, the occult forces, in a clash above my father's heart.

There was an explosion, he said, a quick creation of power. Positive and negative. A flash, a beam. Something like the splutter with which the antenna of a tram meets its overhead wires again, when it is being changed about.

The Troll made a high squealing noise, like a crab being boiled, and began rapidly to dwindle in size. It dropped my father and turned about, and ran wailing, as if it had been terribly burnt, for the window. Its colour waned as its size decreased. It was one of those air-toys now, that expire with a piercing whistle. It scrambled over the window-sill, scarcely larger than a little child, and sagging visibly.

My father leaped out of bed and followed it to the window. He saw it drop on the terrace like a toad, gather itself together, stumble off, staggering and whistling like a bat, down the valley of the Abiskojokk.

My father fainted.

In the morning the manageress said, 'There has been such a terrible tragedy. The poor Dr Professor was found this morning in the lake. The worry about his wife had certainly unhinged his mind.'

A subscription for a wreath was started by the American, to which my father subscribed five shillings; and the body was shipped off next morning, on one of the twelve trains that travel between Upsala and Narvik every day.

1935

RICHARD HUGHES

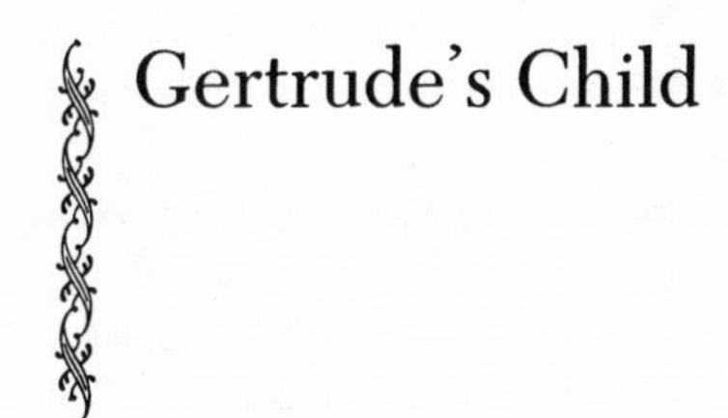

Gertrude's Child

ONE night, Gertrude the wooden doll got furious because the little girl she belonged to was being unkind to her.

'I won't belong to you any more!' said Gertrude: 'I don't want to belong to anyone, only myself.'

So Gertrude ran away. I mean, she ran *right* away—right along the main road out into the world on her own; and the night was dark.

Gertrude was glad she was made of wood and not easily hurt, for her hard little wooden feet went clickety-clop on the hard road without any shoes yet didn't get sore. She was painted with oil paint, too, so the rain just ran off her. Also, wooden dolls don't need any dinner. At dolls' parties they eat what you give them of course, and enjoy it; but in between parties wooden dolls need eat nothing at all.

Gertrude almost felt sorry for the little girl she had run away from, for being made of that soft stuff all children are made of which scratches and bruises so easily, and falls ill . . . But no, for Gertrude was wooden and hard right through, and *couldn't* be sorry for someone who'd been unkind to her. 'I hope she falls down and bleeds!' said Gertrude to herself: 'That's what she deserves!'

Then daylight came, and the sun came out and dried Gertrude, and made the paint on her hard wooden shoulders smell good. She began to sing (quite loud, though her voice was woody).

But all that day Gertrude met no one at all on the road, and began to feel lonely. She thought that it might be nice after all to have a friend—not a soft one of course, but a sensible hard one like herself. So Gertrude made up a story in her head about another wooden doll, and pretended this other wooden doll was walking beside her and talking to her (but it wasn't, of course: she was quite alone really).

When it got dark again Gertrude looked up at the face of the moon, because it was the only face round there to look at. 'You're not much company, Moon!' said Gertrude. 'But that's all right because I don't really *want* company.' (This wasn't quite true, though she wished it was.)

But she had not walked very much farther when she came up behind an old man carrying a load on his back. The old man took her hand, and they walked along together hand-in-hand for a while. But he didn't say anything to her—not a word. 'Very old men never do know what to say to dolls,' Gertrude told herself.

But at last the old man stopped outside a cottage, and then he did speak to her: 'Would you like a little girl of your own?' he asked Gertrude.

'No!' said Gertrude: 'I don't want to belong to any little girl again, ever!'

'You don't understand,' said the old man: '*You* wouldn't belong to *her*, *she* would belong to *you*.'

'I don't see how . . .' said Gertrude.

'Come round to my shop in the morning and I'll explain,' said the old man. 'You see, I sell little girls in my shop. And little boys too, if you'd rather.'

'Who *on earth* wants to buy them?' asked Gertrude, astonished.

'I'll tell you tomorrow,' said the old man. 'Good night for now!'

He went into his cottage and shut the door.

Gertrude walked on. She wondered who he sold little girls to. Suppose she bought one: would a little girl turn out more trouble than she was worth? Certainly no child could be as good company for Gertrude as a real doll would be—but, perhaps dolls couldn't buy *dolls*. . . .

Gertrude had felt very lonely while it was dark, but when morning came she did not feel quite so lonely and so she forgot the idea of buying herself a child. She was thinking of something quite different, when all of a sudden she saw the old man standing at the door of a small road-side shop.

Seeing him made her remember. 'Hullo!' she said: '*Now* tell me who buys your children?'

'Oh, mostly dolls like yourself do,' said the old man: 'Or else other toys. And puppies of course; but puppies don't often have money enough—they spend it too fast. Sometimes I sell one to a pony. But ponies seldom want children very much—they're too happy playing about by themselves. As for kittens . . . I don't really like selling children to kittens, because kittens can be so cruel!'

'What kind of children do you sell?' asked Gertrude.

'I sell all kinds and all ages,' said the old man. 'Fat ones and thin ones, pretty and plain, good ones and naughty ones. You'll see some in the shop-window here, but I've lots more inside.'

Gertrude looked in the window. It was got up to look like a Christmas party. Children of all sizes were sitting about in their very best clothes. Little girls had no creases at all in their pretty dresses: little boys had clean hands, and their hair was all smarmed down and oiled. The children looked stiff and shy, like people having their photographs taken. Not one of them wriggled: being stared at in a window like this made them even too shy to pull the crackers they held in their hands.

When she saw them, a great longing came over Gertrude to have one of her own: 'Oh, I *must* have one of those grand-looking ones!' said Gertrude.

'Come and look at the others inside before you make up your mind,' said the old man, and led her into the shop.

Inside, Gertrude saw piles of children all over everywhere—every kind of child you could think of: as well as English ones there were Spanish and Russian and Dutch, and some Africans, and a special shelf of Chinese ones, and even an Indian pair who were brown.

There was, too, one very shining and precious child in a box by herself, who lay very still. 'That's a *most* beautiful one,' Gertrude whispered.

'Yes, but she wouldn't be any good to you,' said the old man very sadly: 'She doesn't move at all now. Her eyes won't open any more.'

Still, there were plenty of children to choose from. Some were sitting in rows on the shelves (which were too high for them to climb down without help). Because it was the middle of the morning they were all eating biscuits and drinking glasses of milk.

'Can they sing?' Gertrude asked.

'Of course they can sing!' said the old man. He waved his hand, and the children stopped munching their biscuits and sang Gertrude's favourite song.

'Now that you've seen them all you'd better choose one,' said the old man.

'May I feel them first?' asked Gertrude, because she couldn't believe they were real.

'Yes, if your hands are clean,' said the old man.

So Gertrude poked them with her hard wooden fingers and felt their ribs, and the ones who were ticklish giggled and squealed no end. 'Now you've poked them enough,' said the old man. 'Hurry up and choose, before you've made them all spill their milk.'

Gertrude liked nearly all of them. But there was one little girl she was certain sure she would best like to have for her own. This little girl was about six years old, and a bit thin: she had curly yellow hair and a pretty pink dress, and short white socks, and a happy look in her eye.

'If she's naughty,' thought Gertrude, 'I'll smack her and smack her and smack her! That's the only way of making them good,' thought Gertrude (since that was how she'd been treated herself).

But the old man seemed to know what she was thinking, and said to Gertrude: 'If I sell her to you, you must promise to be kind to her! Remember, children aren't hard like you wooden dolls. If you drag *children* by the leg head-down through bushes, they get scratched and bruised. If you drop them from the top of the stairs, they break their necks. If you take their clothes off and leave them out in the cold, they get ill. If you forget to feed them they die, just like animal pets do.'

'Oh, I promise to be kind to her,' said Gertrude: 'But you said about her clothes—do you mean her clothes really take off?'

'Of course they do!' said the old man: 'But remember what I said about being kind to her, and not letting her catch cold.'

'Yes yes!' said Gertrude, even more hurriedly. Quickly she paid the price, then caught her little girl by the arm and ran out of the shop with her.

'Don't forget to give her her dinner!' the old man shouted after them.

'Good-bye! Good-bye, child!' called all the children left on the shelf (all except the very beautiful one in the box whose eyes wouldn't open).

Gertrude was delighted with her child. At first all she wanted to do was to walk up and down the street, leading her child by the hand.

Gertrude saw other dolls and puppies and people like that taking *their* children out for a walk, and she hoped they were jealous of her for having such a pretty one. But soon she got tired of this, and then she began to remember all the many things you have to do when you have a child to look after: all the washing and minding and mending for her, and feeding and cleaning and cooking and catching and combing and teaching and sewing and cuddling and bandaging and reading aloud. You couldn't do all that just wandering around: it meant having a home.

'There's a house just round the corner from here we could live in,' said the child.

'And I'm going to call you Annie,' said Gertrude. 'You can't manage without a name.'

'Good!' said the child, 'I'm Annie, then.' And together they went in through the gate.

Inside the gate, in front of the house there was a beautiful garden, with flowers growing. But round the back there was an orchard, so they went straight there to explore it.

Suddenly Gertrude began wondering if what the old man had said was true and Annie's clothes really did come off: for she feared that they might be sewed to her skin, the way a doll's clothes sometimes are. She took Annie's pretty pink dress off, just to find out. . . . 'I'll remember to put it on her again *for certain*,' thought Gertrude.

Then she went on undressing Annie, dropping her clothes on the ground. 'Stop undressing me, it's c-c-cold!' said Annie.

'Nonsense!' said Gertrude. 'I must just find out if they *all* really unbutton.'

Annie tried to run away, but Gertrude kept catching her and taking off one thing more. Annie was so hard to catch that Gertrude kept dropping the clothes all over everywhere. The sun had gone in now, and it started to snow. 'Oooooo! I'm so *cold*!' shivered Annie.

'Nonsense!' said Gertrude (who never felt cold herself, of course). 'You're not a bit cold really, so don't make a fuss!' And she undressed Annie completely.

When there were no more clothes to take off, Gertrude left Annie and started making snowballs to throw at the birds. Annie sat on the ground all bare, and shivered and howled. Annie couldn't find her clothes to put on again because they were all buried by now in the falling snow.

But at last Gertrude got tired of snowballing the birds. 'Come on in. Don't sit there dawdling around,' she said to Annie. 'Come into the house!'

'Is it dinner-time yet?' asked Annie, her teeth chattering. 'It feels like it must be.'

'*Dinner*?' said Gertrude. 'Yes—later on. But first I must cut your hair.'

So Gertrude took a pair of scissors and began cutting Annie's lovely hair. But she soon found she had cut off too much on one side, which meant she then had to cut more off the other to match; and so it went on, till Annie's head had hardly any hair left. Then Gertrude was sad, because nothing was left to cut.

'Is it dinner-time *now*?' asked Annie, when her hair was all gone.

'No, you bad child! It's your bed-time,' said Gertrude.

It wasn't really anywhere near bed-time, but Gertrude was cross with Annie for looking like a scarecrow with her hair all cut off. Besides, she wanted to be free of the child for a bit, to think her own thoughts. So she filled the bath with warm water, dropped Annie in as quick as she could, and then went away and left her there in the bath!

Gertrude went off to explore the house. It was a wonderful house, with cupboards everywhere—and every sort of thing you could want was there in the cupboards, all ready to hand! Gertrude spent hours and hours exploring the house, and came at last to the kitchen: There on the table was a whole pound of sausages. 'I think I'll have some supper, for once,' said Gertrude to herself. So she cooked the sausages, then sat down in front of the kitchen stove and ate them all up.

As she finished the last of them Gertrude felt sleepy, so went straight upstairs and got into a bed. She had forgotten all about poor Annie, left sitting there in the bath!

It was not till Gertrude was just dropping off to sleep that at last she remembered Annie. 'Poor Annie!' she thought: 'How horrible, having to pass the whole night in the bath!' But it was very, very dark; and Gertrude didn't want to get out of bed. 'Perhaps she *likes* sleeping in baths better than beds,' Gertrude told herself. 'Yes, I'm sure Annie would *rather* stop all night in the bath. . . .'

Yet Gertrude knew in her heart that this couldn't be true. Only a fish (or a mermaid) could like sleeping in water all night. 'Come on, Gertrude!' she said to herself: 'Up you get!'

Gertrude jumped out of bed, and taking two torches (one for each hand) she ran to the bathroom. There was Annie, still sitting in the water, which by now was quite cold. Annie looked very sad, and her teeth were chattering. But Gertrude soon had her out of the water, and dried her with a towel, and carried Annie off with her to bed.

'Darling Annie!' said Gertrude in bed, putting her hard wooden arm round Annie's neck.

'Darling Doll!' said Annie. 'How kind to me you are!'

And so they both fell asleep, with their arms tight round each other.

When they woke in the morning, Annie had a cold in her head. Her nose was red, and dripping. She had no hair, and no clothes. She didn't look pretty now, as she had in the shop.

'I must make you some new clothes, now that you've lost your others,' said Gertrude. So she took the cloth off the table and cut it up and made it into clothes. But they were not very nice clothes, because Gertrude had never learned sewing. Indeed Gertrude had hardly any idea at all about how to make clothes. All the same, Annie seemed just as pleased as if her tablecloth dress was the most beautiful dress in the world.

'What are we going to do today?' asked Annie, admiring her new dress in the mirror.

'We're going to have a party,' said Gertrude: 'It's your birthday today! But first you have to be smacked for losing your clothes.'

It wasn't fair for Annie to be smacked, because it was Gertrude herself who had lost them. But Annie just said, 'I'm sorry, and I won't do it again.'

'"Sorry" is not enough! You'll have to be punished as well,' said Gertrude.

Annie howled when Gertrude smacked her (and so would *you* howl if somebody smacked you with hard wooden hands!). But Gertrude didn't know how much she was hurting, of course.

'Stop crying at once, now!' said Gertrude as soon as the smacking was over. 'It's time to get ready for the party.'

So all the morning they made cakes and baked them, and put icing on top to look like flowers. Then they set out the candles, and hung up red paper streamers to make the room look pretty, and phoned people asking them to come.

At three in the afternoon they heard someone knock on the door. Annie wanted to open it. But Gertrude sent her up to the bathroom to wash her hands. 'Wait in the bathroom till I come up to see you're really clean!' Gertrude called up the stairs after Annie, and went to open the door herself.

It was the first guest come for the party. There on the step stood a big teddy-bear, leading a very small boy by the hand.

'Come in, and be a good boy,' the toy bear whispered to his little boy as Gertrude let them in. 'And remember you mustn't shout for things: you must wait till the cakes are offered you, and say "Thank you".'

'Yes, Uncle Teddy,' said the little boy (but he wasn't really listening).

Next came a rocking-horse. *He* had to be helped up the steps by the three children he had brought with him. One pulled in front and two pushed behind. The children looked very proud to belong to such a fine horse.

Then came a small dumpy doll, dragging with her quite a big lanky schoolgirl. 'Surely you're too old to belong to a doll!' Gertrude burst out when she saw her.

'That's what *I* think,' said the big girl: 'But she has had me since I was tiny, years ago—and now she won't believe I'm almost grown-up!'

'Don't talk so daft, Miss Theodora!' said the doll severely: 'You're no more than a big baby still—and don't you forget it! Now, you behave nicely or I'll punish you!'

The little doll looked so fierce that the big girl was afraid of her: 'S-sorry,' the big girl said, and put her thumb in her mouth.

Next came a puppy, dragging a small boy behind him on a rope. The puppy marched straight in without even a 'How-d'you-do' to Gertrude and jumped up into a chair at the table where the food was spread out. The little boy tried to climb up beside him, but 'Lie down!' barked the puppy to the little boy, 'Or you'll be tied up outside and not have any cakes at all!'

So the little boy crept under the table, and curled up by his master's chair.

Then all the dolls and toys climbed into chairs. But their children were not allowed at table at all! The children had to sit in a row on a bench in the corner: only the puppy's little boy was allowed to stay curled up on the floor, chewing his rope.

'You be good!' shouted all the dolls and toys to the children together. 'Then perhaps we'll let you have some bread-and-butter, if by the time we've finished the cakes we can't eat any more ourselves.'

'I hear you have a child too, haven't you?' the teddy-bear asked Gertrude with his mouth full of chocolate biscuit: 'Where is she?'

Goodness gracious! Once again Gertrude had altogether forgotten Annie, after sending her up to the bathroom to wash her hands!

But then Gertrude looked round the party, and saw all those proud toys with their charming fashion-magazine children, in their very best clothes, sitting so good and neat in the corner; and Gertrude felt ashamed of Annie with her tablecloth dress, and no hair, and her nose all runny. Gertrude couldn't bear all these grand toys to see Annie, or even to know she owned a child who looked so common and shabby.

Then Gertrude told a lie: 'Annie's been naughty,' she said. 'I had to send her to bed for a punishment.' (Poor Annie! And this was *her* birthday party, after all!)

'Never mind,' said the teddy-bear: 'It's probably just as well. If she's a naughty child I'm sure I don't want *my* little boy to meet her. She might make him naughty too.'

'Quite right!' said all the other toys together: 'I'm sure if she's naughty we none of us want *our* children to play with her!'

Gertrude began to feel she wasn't liking this party any longer. . . .

Just then they heard a scream from the garden outside, and the puppy's little boy got up and ran to the window, trailing his rope. 'I say!' he cried out, very excited: 'It's a—'

'Lie down!' barked the puppy. 'And don't speak till you're spoken to!'

'But it's a—'

'Do as you're told!' yapped the puppy: So the little boy became quiet.

'But I saw it too!' said one of the other children: 'It was—'

'Be quiet or I'll rock on you!' said the rocking-horse, furious.

'I *won't* be quiet!' said the child bravely. 'There's a lion in the garden, and he has caught Gertrude's child and is eating her up!'

'Well, well,' said the teddy-bear: 'What a *very* naughty little girl she must be, to need to be eaten by a lion!'

'It sounds horrid and vulgar,' said the prim little doll that the big girl belonged to: 'I'm sure I don't want to hear anything about it till I've finished my tea.'

'Quite right!' said the teddy-bear: 'It's spoiling my appetite too! Let us talk of pleasanter subjects.'

'I think I would like some more ice-cream,' said the rocking-horse: 'If Annie is being eaten by a lion there's no use saving any for *her*.'

But Gertrude sprang to her feet in a blazing rage. 'You brutes and horrors and pigs!' Gertrude cried: 'You're not going just to sit there and *let* her be eaten, are you?'

'It's none of *our* business,' said the party.

But Gertrude didn't wait.

She seized the teapot in one hand and a spoon in the other and jumped out of the window. There was the lion, skulking about in the garden with Annie in his mouth, looking for a comfortable place to lie down and eat her.

'Drop her!' cried Gertrude: 'Put my child down at once, Sir!'

But the lion only growled, and flicked his tail.

Poor Annie was very frightened, and not screaming now. The lion was huge. But Gertrude didn't wait for a moment: she threw the teapot full of hot tea right in his face and sprang at him with the spoon.

The lion roared with pain from the hot tea and dropped Annie—but instead, he seized Gertrude's arm in his teeth.

'Run, Annie!' cried Gertrude: 'Run in the house and be safe!'

'I *won't* run and leave you to be eaten!' cried Annie.

Annie tore up handfuls of snow and started pelting the lion to make him let go of Gertrude's arm, but—scrunch! The lion had bitten Gertrude's arm right off.

'*Ugh!*' growled the lion, 'she's only made of wood after all, and I don't like eating wooden people one bit!'

Meanwhile all the children had jumped out of the window (leaving the toys to finish the ice-cream) and were pelting and booing the lion.

'I suppose they're all made of wood too, and don't taste as nice as they look,' grumbled the lion, trying with his claws to get the wood splinters out of his teeth.

Just then a big ball of snow hit the lion slap in the eye. He dropped Gertrude's arm on the ground, and ran right away, and was gone.

But Gertrude had lost her arm. 'Oh, *poor* Gertrude!' said Annie.

'Never mind,' said Gertrude bravely. 'It doesn't hurt *too* much.'

'How lucky you're made of wood and don't feel things the way we do!' said Annie.

'Y-y-yes,' said Gertrude, trying her best not to cry.

But it was hard for her not to cry, for something strange was happening to Gertrude: never in all her life had she felt so *un*wooden as now! Indeed her gone arm was hurting her horribly—almost as if she wasn't a doll, but a person.

Then the brave little boy who belonged to the rocking-horse picked up Gertrude's chewed arm and examined it. 'I think I can mend this,' he said, 'with my carpentry tools.' So he mended Gertrude's arm and fixed it on again, and the children all cheered.

Then the children went back indoors, because the toys they belonged to (and also the puppy) were calling angrily from the window. But Gertrude and Annie stopped outside together.

'Annie!' said Gertrude; 'Listen. I think it's a stupid idea, dolls *having* to belong to children or children to dolls. Why can't they just be friends?'

'And both look after each other?'

'Yes. Anyway, that's how *we* are going to be from now on,' said Gertrude: 'You won't belong to me and I won't belong to you—not neither. So, now let us start on our travels!'

Then Annie and Gertrude put their arms round each other's waists, and started along that hard black road together. And the curious thing was this: Annie thought Gertrude's arm now felt soft, and warm—almost like the arm of another child: while Gertrude found Annie's arm comforting and *strong*—almost as if it too were a wooden one.

That's all about Gertrude and Annie for now. . . .

1940

JAMES THURBER

The Unicorn in the Garden

ONCE upon a sunny morning a man who sat in a breakfast nook looked up from his scrambled eggs to see a white unicorn with a golden horn quietly cropping the roses in the garden. The man went up to the bedroom where his wife was still asleep and woke her. 'There's a unicorn in the garden,' he said. 'Eating roses.' She opened one unfriendly eye and looked at him. 'The unicorn is a mythical beast,' she said, and turned her back on him. The man walked slowly downstairs and out into the garden. The unicorn was still there; he was now browsing among the tulips. 'Here, unicorn,' said the man, and he pulled up a lily and gave it to him. The unicorn ate it gravely. With a high heart, because there was a unicorn in his garden, the man went upstairs and roused his wife again. 'The unicorn,' he said, 'ate a lily.' His wife sat up in bed and looked at him, coldly. 'You are a booby,' she said, 'and I am going to have you put in the booby-hatch.' The man, who had never liked the words 'booby' and 'booby-hatch,' and who liked them even less on a shining morning when there was a unicorn in the garden, thought for a moment. 'We'll see about that,' he said. He walked over to the door. 'He has a golden horn in the middle of his forehead,' he told her. Then he went back to the garden to watch the unicorn; but the unicorn had gone away. The man sat down among the roses and went to sleep.

As soon as the husband had gone out of the house, the wife got up and dressed as fast as she could. She was very excited and there was a gloat in her eye. She telephoned the police and she telephoned a psychiatrist; she told them to hurry to her house and bring a strait-jacket. When the police and the psychiatrist arrived they sat down in chairs and looked at her, with great interest. 'My husband,' she said, 'saw a unicorn this morning.' The police looked at the psychiatrist and the psychiatrist looked at the police. 'He told me it ate a lily,' she said. The psychiatrist looked at the police and the police looked at the psychiatrist. 'He told me it had a golden horn in the middle of its forehead,' she said. At a solemn signal from the psychiatrist, the police leaped from their chairs and seized the wife. They had a hard time subduing her, for she put up a terrific struggle, but they finally subdued her. Just as they got her into the strait-jacket, the husband came back into the house.

'Did you tell your wife you saw a unicorn?' asked the police. 'Of course not,' said the husband. 'The unicorn is a mythical beast.' 'That's all I wanted to know,' said the psychiatrist. 'Take her away. I'm sorry, sir, but your wife is as crazy as a jay bird.' So they took her away, cursing and screaming, and shut her up in an institution. The husband lived happily ever after.

Moral: Don't count your boobies until they are hatched.

1940

SYLVIA
TOWNSEND
WARNER

Bluebeard's Daughter

EVERY child can tell of his ominous pigmentation, of his ruthless temper, of the fate of his wives and of his own fate, no less bloody than theirs; but—unless it be here and there a Director of Oriental Studies—no one now remembers that Bluebeard had a daughter. Amid so much that is wild and shocking this gentler trait of his character has been overlooked. Perhaps, rather than spoil the symmetry of a bad husband by an admission that he was a good father, historians have suppressed her. I have heard her very existence denied, on the grounds that none of Bluebeard's wives lived long enough to bear him a child. This shows what it is to give a dog a bad name. To his third wife, the mother of Djamileh, Bluebeard was most tenderly devoted, and no shadow of suspicion rested upon her quite natural death in childbed.

From the moment of her birth Djamileh became the apple of Bluebeard's eye. His messengers ransacked Georgia and Circassia to find wet-nurses of unimpeachable health, beauty, and virtue; her infant limbs were washed in nothing but rosewater, and swaddled in Chinese silks. She cut her teeth upon a cabochon emerald engraved with propitious mottoes, and all the nursery vessels, mugs, platters, ewers, basins, and chamber-pots were of white jade. Never was there a more adoring and conscientious father than Bluebeard, and I have sometimes thought that the career of this much-widowed man was inevitably determined by his anxiety to find for Djamileh an ideal stepmother.

Djamileh's childhood was happy, for none of the stepmothers lasted long enough to outwear their good intentions, and every evening, whatever his occupations during the day, Bluebeard came to the nursery for an hour's romp. But three days before her ninth birthday Djamileh was told that her father was dead; and while she was still weeping for her loss she was made to weep even more bitterly by the statement that he was a bad man and that she must not cry for him. Dressed in crape, with the Bluebeard diamonds sparkling like angry tears beneath her veils, and wearing a bandage on her wrist, Fatima came to Djamileh's pavilion and paid off the nurses and governesses. With her came Aunt Ann, and a strange young man whom she was told to call Uncle Selim; and while the nurses lamented and packed and the governesses sulked, swooned, and clapped their hands for sherbet, Djamileh listened to this trio disputing as to what should be done with her.

'For she can't stay here alone,' said Fatima. 'And nothing will induce me to spend another night under this odious roof.'

'Why not send her to school?'

'Or to the Christians?' suggested Selim.

'Perhaps there is some provision for her in the will?'

'Will! Don't tell me that such a monster could make a will, a valid will. Besides, he never made one.'

Fatima stamped her foot, and the diamond necklace sidled on her stormy bosom. Still disputing, they left the room.

That afternoon all the silk carpets and embroidered hangings, all the golden dishes and rock-crystal wine-coolers, together with the family jewels and Bluebeard's unique collection of the Persian erotic poets, were packed up and sent by camel to Selim's residence in Teheran. Thither travelled also Fatima, Ann, Selim, and Djamileh, together with a few selected slaves, Fatima in one litter with Selim riding at her side, doing his best to look stately but not altogether succeeding, since his mount was too big for him, Ann and Djamileh in the other. During the journey Ann said little, for she was engaged in ticking off entries in a large scroll. But once or twice she told Djamileh not to fidget, and to thank her stars that she had kind friends who would provide for her.

As it happened, Djamileh was perfectly well provided for. Bluebeard had made an exemplary and flawless will by which he left all his property to his only daughter and named his solicitor as her guardian until she should marry. No will can please everybody; and there was considerable heartburning when Badruddin removed Djamileh and her belongings from the care of Fatima, Ann, and Selim, persisting to the last filigree egg-cup in his thanks for their kind offices towards the heiress and her inheritance.

Badruddin was a bachelor, and grew remarkably fine jasmines.

Every evening when he came home from his office he filled a green watering-pot and went to see how they had passed the day. In the latticed garden the jasmine bush awaited him like a dumb and exceptionally charming wife. Now he often found Djamileh sitting beneath the bush, pale and silent, as though, in response to being watered so carefully, the jasmine had borne him a daughter.

It would have been well for Djamileh if she had owed her being to such an innocent parentage. But she was Bluebeard's daughter, and all the girl-babies of the neighbourhood cried in terror at her father's name. What was more, the poor girl could not look at herself in the mirror without being reminded of her disgrace. For she had inherited her father's colouring. Her hair was a deep butcher's blue, her eyebrows and eyelashes were blue also. Her complexion was clear and pale, and if some sally of laughter brought a glow to her cheek it was of the usual pink, but the sinister parental pigmentation reasserted itself on her lips, which were deep purple as though stained with eating mulberries; and the inside of her mouth and her tongue were dusky blue like a well-bred chow-dog's. For the rest she was like any other woman, and when she pricked her finger the blood ran scarlet.

Looks so much out of the common, if carried off with sufficient assurance, might be an asset to a modern miss. In Djamileh's time taste was more classical. Blue hair and purple lips, however come by, would have been a serious handicap for any young woman—how much more so, then, for her, in whom they were not only regrettable but scandalous. It was impossible for Bluebeard's badged daughter to be like other girls of her age. The purple mouth seldom smiled; the blue hair, severely braided by day, was often at night wetted with her tears. She might, indeed, have dyed it. But filial devotion forbade. Whatever his faults, Bluebeard had been a good father.

Djamileh had a great deal of proper feeling; it grieved her to think of her father's crimes. But she had also a good deal of natural partiality, and disliked Fatima; and this led her to try to find excuses for his behaviour. No doubt it was wrong, very wrong, to murder so many wives; but Badruddin seemed to think that it was almost as wrong to have married them, at any rate to have married so many of them. Experience, he said, should have taught the deceased that female curiosity is insatiable; it was foolish to go on hoping to find a woman without curiosity. Speaking with gravity, he conjured his ward to struggle, as far as in her lay, with this failing, so natural in her own sex, so displeasing to the other.

Djamileh fastened upon his words. To mark her reprobation of curiosity, the fault which had teased on her father to his ruin, she resolved never to be in the least curious herself. And for three weeks she did not ask a single question. At the end of the third week she fell into a violent fever, and Badruddin, who had been growing more and more disquieted by what appeared to him to be a protracted fit of sulks, sent for a doctoress. The doctoress was baffled by the fever, but did not admit it. What the patient needed, she said, was light but distracting conversation. Mentioning in the course of her chat that she had discovered from the eunuch that the packing-case in the lobby contained a new garden hose, the doctoress had the pleasure of seeing Djamileh make an instant recovery from her fever. Congratulating herself on her skill and on her fee, the old dame went off, leaving Djamileh to realize that it was not enough to refrain from asking questions, some more radical method of combating curiosity must be found. And so when Badruddin, shortly after her recovery, asked her in a laughing way how she would like a husband, she replied seriously that she would prefer a public-school education.

This was not possible. But the indulgent solicitor did what he could to satisfy this odd whim, and Djamileh made such good use of her opportunities that by the time she was fifteen she had spoilt her handwriting, forgotten how to speak French, lost all her former interest in botany, and asked only the most unspeculative questions. Badruddin was displeased. He sighed to think that the intellectual Bluebeard's child should have grown up so dull-witted, and spent more and more time in the company of his jasmines. Possibly, even, he consulted them, for though they were silent they could be expressive. In any case, after a month or so of inquiries, interviews, and drawing up treaties, he told Djamileh that, acting under her father's will, he had made arrangements for her marriage.

Djamileh was sufficiently startled to ask quite a number of questions, and Badruddin congratulated himself on the aptness of his prescription. His choice had fallen upon Prince Kayel Oumarah, a young man of good birth, good looks, and pleasant character, but not very well-to-do. The prince's relations were prepared to overlook Djamileh's origin in consideration of her fortune, which was enormous, and Kayel, who was of a rather sentimental turn of mind, felt that it was an act of chivalry to marry a girl whom other young men might scorn for what was no fault of hers, loved her already for being so much obliged to him, and wrote several ghazals expressing a preference for blue hair.

> 'What wouldn't I do, what wouldn't I do,
> To get at that hair of heavenly blue?'

(the original Persian is, of course, more elegant) sang Kayel under her window. Djamileh thought this harping on her hair not in the best of taste, more especially since Kayel had a robust voice and the whole street might hear him. But it was flattering to have poems written about her (she herself had no turn for poetry), and when she peeped through the lattice she thought that he had a good figure and swayed to and fro with a great deal of feeling. Passion and a good figure can atone for much; and perhaps when they were man and wife he would leave off making personal remarks.

After a. formal introduction, during which Djamileh offered Kayel symbolical sweetmeats and in her confusion ate most of them herself, the young couple were married. And shortly afterwards they left town for the Castle of Shady Transports, the late Bluebeard's country house.

Djamileh had not set eyes on Shady Transports since she was carried away from it in the same litter as Aunt Ann and the inventory. It had been in the charge of a caretaker ever since. But before the wedding Badruddin had spent a few days at the village inn, and under his superintendence the roof had been mended, the gardens trimmed up, all the floors very carefully scrubbed, and a considerable quantity of female attire burned in the stable yard. There was no look of former tragedy about the place when Djamileh and Kayel arrived. The fountain plashed innocently in the forecourt, all the most appropriate flowers in the language of love were bedded out in the parterre, a troop of new slaves, very young and handsomely dressed, stood bowing on either side of the door, and seated on cushions in stiff attitudes of expectation Maya and Moghreb, Djamileh's favourite dolls, held out their jointed arms in welcome.

Tears came into her eyes at this token of Badruddin's understanding heart. She picked up her old friends and kissed first one and then the other, begging their pardon for the long years in which they had suffered neglect. She thought they must have pined, for certainly they weighed much less than of old. Then she recollected that she was grown up, and had a husband.

At the moment he was not to be seen. Still clasping Maya and Moghreb, she went in search of him, and found him in the armoury, standing lost in admiration before a display of swords, daggers, and cutlasses. Djamileh remembered how, as a child, she had been held up to admire, and warned not to touch.

'That one comes from Turkestan,' she said. 'My father could cut off a man's head with it at a single blow.'

Kayel pulled the blade a little way from the sheath. It was speckled with rust, and the edge was blunted.

'We must have them cleaned up,' he said. 'It's a pity to let them get like this, for I've never seen a finer collection.'

'He had a splendid collection of poets, too,' said Djamileh. 'I was too young to read them then, of course, but now that I am married to a poet myself I shall read them all.'

'What a various-minded man!' exclaimed Kayel as he followed her to the library.

It is always a pleasure to explore a fine old rambling country house. Many people whose immediate thoughts would keep them tediously awake slide into a dream by fancying that such a house has—no exact matter how—come into their possession. In fancy they visit it for the first time, they wander from room to room, trying each bed in turn, pulling out the books, opening Indian boxes, meeting themselves in mirrors. . . . All is new to them, and all is theirs.

For Kayel and Djamileh this charming delusion was a matter of fact. Djamileh indeed declared that she remembered Shady Transports from the days of her childhood, and was always sure that she knew what was round the next corner; but really her recollections were so fragmentary that except for the sentiment of the thing she might have been exploring her old home for the first time. As for Kayel, who had spent most of his life in furnished lodgings, the comfort and spaciousness of his wife's palace impressed him even more than he was prepared to admit. Exclaiming with delight, the young couple ransacked the house, or wandered arm in arm through the grounds, discovering fishponds, icehouses, classical grottoes, and rustic bridges. The gardeners heard their laughter among the blossoming thickets, or traced where they had sat by the quantity of cherry-stones.

At last a day came when it seemed that Shady Transports had yielded up to them all its secrets. A sharp thunderstorm had broken up the fine weather. The rain was still falling, and Kayel and Djamileh sat in the western parlour playing chess like an old married couple. The rain had cooled the air, indeed it was quite chilly; and Kayel, who was getting the worst of the game, complained of a draught that blew on his back and distracted him.

'There can't really be a draught, my falcon,' objected Djamileh, 'for draughts don't blow out of solid walls, and there is only a wall behind you.'

'There is a draught,' persisted he. 'I take your pawn. No, wait a moment, I'm not sure that I do. How can I possibly play chess in a whirlwind?'

'Change places,' said his wife, 'and I'll turn the board.'

They did so and continued the game. It was now Djamileh's move; and as she sat gazing at the pieces Kayel fell to studying her intent and unobservant countenance. She was certainly quite pretty, very pretty even, in spite of her colouring. Marriage had improved her, thought he. A large portrait of Bluebeard hung on the wall behind her. Kayel's glance went from living daughter to painted sire, comparing the two physiognomies. Was there a likeness—apart, of course, from the blue hair? Djamileh was said to be the image of her mother; certainly the rather foxlike mask before him, the narrow eyes and pointed chin, bore no resemblance to the prominent eyes and heavy jowl of the portrait. Yet there was a something . . . the pouting lower lip, perhaps, emphasized now by her considering expression. Kayel had another look at the portrait.

'Djamileh! There *is* a draught! I saw the hangings move.' He jumped up and pulled them aside. 'What did I say?' he inquired triumphantly.

'Oh! Another surprise! Oh, haven't I a lovely Jack-in-the-Box house?'

The silken hangings had concealed a massive stone archway, closed by a green baize door.

Kayel nipped his wife's ear affectionately. 'You who remember everything so perfectly—what's behind that door?'

'Rose-petal conserve,' she replied. 'I have just remembered how it used to be brought out from the cupboard when I was good.'

'I don't believe it. I don't believe there's a cupboard, I don't believe you were ever good.'

'Open it and see.'

Beyond the baize door a winding stair led into a small gallery or corridor, on one side of which were windows looking into the park, on the other, doors. It was filled with a green and moving light reflected from the wet foliage outside. They turned to each other with rapture. A secret passage—five doors in a row, five new rooms waiting to be explored! With a dramatic gesture Kayel threw open the first door. A small dark closet was revealed, perfectly empty. A trifle dashed, they opened the next door. Another closet, small, dark, and empty. The third door revealed a third closet, the exact replica of the first and second.

Djamileh began to laugh at her husband's crestfallen air.

'In my day,' she said, 'all these cupboards were full of rose-petal conserve. So now you see how good I was.'

Kayel opened the fourth door.

He was a solemn young man, but now he began to laugh also. Four empty closets, one after another, seemed to these amiable young people the height of humour. They laughed so loudly that they did not hear a low peal of thunder, the last word of the retreating storm. A dove who had her nest in the lime tree outside the window was startled by their laughter or by the thunder; she flew away, looking pale and unreal against the slate-coloured sky. Her flight stirred the branches, which shook off their raindrops, spattering them against the casement.

'Now for the fifth door,' said Kayel.

But the fifth door was locked.

'Djamileh, dear, run and ask the steward for the keys. But don't mention which door we want unfastened. Slaves talk so, they are always imagining mysteries.'

'I am rather tired of empty cupboards, darling. Shall we leave this one for the present? At any rate till after tea? So much emptiness has made me very hungry, I really need my tea.'

'Djamileh, fetch the keys.'

Djamileh was an obedient wife, but she was also a prudent one. When she had found the bunch of keys she looked carefully over those which were unlabelled. They were many, and of all shapes and sizes; but at last she found the key she had been looking for and which she had dreaded to find. It was a small key, made of gold and finely arabesqued; and on it there was a small dark stain that might have been a bloodstain.

She slipped it off the ring and hid it in her dress.

Returning to the gallery, she was rather unpleasantly struck by Kayel's expression. She could never have believed that his open countenance could wear such a look of cupidity or that his eyes could become so beady. Hearing her step, he started violently, as though roused from profound absorption.

'There you are! What an age you have been—darling! Let's see now. Icehouse, Stillroom, Butler's Pantry, Wine-cellar, Family Vault . . . I wonder if this is it?'

He tried key after key, but none of them fitted. He tried them all over again, upside-down or widdershins. But still they did not fit. So then he took out his pocket-knife, and tried to pick the lock. This also was useless.

'Eblis take this lock!' he exclaimed. And suddenly losing his temper, he began to kick and batter at the door. As he did so there was a little click; and one of the panels of the door fell open upon a hinge, and disclosed a piece of parchment, framed and glazed, on which was an inscription in ancient Sanskrit characters.

'What the . . . Here, I can't make this out.'

Djamileh, who was better educated than her husband in such useless studies as calligraphy, examined the parchment, and read aloud: 'CURIOSITY KILLED THE CAT.'

Against her bosom she felt the little gold key sidle, and she had the unpleasant sensation which country language calls: 'The grey goose walking over your grave.'

'I think,' she said gently, 'I think, dear husband, we had better leave this door alone.'

Kayel scratched his head and looked at the door.

'Are you sure that's what it means? Perhaps you didn't read it right.'

'I am quite sure that is what it means.'

'But, Djamileh, I do want to open the door.'

'So do I, dear. But under the circumstances we had better not do anything of the sort. The doors in this house are rather queer sometimes. My poor father . . . my poor stepmothers . . .'

'I wonder,' mused Kayel, 'if we could train a cat to turn the lock and go in first.'

'Even if we could, which I doubt, I don't think that would be at all fair to the cat. No, Kayel, I am sure we should agree to leave this door alone.'

'It's not that I am in the least inquisitive,' said Kayel, 'for I am not. But as master of the house I really think it my duty to know what's inside this cupboard. It might be firearms, for instance, or poison, which might get into the wrong hands. One has a certain responsibility, hang it!'

'Yes, of course. But all the same I feel sure we should leave the door alone.'

'Besides, I have you to consider, Djamileh. As a husband, you must be my first consideration. Now you may not want to open the door just now; but suppose, later on, when you were going to have a baby, you developed one of those strange yearnings that women at such times are subject to; and suppose it took the form of longing to know what was behind this door. It might be very bad for you, Djamileh, it might imperil your health, besides birth-marking the baby. No! It's too grave a risk. We had much better open the door immediately.'

And he began to worry the lock again with his pen-knife.

'Kayel, please don't. *Please* don't. I implore you, I have a feeling—'

'Nonsense. Women always have feelings.'

'—as though I were going to be sick. In fact, I am sure I am going to be sick.'

'Well, run off and be sick, then. No doubt it was the thunderstorm, and all those strawberries.'

'I can't run off, Kayel. I don't feel well enough to walk; you must carry me. Kayel!'—she laid her head insistently on his chest—'Kayel! I felt sick this morning, too.'

And she laid her limp weight against him so firmly that with a sigh he picked her up and carried her down the corridor.

Laid on the sofa, she still kept a firm hold on his wrist, and groaned whenever he tried to detach himself. At last, making the best of a bad job, he resigned himself, and spent the rest of the day reading aloud to her from the erotic Persian poets. But he did not read with his usual fervour; the lyrics, as he rendered them, might as well have been genealogies. And Djamileh, listening with closed eyes, debated within herself why Kayel should be so cross. Was it just the locked closet? Was it, could it be, that he was displeased by the idea of a baby with Bluebeard blood? This second possibility was highly distressing to her, and she wished, more and more fervently, as she lay on the sofa keeping up a pretence of delicate health and disciplining her healthy appetite to a little bouillon and some plain sherbet, that she had hit upon a pretext with fewer consequences entailed.

It seemed to her that they were probably estranged for ever. So it was a great relief to be awakened in the middle of the night by Kayel's usual affable tones, even though the words were:

'Djamileh, I believe I've got it! All we have to do is to get a stonemason, and a ladder, and knock a hole in the wall. Then we can look in from outside. No possible harm in that.'

All the next day and the day after, Kayel perambulated the west wing of Shady Transports with his stonemasons, directing them where to knock holes in the walls; for it had been explained to the slaves that he intended to bring the house up to date by throwing out a few bow-windows. But not one of these perspectives (the walls of Shady Transports were exceedingly massy) afforded a view into the locked closet. While these operations were going on he insisted that Djamileh should remain at his side. It was essential, he said, that she should appear interested in the improvements, because of the slaves. All this while she was carrying about that key on her person, and debating whether she should throw it away, in case Kayel, by getting possession of it, should endanger his life, or whether she should keep it and use it herself the moment he was safely out of the way.

Jaded in nerves and body, at the close of the second day they had a violent quarrel. It purported to be about the best method of pruning acacias, but while they were hurrying from sarcasm to acrimony, from acrimony to abuse, from abuse to fisticuffs, they were perfectly aware that in truth they were quarrelling as to which of them should first get at that closet.

'Laterals! Laterals!' exclaimed Djamileh. 'You know no more of pruning than you know of dressmaking. That's right! Tear out my hair, do!'

'No, thank you.' Kayel folded his arms across his chest. 'I have no use for *blue hair*.'

Pierced by this taunt, Djamileh burst into tears. The soft-hearted Kayel felt that he had gone too far, and made several handsome apologies for the remark; but it seemed likely his apologies would be in vain, for Djamileh only came out of her tears to ride off on a high horse.

'No, Kayel,' she said, putting aside his hand, and speaking with exasperating nobility and gentleness. 'No, no, it is useless, do not let us deceive ourselves any longer. I do not blame you; your feeling is natural and one should never blame people for natural feelings.'

'Then why have you been blaming me all this time for a little natural curiosity?'

Djamileh swept on.

'And how could you possibly have felt anything but aversion for one in whose veins so blatantly runs the blood of the Bluebeards, for one whose hair, whose lips, stigmatize her as the child of an unfortunate monster? I do not blame you, Kayel. I blame myself, for fancying you could ever love me. But I will make you the only amends in my power. I will leave you.'

A light quickened in Kayel's eye. So he thought she would leave him at Shady Transports, did he?

'Tomorrow we will go *together* to Badruddin. He arranged our marriage, he had better see about our divorce.'

Flushed with temper, glittering with tears, she threw herself into his willing arms. They were still in all the raptures of sentiment and first love, and in the even more enthralling raptures of sentiment and first grief, when they set out for Teheran. Absorbed in gazing into each other's eyes and wiping away each other's tears with pink silk handkerchiefs, they did not notice that a drove of stampeding camels was approaching their palanquin; and it was with the greatest surprise and bewilderment that they found themselves tossed over a precipice.

When Djamileh recovered her senses she found herself lying in a narrow green pasture, beside a water-course. Some fine broad-tailed sheep were cropping the herbage around, and an aged shepherdess was bathing her forehead and slapping her hands.

'How did I come here?' she inquired.

'I really cannot tell you,' answered the shepherdess. 'All I know is that about half an hour ago you, and a handsome young man, and a coachman, and a quantity of silk cushions and chicken sandwiches appeared, as it were from heaven, and fell amongst us and our sheep. Perhaps as you are feeling better you would like one of the sandwiches?'

'Where is that young man? He is not dead?'

'Not at all. A little bruised, but nothing worse. He recovered before you, and feeling rather shaken he went off with the shepherds to have a drink at the inn. The coachman went with them.'

Djamileh ate another sandwich, brooding on Kayel's heartlessness.

'Listen,' she said, raising herself on one elbow. 'I have not time to tell you the whole of my history, which is long and complicated with unheard-of misfortunes. Suffice it to say that I am young, beautiful, wealthy, well born, and accomplished, and the child of doting and distinguished parents. At their death I fell into the hands of an unscrupulous solicitor who, entirely against my will, married me to that young man you have seen. We had not been married for a day before he showed himself a monster of jealousy; and though my conduct has been unspotted as the snow he has continually belaboured me with threats and reproaches, and now has determined to shut me up, for ever, in a hermitage on the Caucasus mountains, inherited from a woman-hating uncle (the whole family are very queer). We were on our way thither when, by the interposition of my good genius, the palanquin overturned, and we arrived among your flocks as we did.'

'Indeed,' replied the aged shepherdess. 'He said nothing of all that. But I do not doubt it. Men are a cruel and fantastic race. I too have lived a life chequered with many strange adventures and unmerited misfortunes. I was born in India, the child of a virtuous Brahmin and of a mother who had, before my birth, graced the world with eleven daughters, each lovelier than the last. In the opinion of many well-qualified persons, I, the youngest of her children, was even fairer—'

'I can well believe it,' said Djamileh. 'But, venerable Aunt, my misfortunes compel me to postpone the pleasure of hearing your story until a more suitable moment. It is, as you will see, essential that I should seize this chance of escaping from my tyrant. Here is a purse. I shall be everlastingly obliged if you will conduct me to the nearest livery-stables where I can hire a small chariot and swift horses.'

Though bruised and scratched Djamileh was not much the worse for her sudden descent into the valley, and following the old shepherdess, who was as nimble as a goat, she scrambled up the precipice, and soon found herself in a hired chariot, driving at full speed towards the Castle of Shady Transports, clutching in her hot hand the key of the locked closet. Her impatience was indescribable, and as for her scruples and her good principles, they had vanished as though they had never been. Whether it was a slight concussion, or pique at hearing that Kayel had left her in order to go off and drink with vulgar shepherds, I do not pretend to say. But in any case, Djamileh had now but one thought, and that was to gratify her curiosity as soon as possible.

Bundling up a pretext of having forgotten her jewellery, she hurried past the house steward and the slaves, refusing refreshment and not listening to a word they said. She ran to the west parlour, threw aside the embroidered hangings, opened the green baize door, flew up the winding stair and along the gallery.

But the door of the fifth closet had been burst open.

It gave upon a sumptuous but dusky vacancy, an underground saloon of great size, walled with mosaics and inadequately lit by seven vast rubies hanging from the ceiling. A flight of marble steps led down to this apartment, and at the foot of the steps lay Kayel, groaning piteously.

'Thank heaven you've come! I've been here for the last half-hour, shouting at the top of my voice, and not one of those accursed slaves has come near me.'

'Oh, Kayel, are you badly hurt?'

'Hurt? I should think I've broken every bone in my body, and I know I've broken my collar-bone. I had to smash that door in, and it gave suddenly, and I pitched all the way down these steps. My second fall today. Oh!'

As she leaned over him the little golden key, forgotten and useless now, slid from her hand.

'My God, Djamileh! You've had that key all this time. And so *that* was why you came back?'

'Yes, Kayel. I came back to open the door. But you got here before me.'

And while that parry still held him she hastened to add:

'We have both behaved so shockingly that I don't think either of us had better reproach the other. So now let us see about your fracture.'

Not till the collar-bone was mending nicely; not till the coverlet which Djamileh had begun to knit as she sat by her husband's bedside, since knitting is always so soothing to invalids, was nearly finished; not till they had solved the last of the acrostics sent to them by a sympathizing Badruddin, did they mention the affair of the closet.

'How could I have the heart to leave you—you, looking so pale, and so appealing?' said Kayel suddenly.

'And the lies I told about you, Kayel, the moment I came to . . . the things I said, the way I took away your character!'

'We must have been mad.'

'We were suffering from curiosity. That was all, but it was quite enough.'

'How terrible curiosity is, Djamileh! Fiercer than lust, more ruthless than avarice. . . .'

'Insatiable as man-eating tigers. . . .'

'Insistent as that itching-powder one buys at low French fairs. . . . O Djamileh, let us vow never to feel curiosity again!'

'I made that vow long ago. You have seen what good it was.'

They meditated, gazing into each other's eyes.

'It seems to me, my husband, that we should be less inquisitive if we had more to do. I think we should give up all our money, live in a village, and work all day in the fields.'

'That only shows, my dearest, that you have always lived in a town. The people who work all day in the fields will sit up all night in the hopes of discovering if their neighbour's cat has littered brindled or tortoiseshell kittens.'

They continued to interrogate each other's eyes.

'A man through whose garden flowed a violent water-course,' said Djamileh, 'complained one day to the stream: "O Stream, you have washed away my hollyhocks, swept off my artichokes, undermined my banks, flooded my bowling-green, and drowned my youngest son, the garland of my grey head. I wish, O Stream, that you would have the kindness to flow elsewhere." "That cannot be," replied the stream, "since Allah has bidden me to flow where I do. But if you were to erect a mill on your property, perhaps you would admit that I have my uses." In other words, Kayel, it seems to me that, since we cannot do away with our curiosity, we had best sublimate it, and take up the study of a science.'

'Let it be astronomy,' answered Kayel. 'Of all sciences, it is the one least likely to intervene in our private life.'

To this day, though Bluebeard's daughter is forgotten, the wife of Kayel the astronomer is held in remembrance. It was she whose sympathetic collaboration supported him through his researches into the Saturnian rings, it was she who worked out the mathematical calculations which enabled him to prove that the lost Pleiad would reappear in the year 1963. As time went on, and her grandchildren came clustering round the telescope, Djamileh's blue hair became silver; but to the day of her death her arched blue brows gave an appearance of alertness to her wrinkled countenance, and her teeth, glistening and perfect as in her girlhood, were shown off to the best advantage by the lining of her mouth, duskily blue as that of a well-bred chow-dog's.

1940

JOHN
COLLIER

The Chaser

ALAN Austen, as nervous as a kitten, went up certain dark and creaky stairs in the neighbourhood of Pell Street, and peered about for a long time on the dim landing before he found the name he wanted written obscurely on one of the doors.

He pushed open this door, as he had been told to do, and found himself in a tiny room, which contained no furniture but a plain kitchen table, a rocking-chair, and an ordinary chair. On one of the dirty buff-coloured walls were a couple of shelves, containing in all perhaps a dozen bottles and jars.

An old man sat in the rocking-chair, reading a newspaper. Alan, without a word, handed him the card he had been given. 'Sit down, Mr Austen,' said the old man very politely. 'I am glad to make your acquaintance.'

'Is it true,' asked Alan, 'that you have a certain mixture that has—er—quite extraordinary effects?'

'My dear sir,' replied the old man, 'my stock in trade is not very large—I don't deal in laxatives and teething mixtures—but such as it is, it is varied. I think nothing I sell has effects which could be precisely described as ordinary.'

'Well, the fact is—' began Alan.

'Here, for example,' interrupted the old man, reaching for a bottle from the shelf. 'Here is a liquid as colourless as water, almost tasteless, quite imperceptible in coffee, milk, wine, or any other beverage. It is also quite imperceptible to any known method of autopsy.'

'Do you mean it is a poison?' cried Alan, very much horrified.

'Call it a glove-cleaner if you like,' said the old man indifferently. 'Maybe it will clean gloves. I have never tried. One might call it a life-cleaner. Lives need cleaning sometimes.'

'I want nothing of that sort,' said Alan.

'Probably it is just as well,' said the old man. 'Do you know the price of this? For one teaspoonful, which is sufficient, I ask five thousand dollars. Never less. Not a penny less.'

'I hope all your mixtures are not as expensive,' said Alan apprehensively.

'Oh dear, no,' said the old man. 'It would be no good charging that sort of price for a love potion, for example. Young people who need a love potion very seldom have five thousand dollars. Otherwise they would not need a love potion.'

'I am glad to hear that,' said Alan.

'I look at it like this,' said the old man. 'Please a customer with one article, and he will come back when he needs another. Even if it *is* more costly. He will save up for it, if necessary.'

'So,' said Alan, 'you really do sell love potions?'

'If I did not sell love potions,' said the old man, reaching for another bottle, 'I should not have mentioned the other matter to you. It is only when one is in a position to oblige that one can afford to be so confidential.'

'And these potions,' said Alan. 'They are not just—just—er—'

'Oh, no,' said the old man. 'Their effects are permanent, and extend far beyond the mere casual impulse. But they include it. Oh, yes, they include it. Bountifully, insistently. Everlastingly.'

'Dear me!' said Alan, attempting a look of scientific detachment. 'How very interesting!'

'But consider the spiritual side,' said the old man.

'I do, indeed,' said Alan.

'For indifference,' said the old man, 'they substitute devotion. For scorn, adoration. Give one tiny measure of this to the young lady—its flavour is imperceptible in orange juice, soup, or cocktails—and however gay and giddy she is, she will change altogether. She will want nothing but solitude, and you.'

'I can hardly believe it,' said Alan. 'She is so fond of parties.'

'She will not like them any more,' said the old man. 'She will be afraid of the pretty girls you may meet.'

'She will actually be jealous?' cried Alan in a rapture. 'Of me?'

'Yes, she will want to be everything to you.'

'She is, already. Only she doesn't care about it.'

'She will, when she has taken this. She will care intensely. You will be her sole interest in life.'

'Wonderful!' cried Alan.

'She will want to know all you do,' said the old man. 'All that has happened to you during the day. Every word of it. She will want to know what you are thinking about, why you smile suddenly, why you are looking sad.'

'That is love!' cried Alan.

'Yes,' said the old man. 'How carefully she will look after you! She will never allow you to be tired, to sit in a draught, to neglect your food. If you are an hour late, she will be terrified. She will think you are killed, or that some siren has caught you.'

'I can hardly imagine Diana like that!' cried Alan, overwhelmed with joy.

'You will not have to use your imagination,' said the old man. 'And, by the way, since there are always sirens, if by any chance you *should*, later on, slip a little, you need not worry. She will forgive you, in the end. She will be terribly hurt, of course, but she will forgive you—in the end.'

'That will not happen,' said Alan fervently.

'Of course not,' said the old man. 'But, if it did, you need not worry. She would never divorce you. Oh, no! And, of course, she herself will never give you the least, the very least, grounds for—uneasiness.'

'And how much,' said Alan, 'is this wonderful mixture?'

'It is not as dear,' said the old man, 'as the glove-cleaner, or life-cleaner, as I sometimes call it. No. That is five thousand dollars, never a penny less. One has to be older than you are, to indulge in that sort of thing. One has to save up for it.'

'But the love potion?' said Alan.

'Oh, that,' said the old man, opening the drawer in the kitchen table, and taking out a tiny, rather dirty-looking phial. 'That is just a dollar.'

'I can't tell you how grateful I am,' said Alan, watching him fill it.

'I like to oblige,' said the old man. 'Then customers come back, later in life, when they are rather better off, and want more expensive things. Here you are. You will find it very effective.'

'Thank you again,' said Alan. 'Good-bye.'

'*Au revoir*,' said the old man.

1941

PHILIP K. DICK

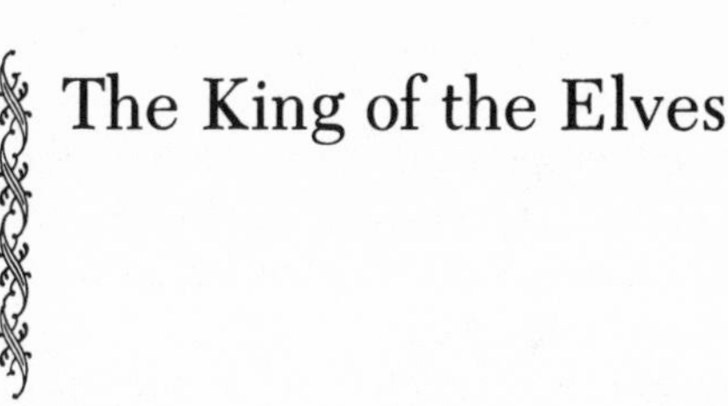

The King of the Elves

It was raining and getting dark. Sheets of water blew along the row of pumps at the edge of the filling station; the tree across the highway bent against the wind.

Shadrach Jones stood just inside the doorway of the little building, leaning against an oil drum. The door was open and gusts of rain blew in onto the wood floor. It was late; the sun had set, and the air was turning cold. Shadrach reached into his coat and brought out a cigar. He bit the end off it and lit it carefully, turning away from the door. In the gloom, the cigar burst into life, warm and glowing. Shadrach took a deep draw. He buttoned his coat around him and stepped out onto the pavement.

'Darn,' he said. 'What a night!' Rain buffeted him, wind blew at him. He looked up and down the highway, squinting. There were no cars in sight. He shook his head, locked up the gasoline pumps.

He went back into the building and pulled the door shut behind him. He opened the cash register and counted the money he'd taken in during the day. It was not much.

Not much, but enough for one old man. Enough to buy him tobacco and firewood and magazines, so that he could be comfortable as he waited for the occasional cars to come by. Not very many cars came along the highway any more. The highway had begun to fall into disrepair; there were many cracks in its dry, rough surface, and most cars preferred to take the big state highway that ran beyond the hills. There was nothing in Derryville to attract them, to make them turn toward it. Derryville was a small town, too small to bring in any of the major industries, too small to be very important to anyone. Sometimes hours went by without—

Shadrach tensed. His fingers closed over the money. From outside came a sound, the melodic ring of the signal wire stretched along the pavement.

Dinggg!

Shadrach dropped the money into the till and pushed the drawer closed. He stood up slowly and walked toward the door, listening. At the door, he snapped off the light and waited in the darkness, staring out.

He could see no car there. The rain was pouring down, swirling with the wind; clouds of mist moved along the road. And something was standing beside the pumps.

He opened the door and stepped out. At first, his eyes could make nothing out. Then the old man swallowed uneasily.

Two tiny figures stood in the rain, holding a kind of platform between them. Once, they might have been gaily dressed in bright garments, but now their clothes hung limp and sodden, dripping in the rain. They glanced half-heartedly at Shadrach. Water streaked their tiny faces, great drops of water. Their robes blew about them with the wind, lashing and swirling.

On the platform, something stirred. A small head turned wearily, peering at Shadrach. In the dim light, a rain-streaked helmet glinted dully.

'Who are you?' Shadrach said.

The figure on the platform raised itself up. 'I'm the King of the Elves and I'm wet.'

Shadrach stared in astonishment.

'That's right,' one of the bearers said. 'We're all wet.'

A small group of Elves came straggling up, gathering around their king. They huddled together forlornly, silently.

'The King of the Elves,' Shadrach repeated. 'Well, I'll be darned.'

Could it be true? They were very small, all right, and their dripping clothes were strange and oddly colored.

But *Elves*?

'I'll be darned. Well, whatever you are, you shouldn't be out on a night like this.'

'Of course not,' the king murmured. 'No fault of our own. No fault . . .' His voice trailed off into a choking cough. The Elf soldiers peered anxiously at the platform.

'Maybe you better bring him inside,' Shadrach said. 'My place is up the road. He shouldn't be out in the rain.'

'Do you think we like being out on a night like this?' one of the bearers muttered. 'Which way is it? Direct us.'

Shadrach pointed up the road. 'Over there. Just follow me. I'll get a fire going.'

He went down the road, feeling his way onto the first of the flat stone steps that he and Phineas Judd had laid during the summer. At the top of the steps, he looked back. The platform was coming slowly along, swaying a little from side to side. Behind it, the Elf soldiers picked their way, a tiny column of silent dripping creatures, unhappy and cold.

'I'll get the fire started,' Shadrach said. He hurried them into the house.

Wearily, the Elf King lay back against the pillow. After sipping hot chocolate, he had relaxed and his heavy breathing sounded suspiciously like a snore.

Shadrach shifted in discomfort.

'I'm sorry,' the Elf King said suddenly, opening his eyes. He rubbed his forehead. 'I must have drifted off. Where was I?'

'You should retire, Your Majesty,' one of the soldiers said sleepily, 'It is late and these are hard times.'

'True,' the Elf King said, nodding. 'Very true.' He looked up at the towering figure of Shadrach, standing before the fireplace, a glass of beer in his hand. 'Mortal, we thank you for your hospitality. Normally, we do not impose on human beings.'

'It's those Trolls,' another of the soldiers said, curled up on a cushion of the couch.

'Right,' another soldier agreed. He sat up, groping for his sword. 'Those reeking Trolls, digging and croaking—'

'You see,' the Elf King went on, 'as our party was crossing from the Great Low Steps toward the Castle, where it lies in the hollow of the Towering Mountains—'

'You mean Sugar Ridge,' Shadrach supplied helpfully.

'The Towering Mountains. Slowly we made our way. A rain storm came up. We became confused. All at once a group of Trolls appeared, crashing through the underbrush. We left the woods and sought safety on the Endless Path—'

'The highway. Route Twenty.'

'So that is why we're here.' The Elf King paused a moment. 'Harder and harder it rained. The wind blew around us, cold and bitter. For an endless time we toiled along. We had no idea where we were going or what would become of us.'

The Elf King looked up at Shadrach. 'We knew only this: Behind us, the Trolls were coming, creeping through the woods, marching through the rain, crushing everything before them.'

He put his hand to his mouth and coughed, bending forward. All the Elves waited anxiously until he was done. He straightened up.

'It was kind of you to allow us to come inside. We will not trouble you for long. It is not the custom of the Elves—'

Again he coughed, covering his face with his hand. The Elves drew toward him apprehensively. At last the king stirred. He sighed.

'What's the matter?' Shadrach asked. He went over and took the cup of chocolate from the fragile hand. The Elf King lay back, his eyes shut.

'He has to rest,' one of the soldiers said. 'Where's your room? The sleeping room.'

'Upstairs,' Shadrach said. 'I'll show you where.'

Late that night, Shadrach sat by himself in the dark, deserted living room, deep in meditation. The Elves were asleep above him, upstairs in the bedroom, the Elf King in the bed, the others curled up together on the rug.

The house was silent. Outside, the rain poured down endlessly, blowing against the house. Shadrach could hear the tree branches slapping in the wind. He clasped and unclasped his hands. What a strange business it was—all these Elves, with their old, sick king, their piping voices. How anxious and peevish they were!

But pathetic, too; so small and wet, with water dripping down from them, and all their gay robes limp and soggy.

The Trolls—what were they like? Unpleasant and not very clean. Something about digging, breaking and pushing through the woods . . .

Suddenly, Shadrach laughed in embarrassment. What was the matter with him, believing all this? He put his cigar out angrily, his ears red. What was going on? What kind of joke was this?

Elves? Shadrach grunted in indignation. Elves in Derryville? In the middle of Colorado? Maybe there were Elves in Europe. Maybe in Ireland. He had heard of that. But here? Upstairs in his own house, sleeping in his own bed?

'I've heard just about enough of this,' he said. 'I'm not an idiot, you know.'

He turned toward the stairs, feeling for the banister in the gloom. He began to climb.

Above him, a light went on abruptly. A door opened.

Two Elves came slowly out onto the landing. They looked down at him. Shadrach halted halfway up the stairs. Something on their faces made him stop.

'What's the matter?' he asked hesitantly.

They did not answer. The house was turning cold, cold and dark, with the chill of the rain outside and the chill of the unknown inside.

'What is it?' he said again. 'What's the matter?'

'The King is dead,' one of the Elves said. 'He died a few moments ago.'

Shadrach stared up, wide-eyed. 'He did? But—'

'He was very cold and very tired.' The Elves turned away, going back into the room, slowly and quietly shutting the door.

Shadrach stood, his fingers on the banister, hard, lean fingers, strong and thin.

He nodded his head blankly.

'I see,' he said to the closed door. 'He's dead.'

The Elf soldiers stood around him in a solemn circle. The living room was bright with sunlight, the cold white glare of early morning.

'But wait,' Shadrach said. He plucked at his necktie. 'I have to get to the filling station. Can't you talk to me when I come home?'

The faces of the Elf soldiers were serious and concerned.

'Listen,' one of them said. 'Please hear us out. It is very important to us.'

Shadrach looked past them. Through the window he saw the highway, steaming in the heat of day, and down a little way was the gas station, glittering brightly. And even as he watched, a car came up to it and honked thinly, impatiently. When nobody came out of the station, the car drove off again down the road.

'We beg you,' a soldier said.

Shadrach looked down at the ring around him, the anxious faces, scored with concern and trouble. Strangely, he had always thought of Elves as carefree beings, flitting without worry or sense—

'Go ahead,' he said. 'I'm listening.' He went over to the big chair and sat down. The Elves came up around him. They conversed among themselves for a moment, whispering, murmuring distantly. Then they turned toward Shadrach.

The old man waited, his arms folded.

'We cannot be without a king,' one of the soldiers said. 'We could not survive. Not these days.'

'The Trolls,' another added. 'They multiply very fast. They are terrible beasts. They're heavy and ponderous, crude, bad-smelling—'

'The odor of them is awful. They come up from the dark wet places, under the earth, where the blind, groping plants feed in silence, far below the surface, far from the sun.'

'Well, you ought to elect a king, then,' Shadrach suggested. 'I don't see any problem there.'

'We do not elect the King of the Elves,' a soldier said. 'The old king must name his successor.'

'Oh,' Shadrach replied. 'Well, there's nothing wrong with that method.'

'As our old king lay dying, a few distant words came forth from his lips,' a soldier said. 'We bent closer, frightened and unhappy, listening.'

'Important, all right,' agreed Shadrach. 'Not something you'd want to miss.'

'He spoke the name of him who will lead us.'

'Good. You caught it, then. Well, where's the difficulty?'

'The name he spoke was—was your name.'

Shadrach stared. '*Mine?*'

'The dying king said: 'Make him, the towering mortal, your king. Many things will come if he leads the Elves into battle against the Trolls. I see the rising once again of the Elf Empire, as it was in the old days, as it was before—'

'Me!' Shadrach leaped up. 'Me? King of the Elves?'

Shadrach walked about the room, his hands in his pockets. 'Me, Shadrach Jones, King of the Elves.' He grinned a little. 'I sure never thought of it before.'

He went to the mirror over the fireplace and studied himself. He saw his thin, graying hair, his bright eyes, dark skin, his big Adam's apple.

'King of the Elves,' he said. 'King of the Elves. Wait till Phineas Judd hears about this. Wait till I tell him!'

Phineas Judd would certainly be surprised!

Above the filling station, the sun shone, high in the clear blue sky.

Phineas Judd sat playing with the accelerator of his old Ford truck. The motor raced and slowed. Phineas reached over and turned the ignition key off, then rolled the window all the way down.

'What did you say?' he asked. He took off his glasses and began to polish them, steel rims between slender, deft fingers that were patient from years of practice. He restored his glasses to his nose and smoothed what remained of his hair into place.

'What was it, Shadrach?' he said. 'Let's hear that again.'

'I'm King of the Elves,' Shadrach repeated. He changed position, bringing his other foot up on the running board. 'Who would have thought it? Me, Shadrach Jones, King of the Elves.'

Phineas gazed at him. 'How long have you been—King of the Elves, Shadrach?'

'Since the night before last.'

'I see. The night before last.' Phineas nodded. 'I see. And what, may I ask, occurred the night before last?'

'The Elves came to my house. When the old king died, he told them that—'

A truck came rumbling up and the driver leaped out. 'Water!' he said. 'Where the hell is the hose?'

Shadrach turned reluctantly. 'I'll get it.' He turned back to Phineas. 'Maybe I can talk to you tonight when you come back from town. I want to tell you the rest. It's very interesting.'

'Sure,' Phineas said, starting up his little truck. 'Sure, Shadrach. I'm very interested to hear.'

He drove off down the road.

Later in the day, Dan Green ran his flivver up to the filling station.

'Hey, Shadrach,' he called. 'Come over here! I want to ask you something.'

Shadrach came out of the little house, holding a waste-rag in his hand.

'What is it?'

'Come here.' Dan leaned out the window, a wide grin on his face, splitting his face from ear to ear. 'Let me ask you something, will you?'

'Sure.'

'Is it true? Are you really the King of the Elves?'

Shadrach flushed a little. 'I guess I am,' he admitted, looking away. 'That's what I am, all right.'

Dan's grin faded. 'Hey, you trying to kid me? What's the gag?'

Shadrach became angry. 'What do you mean? Sure, I'm the King of the Elves. And anyone who says I'm not—'

'All right, Shadrach,' Dan said, starting up the flivver quickly. 'Don't get mad. I was just wondering.'

Shadrach looked very strange.

'All right,' Dan said. 'You don't hear me arguing, do you?'

By the end of the day, everyone around knew about Shadrach and how he had suddenly become the King of the Elves. Pop Richey, who ran the Lucky Store in Derryville, claimed Shadrach was doing it to drum up trade for the filling station.

'He's a smart old fellow,' Pop said. 'Not very many cars go along there any more. He knows what he's doing.'

'I don't know,' Dan Green disagreed. 'You should hear him, I think he really believes it.'

'King of the Elves?' They all began to laugh. 'Wonder what he'll say next.'

Phineas Judd pondered. 'I've known Shadrach for years. I can't figure it out.' He frowned, his face wrinkled and disapproving. 'I don't like it.'

Dan looked at him. 'Then you think he believes it?'

'Sure,' Phineas said. 'Maybe I'm wrong, but I really think he does.'

'But how could he believe it?' Pop asked. 'Shadrach is no fool. He's been in business for a long time. He must be getting something out of it, the way I see it. But what, if it isn't to build up the filling station?'

'Why, don't you know what he's getting?' Dan said, grinning. His gold tooth shone.

'What?' Pop demanded.

'He's got a whole kingdom to himself, that's what—to do with like he wants. How would you like that, Pop? Wouldn't you like to be King of the Elves and not have to run this old store any more?'

'There isn't anything wrong with my store,' Pop said. 'I ain't ashamed to run it. Better than being a clothing salesman.'

Dan flushed, 'Nothing wrong with that, either.' He looked at Phineas. 'Isn't that right? Nothing wrong with selling clothes, is there, Phineas?'

Phineas was staring down at the floor. He glanced up. 'What? What was that?'

'What you thinking about?' Pop wanted to know. 'You look worried.'

'I'm worried about Shadrach,' Phineas said. 'He's getting old. Sitting out there by himself all the time, in the cold weather, with the rain water running over the floor—it blows something awful in the winter, along the highway—'

'Then you *do* think he believes it?' Dan persisted. 'You *don't* think he's getting something out of it?'

Phineas shook his head absently and did not answer.

The laughter died down. They all looked at one another.

That night, as Shadrach was locking up the filling station, a small figure came toward him from the darkness.

'Hey!' Shadrach called out. 'Who are you?'

An Elf soldier came into the light, blinking. He was dressed in a little gray robe, buckled at the waist with a band of silver. On his feet were little leather boots. He carried a short sword at his side.

'I have a serious message for you,' the Elf said. 'Now, where did I put it?'

He searched his robe while Shadrach waited. The Elf brought out a tiny scroll and unfastened it, breaking the wax expertly. He handed it to Shadrach.

'What's it say?' Shadrach asked. He bent over, his eyes close to the vellum. 'I don't have my glasses with me. Can't quite make out these little letters.'

'The Trolls are moving. They've heard that the old king is dead, and they're rising, in all the hills and valleys around. They will try to break the Elf Kingdom into fragments, scatter the Elves—'

'I see,' Shadrach said. 'Before your new king can really get started.'

'That's right.' The Elf soldier nodded. 'This is a crucial moment for the Elves. For centuries, our existence has been precarious. There are so many Trolls, and Elves are very frail and often take sick—'

'Well, what should I do? Are there any suggestions?'

'You're supposed to meet with us under the Great Oak tonight. We'll take you into the Elf Kingdom, and you and your staff will plan and map the defense of the Kingdom.'

'What?' Shadrach looked uncomfortable. 'But I haven't eaten dinner. And my gas station—tomorrow is Saturday, and a lot of cars—'

'But you are King of the Elves,' the soldier said.

Shadrach put his hand to his chin and rubbed it slowly.

'That's right,' he replied. 'I am, ain't I?'

The Elf soldier bowed.

'I wish I'd known this sort of thing was going to happen,' Shadrach said. 'I didn't suppose being King of the Elves—'

He broke off, hoping for an interruption. The Elf soldier watched him calmly, without expression.

'Maybe you ought to have someone else as your king,' Shadrach decided. 'I don't know very much about war and things like that, fighting and all that sort of business.' He paused, shrugged his shoulders. 'It's nothing I've ever mixed in. They don't have wars here in Colorado. I mean they don't have wars between human beings.'

Still the Elf soldier remained silent.

'Why was I picked?' Shadrach went on helplessly, twisting his hands. 'I don't know anything about it. What made him go and pick me? Why didn't he pick somebody else?'

'He trusted you,' the Elf said. 'You brought him inside your house, out of the rain. He knew that you expected nothing for it, that there was nothing you wanted. He had known few who gave and asked nothing back.'

'Oh.' Shadrach thought it over. At last he looked up. 'But what about my gas station? And my house? And what will they say, Dan Green and Pop down at the store—'

The Elf soldier moved away, out of the light. 'I have to go. It's getting late, and at night the Trolls come out. I don't want to be too far away from the others.'

'Sure,' Shadrach said.

'The Trolls are afraid of nothing, now that the old king is dead. They forage everywhere. No one is safe.'

'Where did you say the meeting is to be? And what time?'

'At the Great Oak. When the moon sets tonight, just as it leaves the sky.'

'I'll be there, I guess,' Shadrach said. 'I suppose you're right. The King of the Elves can't afford to let his kingdom down when it needs him most.'

He looked around, but the Elf soldier was already gone.

Shadrach walked up the highway, his mind full of doubts and wonderings. When he came to the first of the flat stone steps, he stopped.

'And the old oak tree is on Phineas's farm! What'll Phineas say?'

But he was the Elf King and the Trolls were moving in the hills. Shadrach stood listening to the rustle of the wind as it moved through the trees beyond the highway, and along the far slopes and hills.

Trolls? Were there really Trolls there, rising up, bold and confident in the darkness of the night, afraid of nothing, afraid of no one?

And this business of being Elf King . . .

Shadrach went on up the steps, his lips pressed tight. When he reached the top of the stone steps, the last rays of sunlight had already faded. It was night.

Phineas Judd stared out the window. He swore and shook his head. Then he went quickly to the door and ran out onto the porch. In the cold moonlight a dim figure was walking slowly across the lower field, coming toward the house along the cow trail.

'Shadrach!' Phineas cried. 'What's wrong? What are you doing out this time of night?'

Shadrach stopped and put his fists stubbornly on his hips.

'You go back home,' Phineas said. 'What's got into you?'

'I'm sorry, Phineas,' Shadrach answered. 'I'm sorry I have to go over your land. But I have to meet somebody at the old oak tree.'

'At this time of night?'

Shadrach bowed his head.

'What's the matter with you, Shadrach? Who in the world you going to meet in the middle of the night on my farm?'

'I have to meet with the Elves. We're going to plan out the war with the Trolls.'

'Well, I'll be damned,' Phineas Judd said. He went back inside the house and slammed the door. For a long time he stood thinking. Then he went back out on the porch again. 'What did you say you were doing? You don't have to tell me, of course, but I just—'

'I have to meet the Elves at the old oak tree. We must have a general council of war against the Trolls.'

'Yes, indeed. The Trolls. Have to watch for the Trolls all the time.'

'Trolls are everywhere,' Shadrach stated, nodding his head. 'I never realized it before. You can't forget them or ignore them. They never forget you. They're always planning, watching you—'

Phineas gaped at him, speechless.

'Oh, by the way,' Shadrach said. 'I may be gone for some time. It depends on how long this business is going to take. I haven't had much experience in fighting Trolls, so I'm not sure. But I wonder if you'd mind looking after the gas station for me, about twice a day, maybe once in the morning and once at night, to make sure no one's broken in or anything like that.'

'You're going away?' Phineas came quickly down the stairs. 'What's all this about Trolls? Why are you going?'

Shadrach patiently repeated what he had said.

'But what for?'

'Because I'm the Elf King. I have to lead them.'

There was silence. 'I see,' Phineas said, at last. 'That's right, you *did* mention it before, didn't you? But, Shadrach, why don't you come inside for a while and you can tell me about the Trolls and drink some coffee and—'

'Coffee?' Shadrach looked up at the pale moon above him, the moon and the bleak sky. The world was still and dead and the night was very cold and the moon would not be setting for some time.

Shadrach shivered.

'It's a cold night,' Phineas urged. 'Too cold to be out. Come on in—'

'I guess I have a little time,' Shadrach admitted. 'A cup of coffee wouldn't do any harm. But I can't stay very long . . .'

Shadrach stretched his legs out and sighed. 'This coffee sure tastes good, Phineas.'

Phineas sipped a little and put his cup down. The living room was quiet and warm. It was a very neat little living room with solemn pictures on the walls, gray uninteresting pictures that minded their own business. In the corner was a small reed organ with sheet music carefully arranged on top of it.

Shadrach noticed the organ and smiled. 'You still play, Phineas?'

'Not much any more. The bellows don't work right. One of them won't come back up.'

'I suppose I could fix it sometime. If I'm around, I mean.'

'That would be fine,' Phineas said. 'I was thinking of asking you.'

'Remember how you used to play "Vilia" and Dan Green came up with that lady who worked for Pop during the summer? The one who wanted to open a pottery shop?'

'I sure do,' Phineas said.

Presently, Shadrach set down his coffee cup and shifted in his chair.

'You want more coffee?' Phineas asked quickly. He stood up. 'A little more?'

'Maybe a little. But I have to be going pretty soon.'

'It's a bad night to be outside.'

Shadrach looked through the window. It was darker; the moon had almost gone down. The fields were stark. Shadrach shivered. 'I wouldn't disagree with you,' he said.

Phineas turned eagerly. 'Look, Shadrach. You go on home where it's warm. You can come out and fight Trolls some other night. There'll always be Trolls. You said so yourself. Plenty of time to do that later, when the weather's better. When it's not so cold.'

Shadrach rubbed his forehead wearily. 'You know, it all seems like some sort of a crazy dream. When did I start talking about Elves and Trolls? When did it all begin?' His voice trailed off. 'Thank you for the coffee.' He got slowly to his feet. 'It warmed me up a lot. And I appreciated the talk. Like old times, you and me sitting here the way we used to.'

'Are you going?' Phineas hesitated. '*Home?*'

'I think I better. It's late.'

Phineas got quickly to his feet. He led Shadrach to the door, one arm around his shoulder.

'All right, Shadrach, you go on home. Take a good hot bath before you go to bed. It'll fix you up. And maybe just a little snort of brandy to warm the blood.'

Phineas opened the front door and they went slowly down the porch steps, onto the cold, dark ground.

'Yes, I guess I'll be going,' Shadrach said.'Good night—'

'You go on home.' Phineas patted him on the arm. 'You run along home and take a good hot bath. And then go straight to bed.'

'That's a good idea. Thank you, Phineas. I appreciate your kindness.' Shadrach looked down at Phineas's hand on his arm. He had not been that close to Phineas for years.

Shadrach contemplated the hand. He wrinkled his brow, puzzled.

Phineas's hand was huge and rough and his arms were short. His fingers were blunt; his nails broken and cracked. Almost black, or so it seemed in the moonlight.

Shadrach looked up at Phineas. 'Strange,' he murmured.

'What's strange, Shadrach?'

In the moonlight, Phineas's face seemed oddly heavy and brutal. Shadrach had never noticed before how the jaw bulged, what a great protruding jaw it was. The skin was yellow and coarse, like parchment. Behind the glasses, the eyes were like two stones, cold and lifeless. The ears were immense, the hair stringy and matted.

Odd that he never noticed before. But he had never seen Phineas in the moonlight.

Shadrach stepped away, studying his old friend. From a few feet off, Phineas Judd seemed unusually short and squat. His legs were slightly bowed. His feet were enormous. And there was something else—

'What is it?' Phineas demanded, beginning to grow suspicious. 'Is there something wrong?'

Something was completely wrong. And he had never noticed it, not in all the years they had been friends. All around Phineas Judd was an odor, a faint, pungent stench of rot, of decaying flesh, damp and moldy.

Shadrach glanced slowly about him. 'Something wrong?' he echoed. 'No, I wouldn't say that.'

By the side of the house was an old rain barrel, half fallen apart. Shadrach walked over to it.

'No, Phineas. I wouldn't say there's something wrong.'

'What are you doing?'

'Me?' Shadrach took hold of one of the barrel staves and pulled it loose. He walked back to Phineas, carrying the barrel stave carefully. 'I'm King of the Elves. Who—or what—are you?'

Phineas roared and attacked with his great murderous shovel hands.

Shadrach smashed him over the head with the barrel stave. Phineas bellowed with rage and pain.

At the shattering sound, there was a clatter and from underneath the house came a furious horde of bounding, leaping creatures, dark bent-over things, their bodies heavy and squat, their feet and heads immense. Shadrach took one look at the flood of dark creatures pouring out from Phineas's basement. He knew what they were.

'Help!' Shadrach shouted. 'Trolls! Help!'

The trolls were all around him, grabbing hold of him, tugging at him, climbing up him, pummeling his face and body.

Shadrach fell to with the barrel stave, swung again and again, kicking Trolls with his feet, whacking them with the barrel stave. There seemed to be hundreds of them. More and more poured out from under Phineas's house, a surging black tide of pot-shaped creatures, their great eyes and teeth gleaming in the moonlight.

'Help!' Shadrach cried again, more feebly now. He was getting winded. His heart labored painfully. A Troll bit his wrist, clinging to his arm. Shadrach flung it away, pulling loose from the horde clutching his trouser legs, the barrel stave rising and falling.

One of the Trolls caught hold of the stave. A whole group of them helped, wrenching furiously, trying to pull it away. Shadrach hung on desperately. Trolls were all over him, on his shoulders, clinging to his coat, riding his arms, his legs, pulling his hair—

He heard a high-pitched clarion call from a long way off, the sound of some distant golden trumpet, echoing in the hills.

The Trolls suddenly stopped attacking. One of them dropped off Shadrach's neck. Another let go of his arm.

The call came again, this time more loudly.

'Elves!' a Troll rasped. He turned and moved toward the sound, grinding his teeth and spitting with fury.

'Elves!'

The Trolls swarmed forward, a growing wave of gnashing teeth and nails, pushing furiously toward the Elf columns. The Elves broke formation and joined battle, shouting with wild joy in their shrill, piping voices. The tide of Trolls rushed against them, Troll against Elf, shovel nails against golden sword, biting jaw against dagger.

'Kill the Elves!'

'Death to the Trolls!'

'Onward!'

'Forward!'

Shadrach fought desperately with the Trolls that were still clinging to him. He was exhausted, panting and gasping for breath. Blindly, he whacked on and on, kicking and jumping, throwing Trolls away from him, through the air and across the ground.

How long the battle raged, Shadrach never knew. He was lost in a sea of dark bodies, round and evil-smelling, clinging to him, tearing, biting, fastened to his nose and hair and fingers. He fought silently, grimly.

All around him, the Elf legions clashed with the Troll horde, little groups of struggling warriors on all sides.

Suddenly Shadrach stopped fighting. He raised his head, looking uncertainly around him. Nothing moved. Everything was silent. The fighting had ceased.

A few Trolls still clung to his arms and legs. Shadrach whacked one with the barrel stave. It howled and dropped to the ground. He staggered back, struggling with the last Troll, who hung tenaciously to his arm.

'Now you!' Shadrach gasped. He pried the Troll loose and flung it into the air. The Troll fell to the ground and scuttled off into the night.

There was nothing more. No Troll moved anywhere. All was silent across the bleak moon-swept fields.

Shadrach sank down on a stone. His chest rose and fell painfully. Red specks swam before his eyes. Weakly, he got out his pocket handkerchief and wiped his neck and face. He closed his eyes, shaking his head from side to side.

When he opened his eyes again, the Elves were coming toward him, gathering their legion together again. The Elves were disheveled and bruised. Their golden armor was gashed and torn. Their helmets were bent or missing. Most of their scarlet plumes were gone. Those that still remained were drooping and broken.

But the battle was over. The war was won. The Troll hordes had been put to flight.

Shadrach got slowly to his feet. The Elf warriors stood around him in a circle, gazing up at him with silent respect. One of them helped steady him as he put his handkerchief away in his pocket.

'Thank you,' Shadrach murmured. 'Thank you very much.'

'The Trolls have been defeated,' an Elf stated, still awed by what had happened.

Shadrach gazed around at the Elves. There were many of them, more than he had ever seen before. All the Elves had turned out for the battle. They were grim-faced, stern with the seriousness of the moment, weary from the terrible struggle.

'Yes, they're gone, all right,' Shadrach said. He was beginning to get his breath. 'That was a close call. I'm glad you fellows came when you did. I was just about finished, fighting them all by myself.'

'All alone, the King of the Elves held off the entire Troll army,' an Elf announced shrilly.

'Eh?' Shadrach said, taken aback. Then he smiled. 'That's true, I *did* fight them alone for a while. I *did* hold off the Trolls all by myself. The whole darn Troll army.'

'There is more,' an Elf said.

Shadrach blinked. 'More?'

'Look over here, O King, mightiest of all the Elves. This way. To the right.'

The Elves led Shadrach over.

'What is it?' Shadrach murmured, seeing nothing at first. He gazed down, trying to pierce the darkness. 'Could we have a torch over here?'

Some Elves brought little pine torches.

There, on the frozen ground, lay Phineas Judd, on his back. His eyes were blank and staring, his mouth half open. He did not move. His body was cold and stiff.

'He is dead,' an Elf said solemnly.

Shadrach gulped in sudden alarm. Cold sweat stood out abruptly on his forehead. 'My gosh! My old friend! What have I done?'

'You have slain the Great Troll.'

Shadrach paused.

'I *what?*'

'You have slain the Great Troll, leader of all the Trolls.'

'This has never happened before,' another Elf exclaimed excitedly. 'The Great Troll has lived for centuries. Nobody imagined he could die. This is our most historic moment.'

All the Elves gazed down at the silent form with awe, awe mixed with more than a little fear.

'Oh, go on!' Shadrach said. 'That's just Phineas Judd.'

But as he spoke, a chill moved up his spine. He remembered what he had seen a little while before, as he stood close by Phineas, as the dying moonlight crossed his old friend's face.

'Look.' One of the Elves bent over and unfastened Phineas's blue-serge vest. He pushed the coat and vest aside. 'See?'

Shadrach bent down to look.

He gasped.

Underneath Phineas Judd's blue-serge vest was a suit of mail, an encrusted mesh of ancient, rusting iron, fastened tightly around the squat body. On the mail stood an engraved insignia, dark and time-worn, embedded with dirt and rust. A moldering half-obliterated emblem. The emblem of a crossed owl leg and toadstool.

The emblem of the Great Troll.

'Golly,' Shadrach said. 'And *I* killed him.'

For a long time he gazed silently down. Then, slowly, realization began to grow in him. He straightened up, a smile forming on his face.

'What is it, O King?' an Elf piped.

'I just thought of something,' Shadrach said. 'I just realized that—that since the Great Troll is dead and the Troll army has been put to flight—'

He broke off. All the Elves were waiting.

'I thought maybe I—that is, maybe if you don't need me any more—'

The Elves listened respectfully. 'What is it, Mighty King? Go on.'

'I thought maybe now I could go back to the filling station and not be king any more.' Shadrach glanced hopefully around at them. 'Do you think so? With the war over and all. With him dead. What do you say?'

For a time, the Elves were silent. They gazed unhappily down at the ground. None of them said anything. At last they began moving away, collecting their banners and pennants.

'Yes, you may go back,' an Elf said quietly. 'The war is over. The

Trolls have been defeated. You may return to your filling station, if that is what you want.'

A flood of relief swept over Shadrach. He straightened up, grinning from ear to ear. 'Thanks! That's fine. That's really fine. That's the best news I've heard in my life.'

He moved away from the Elves, rubbing his hands together and blowing on them.

'Thanks an awful lot.' He grinned around at the silent Elves. 'Well, I guess I'll be running along, then. It's late. Late and cold. It's been a hard night. I'll—I'll see you around.'

The Elves nodded silently.

'Fine. Well, good night.' Shadrach turned and started along the path. He stopped for a moment, waving back at the Elves. 'It was quite a battle, wasn't it? We really licked them.' He hurried on along the path. Once again he stopped, looking back and waving. 'Sure glad I could help out. Well, good night!'

One or two of the Elves waved, but none of them said anything.

Shadrach Jones walked slowly toward his place. He could see it from the rise, the highway that few cars traveled, the filling station falling to ruin, the house that might not last as long as himself, and not enough money coming in to repair them or buy a better location.

He turned around and went back.

The Elves were still gathered there in the silence of the night. They had not moved away.

'I was hoping you hadn't gone,' Shadrach said, relieved.

'And we were hoping you would not leave,' said a soldier.

Shadrach kicked a stone. It bounced through the tight silence and stopped. The Elves were still watching him.

'Leave?' Shadrach asked. 'And me King of the Elves?'

'Then you will remain our king?' an Elf cried.

'It's a hard thing for a man of my age to change. To stop selling gasoline and suddenly be a king. It scared me for a while. But it doesn't any more.'

'You will? You *will*?'

'Sure,' said Shadrach Jones.

The little circle of Elf torches closed in joyously. In their light, he saw a platform like the one that had carried the old King of the Elves. But this one was much larger, big enough to hold a man, and dozens of the soldiers waited with proud shoulders under the shafts.

A soldier gave him a happy bow. 'For you, Sire.'

Shadrach climbed aboard. It was less comfortable than walking, but he knew this was how they wanted to take him to the Kingdom of the Elves.

1953

NAOMI MITCHISON

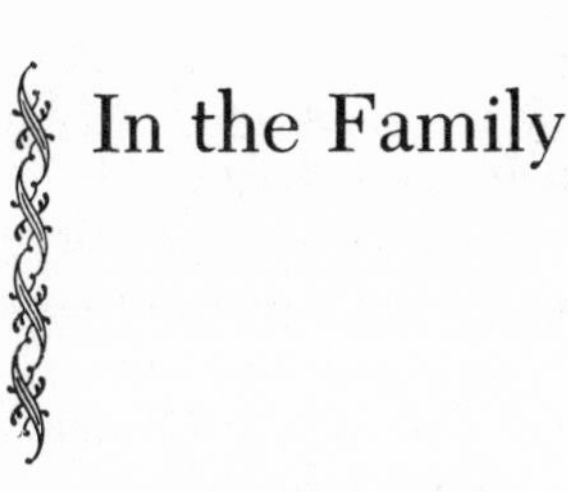

In the Family

It was in the family to be seeing things that are not meant to be seen. And it was not nice for them, not at all. They could have done without seeing the most of what they saw. Mostly all of them had seen the lights, one way or another, and his mother had seen the funeral itself coming round the head of the glen, and fine she knew the bearers, and the corpse she could guess, and sure enough that was how it was before the week was out, but she herself was sick in her bed for the whole day after it.

If you go far enough back, there was his grandfather's uncle that was a great piper, and he was coming back over the hills from a wedding at the far side. It was a good-going wedding, the way they were in those days, and it had lasted all of four nights and they would not have been sparing of the whisky for their pipers. And as he was coming back by Knocnashee, the fairies came out from the hill and asked would he come in a whilie and pipe for them. And he would have done it sure enough, for the whisky had made him as bold as a robin, but he was just too drunk to do justice to his own pipes, so he said he would come to them another day. He never did, and indeed, he died in his bed. But he was in a terrible fear all his life afterwards in case they would mind on what he had said. And there was a kind of fear on all the family, in case it might be remembered on one of them, and he would need to keep the promise of the old piper.

His father had a young sister, Janet, a terrible bonny lassie she was, with a high colour on her. But there was a thing went wrong and the doctor could not say right what it was, so he was for sending her to the hospital in Glasgow. Well then, the evening before she was to go, she was walking up by the old well, and who should she see but the fairy woman. And the fairy woman warned her terrible not to go to the hospital, walking beside her all the while without movement of her feet, the way it is with the fairy folk. And the lassie came back and she was gey put about and she said she would not go. So they sent for the doctor to tell him the thing, but he was wild affronted and said he had it all arranged and go she must or the great doctors in Glasgow would have his head off. So the lassie grat and her mother grat, but the end of it was they could not go against the doctor, and the poor lassie went to the hospital and died there.

So the rest of them said to themselves that if ever they were to get a warning they would bide by it. But for long enough it did not come. His father, Donald MacMillan, had a wee glimpse of the fairies one time; he was under-keeper at the Castle and he was coming back in the early morning from setting his traps on the hill and as he came down by the back road the fairies were riding round the Castle, wee dark folk on white ponies with a glitter of gold on the bridle. But he held his tongue about it, for it is not an under-keeper's place to be seeing the fairies belonging to the Castle ones. And he did not like it, no, not at all, and after that he volunteered for the Army, and it was ten years before he came back.

But Angus himself had seen not a thing. And he was apt to say there was nothing at all in it, and when his sister Effie came back from a dance one night in a terrible fright and saying that a ghost had walked with her all the way from the cross-roads, he was not believing her. For the thing is this. You may know well enough that something has happened. But, gin the Kirk is against it, and the schools, and the newspapers and the wireless forby, you will find it hard enough to believe your own eyes and ears. And the other thing is, that, with yon kind of sights, it is never the same twice. I will see a thing one way and you will see it another, but maybe it is the same in itself. Yet it might be the Ministers are in the right of it and we should not be speaking of such things at all.

But, however that may be, Angus was a well-doing laddie. He was driving one of the forestry lorries and he cared for his lorry the same as it was a horse. He was for ever cleaning and oiling it and putting in a wee thing here or there and never once did he have any accident on the road. And he was a church member and in the choir and he was not drinking scarcely at all except maybe now and then, and he was going with a lassie in the choir too, and she was in the alto line and he was in the tenor. And not one of the family had seen an unchancy thing for years now, and his sister Effie married and with three bairns and indeed she was inclined to laugh at the story of her own ghost now. For it had mistaken her surely, the poor thing, since all it had been after from her was the lend of a horse and cart, and it turned out afterwards that it was an old soul from one of the crofts at the head of the glen and he had died more than a year back and terrible put about over this same thing, and so he had needed to come back.

He had the Gaelic, had Angus, not so as to read excepting it might be some easy kind of song, but well enough to speak a bit and to know what was said to him. This was mostly because his father and mother used to be speaking the Gaelic across the table the way the bairns would not be understanding them, and Angus was that angry at this when he was a wee fellow that he started to learn the Gaelic out of pure devilment to know what his elders were after saying.

Well, the choir had been practising all winter and the time came they were to go over and give a concert in aid of the church funds at Auchandrum, and there was to be a dance after. They would hire a bus and it would be just fine, with all the folk singing away to pass the time on the road and going back with the grand chances there would be in the dark with the lassies, and indeed Angus was mostly sure that Peigi MacLean—for that was his lassie's name—would be sitting beside him at the back, though she would not promise.

Then it came on to two days before, and Angus had finished up and parked his lorry at the saw mill with a tarpaulin over her, and he was taking the short cut home through the larches and by the old well that nobody went to now, since there was a fine new County Council water supply at that side of the village and some of the houses with bathrooms even. There was a woman sitting by the side of the well and at first he thought she was a summer visitor and maybe a painter or that, since she had a long green cloak of an old-fashioned kind that our own women-folk would never be wearing. But as he came nearer she stood, and he saw that she was going to speak to himself and a kind of uneasiness came at him and that got suddenly worse when she spoke to him by name and that in the Gaelic. So he got a grip on the monkey wrench that he was taking back to the house and he answered her and she began to walk by his side and he saw that there was no movement of her feet under the green cloak and for a while he could not anyways listen to what she was saying, the way the blood was pouring in spate through him and the sweat standing out on his forehead and the greatest fear at him lest she should reach out her hand and touch his own.

But after a while it came to him that she was speaking about the choir and their journey to Auchandrum and she was warning him, warning him, that he must not, above all, let Peigi go on the bus, or there would be no more happiness for him all his life long. And he tried wild to speak, but he could not get his mouth round the words for a time. At last he said in the Gaelic: 'How are you telling me this, woman of the Sidhe?'

She said: 'It is because of the friendship between ourselves and your grandfather's uncle that was a good piper and a kindly man, drunk or sober, and I am telling you for your great good, Angus, son of Donald.'

So he said: 'How will I know if it is true?'

She smiled, and some way it was a sweet smile, but far off, since she had the face and body of a young lassie, yet her look was that of an old, old woman, beyond love or hate. She said: 'I will show a sign on the thing you hold in your hand, and believe you me, Angus!'

And he looked full into her eyes, for the fear was beginning to leave him, but as he did that he began to see the young larches behind her on the hillside, and in a short while only it was clear through her he was looking and nothing at all to show where she had been, only the sweat cold on his face. So he went home and said nothing, and his mother brought in the tea, and suddenly she laughed and asked him what kind of job at all had he been doing on his lorry, for there was the monkey wrench lying over on the press, and it wreathed round with the bonny wild honeysuckle.

Well, he thought and he better thought, and he worked the talk round until his mother began to speak of her good-sister Janet, who had not been let to take the warning, but had died in the Glasgow hospital, and he thought it would be a wild thing if the like of thon were to happen to his own lassie through his fault. So when he had washed he went over to the choir practice and there sure enough was Peigi. So on the way back he was speaking to her, begging her not to go to the concert. But she was not listening to him at all, for she thought he was on for some kind of devilment and she said good-night to him, kind of sharp.

Well, the next day he was hashing and fashing away at it, and in the evening again he went over to Peigi's folks' house and he got the talk round to warnings, so that she would be prepared, and then, as she was seeing him to the gate, he told her how it had been. Well, she was terrible put about and at first she was not believing him, and then she was, but she had her dance dress washed and ironed and what would the rest of the lasses say? And there could be nothing in it and nobody believed in such blethers nowadays. And the more Angus pled and swat, the more reasons Peigi was finding not to be heeding him at all and the more she needed to say to herself that he had maybe been drinking and if she were to do what he asked, all the folk would be speaking of it and she was not to be made a clown of by Angus or any other lad, and it would be a great dance after the concert and was Angus not coming to be her partner?

'If you go, I will go, Peigi,' said Angus, 'for if there is a danger coming to you, I would soonest be there.'

So he went to his bed and he tried to tell himself that none of this was real; but it was beyond him not to believe, and indeed in his sleeping he was seeing the fairy woman just as plain, and she speaking to him again. And in the morning it was on him wild to keep Peigi from the concert, yet all the forenoon he could not think of a way. But towards four he was taking back the lorry empty to the sawmill and he saw Peigi on the road going for a message for her mother. He cried on her to come up beside him and he would give her a lift along, and up she came. But when he got to the cross-roads he put on speed and swung his lorry round and up the glen. 'What in God's name are you up to, Angus!' said Peigi and held on tight.

'I am taking you away from the concert,' said Angus, and he did not look at her, but only at the road ahead of him.

'Well then, I will scream!' said Peigi.

'It will not help you any,' said Angus, 'and maybe least said is soonest mended. And indeed I am terrible sorry, Peigi, but it is on me to do this. And you had best not try to snatch at the steering wheel, Peigi, for I would need to hit you and I amna wanting to.'

So Peigi sat as far as she could from him in the cab of the lorry, and they never passed another car, but only an old farm cart or so, and Angus hooted and put on all speed and there was nothing she could do, and by and by she cried a bittie because she was thinking of the dance dress and the nice evening she would be missing, and what in all the world could she say to the rest of the choir?

But Angus was watching his petrol gauge. He was ten miles out of the village now, and he swung round onto a side road that went up past the common grazing towards some old quarries. There would be no chance of Peigi hailing a car and getting a lift back on a road the like of this. And they went bumping up along the old tracks and it was near six and the bus would be starting for Auchandrum in half an hour. At last he stopped the lorry at the mouth of one of the quarries and he said: 'I am that spited, Peigi, but I know well I am in the right of it.'

Peigi said nothing and when he tried to come near her she snatched herself away. So there they sat and no pleasure in it for either of them, but well they knew what would be said of them in the village. And forby that, Angus was thinking how it would be when he did not bring his lorry back to the sawmill. He liked his job with the forestry fine, but this way he would be leaving it with no good character. And the more he thought, the blacker things looked ahead of him, and sudden he began to wonder if the fairy woman had played a trick on him. And it was a terrible thought, yon.

So he started up his engine again and he backed and turned and came down cannily in second and Peigi beside him as cross as a sack of weasels. 'Will you no' speak to me at all, Peigi?' he asked, as they came out into the glen road and down towards the village.

'I will never speak to you again!' she said, and a terrible hurt feeling came over his heart and he hated the fairy woman and the warning and all the two days and nights of it. So that way they came down to the village. 'Do you not take me to the house!' said Peigi suddenly, 'leave you me here and maybe—och, maybe they'll not know and there could be time to think of a thing to say!'

He stopped the lorry and jumped out to help her down, but she had jumped clear, and when he came to her she gave a cry and started to run along the road. So he did not follow until she was out of sight and he knew she would take the back road by old Donnie's hen house.

As he came down to the post office he saw a ring of folk round the telephone kiosk and his own brother cried on him to stop and he jumped out. 'Is it yourself, Angus!' his brother shouted and seized him by the two hands and there were tears streaming out of his eyes and then he said: 'Did Peigi go with the choir?'

'She did not,' said Angus and there was her father shaking him by the hands as well, and it all came out there had been an accident to the bus and three or four folk hurt bad and mostly all cut with the glass and they had just heard it on the telephone, but they werna just sure who was on it. And Angus found himself going off into a wild silly kind of daft laughing and he could not stop himself any. So he got back into the lorry and drove off to the saw mill and parked her, and then he started to clean her and polish her, for he did not know what would the forester say, and he might be getting his books and he might never be back in the cab of his lorry.

Sure enough, while he was at it, up came the forester, for he had been terrible put about when Angus MacMillan never brought in the lorry, and he mostly so dependable, and the forester had his own bosses over him. 'So this is you at last!' said the forester. 'And what have you to say?'

But Angus found no words for he could not begin to speak of a fairy woman. At last he kicked at the tyre of the lorry and said: 'I am terrible sorry, Duncan, and I have not hurt her any, and—and I will pay for the petrol.'

'I will need to think what to do about this,' said the forester, 'for I cannot have such a carry-on with my lorries. I will see you in the morning.' So Angus said never a word, but home he went, and his own folks speaking about nothing but the accident and the lucky he was to be out of it. They knew that Peigi MacLean had not been in it, but they did not know yet that she had been with him. One of his chums had a leg broken, and another was bleeding from his inside and the doctor up with him, and Peigi's young sister had a terrible nasty cut on her head and it was a wonder nobody was killed. If Peigi had been sitting beside her sister on the seat nearest the window the glass would have broken all over her face. Angus went off to his bed and he was terrible tired, but some way he could not sleep. And at least, he said to himself, it was a true warning, and I took it, and the fairy has not tricked me. So he got out of his bed and felt for his flash, and he walked around in his shirt looking for any one thing that was worth giving as a gift. And there seemed to be nothing. For what would a fairy do with the set of cuff links he won at the Nursing Association raffle or the printed letter that had come with his Welcome Home money or the bottle of port he was saving up for Effie's new bairn's christening, or his good boots, even? But at last he came on a thing and it was a wee kind of medal that his father had worn at his watch chain and his father before him, and it might have belonged to the piper for that matter. He did not know was it silver or not and he gave it a bit rub on his sleeve. And he went out in his bare feet, and his shirt blowing round his legs, and up to the well, and the queer thing was he kind of half wanted to see the fairy woman, but there was no breath at all of her. So he threw the wee medal into the well and he stood and he said a kind of half prayer, and he came stumbling back down the path with the stones sore on his feet, but after that he slept as quiet as a herring in a barrel.

But the next morning he went down to the saw mill and the forester sent for him to his wee office with the papers and the telephone and the bundles of axes and saw blades. But still he could say nothing, and after a time the forester said in an angry kind of way that if he had no excuse then he was sacked. But at that Angus had to speak and out it came, the whole story, and half-way through the forester got up and shut the door of his office. At the end he said: 'There is folks that would shut you up in the asylum for the like of yon, Angus.'

'But it is not touched I am!' said Angus anxiously.

'I know plenty that would think it,' said the forester. 'So it was a green cloak she was wearing, yon one? Aye, aye. It is fortunate altogether that you have the Gaelic. But I am wondering wild what would the Minister say.'

'You will not tell him, Duncan!' said Angus. But he knew that the Minister was no great favourite with the forester, on account that he had complained that the noise of the saw-mill was stopping him from composing his sermons, and everyone knowing that the half of them came out of books.

The forester said nothing to this, and Angus stood first on one foot and then on the other. At last the forester said. 'Are you for marrying Peigi?'

'I am, surely,' said Angus, 'but I am no' just so sure of herself. I amna sure if she will speak to me, even.'

'Well,' said the forester, 'you had better be asking her and that way you will keep this story in the one family and maybe it will stay quietest so, for it doesna do to be speaking on such things.'

'I just darena ask her and that is the truth,' said Angus.

'Well,' said the forester, 'if you are marrying Peigi MacLean I will overlook this and keep you on, but if you havena the courage, then you can take your books and off with you and best if you go away out of the place altogether.' And Angus went stumbling out the door of the wee office, and the forester took two new axe heads out of the store and checked them off.

Word had got about of how Angus and Peigi had passed the evening of the concert, and Angus getting plenty from his mates of what could be done with a nice lassie in the cab of a lorry or in the back of it even, and they all thinking it was for badness he had taken her away and kind of half proud of him. But the accident to the bus had more to it, so the talk went over to that, and glad enough was Angus.

After his tea he went over to the MacLean's house, and there was father MacLean staking his peas, and Angus asked after Peigi's sister, and the district nurse was in with her, and then he spoke of the great growth on the peas. At last old MacLean said: 'What is this I hear about yourself and Peigi, Angus MacMillan?'

Angus said: 'It is true enough she came for a drive in my lorry, but it wasna for anything bad.'

'You are telling me that,' said old MacLean and he began to fill his pipe, 'but you were away for three hours and what could you find to be speaking about? So I must ask you, what were you doing?'

'I was driving the lorry mostly,' said Angus, 'and Peigi wasna speaking to me, she was that cross.'

'Then I will ask you, why did you drive the lorry with my daughter in it?'

Then Angus cleared his throat and said in a kind of loud voice: 'I had a warning that Peigi was not to go with the choir.'

Mr MacLean said nothing for a time, but scraped his boot on the edge of a spade and puffed away at his pipe, and at last he said: 'I was hearing just that. And she would have been killed, likely, if she had been in the bus.'

'I was thinking so,' said Angus in a half whisper. And then the District Nurse came out the house and said cheerily that the lassie was going on fine, and now she must be off to the others and what a carry-on she was having with them all. And she started her wee car and drove off.

'Well, well,' said Mr MacLean, 'it is in your family to be seeing things. And things are mostly unchancy. But this time it was the great chance for my own lassie. Aye, aye. You will be wanting to see Peigi, likely?'

'It is what I would like most in the world,' said Angus, 'but—'

'Well, come you in,' said Mr MacLean, 'and we will say no more over this matter of the warning.' They went into the house together. Peigi was there, washing her sister's frock that had been spoilt a bit. Her father said: 'I will go through and see your sister.' Peigi went on slapping and scrubbing away at the dress, letting on she did not see there was a soul there.

After a time Angus said: 'Are you speaking to me, Peigi?'

'Ach well—' she said, 'I might.'

'Duncan was at me over taking the lorry,' said Angus.

'Was he now? And I am sure he was quite right. You should not be taking things without leave.'

'He spoke the way he might be giving me my books, Peigi.'

Peigi let the frock fall back into the lather. 'The dirty clown! How would he do that on you?'

'Well,' said Angus, 'he was only kind of half believing me, maybe. If he were to believe me right I am thinking there would be nothing said. Ach, Peigi, he would believe me if he thought I had done it because I loved you true! He was asking were you and I to be married, Peigi. But I was saying I daredna ask you.'

'How?' she said.

'Och, well, if you hadna been speaking to me, it would have been kind of difficult, Peigi.'

Peigi was flicking the suds off her fingers. She said into the air: 'Isn't it wild now, to think of the nice time we could have been having on the hills and we just ourselves, if only you had explained the thing right to me at the first.'

'Indeed and I did my best!' said Angus.

'Aye, but I was not knowing then that the bus would have this accident, and how would anyone with a grain of sense believe you,

Angus? But we will not speak any more of this warning nor anything to do with it, because I am thinking we will have plenty else to speak about.'

1957

BERNARD MALAMUD

The Jewbird

THE window was open so the skinny bird flew in. Flappity-flap with its frazzled black wings. That's how it goes. It's open, you're in. Closed, you're out and that's your fate. The bird wearily flapped through the open kitchen window of Harry Cohen's top-floor apartment on First Avenue near the lower East River. On a rod on the wall hung an escaped canary cage, its door wide open, but this black-type longbeaked bird—its ruffled head and small dull eyes, crossed a little, making it look like a dissipated crow—landed if not smack on Cohen's thick lamb chop, at least on the table, close by. The frozen foods salesman was sitting at supper with his wife and young son on a hot August evening a year ago. Cohen, a heavy man with hairy chest and beefy shorts; Edie, in skinny yellow shorts and red halter; and their ten-year-old Morris (after her father)—Maurie, they called him, a nice kid though not overly bright—were all in the city after two weeks out, because Cohen's mother was dying. They had been enjoying Kingston, New York, but drove back when Mama got sick in her flat in the Bronx.

'Right on the table,' said Cohen, putting down his beer glass and swatting at the bird. 'Son of a bitch.'

'Harry, take care with your language,' Edie said, looking at Maurie, who watched every move.

The bird cawed hoarsely and with a flap of its bedraggled wings—feathers tufted this way and that—rose heavily to the top of the open kitchen door, where it perched staring down.

'Gevalt, a pogrom!'

'It's a talking bird,' said Edie in astonishment.

'In Jewish,' said Maurie.

'Wise guy,' muttered Cohen. He gnawed on his chop, then put down the bone. 'So if you can talk, say what's your business. What do you want here?'

'If you can't spare a lamb chop,' said the bird, 'I'll settle for a piece of herring with a crust of bread. You can't live on your nerve forever.'

'This ain't a restaurant,' Cohen replied. 'All I'm asking is what brings you to this address?'

'The window was open,' the bird sighed; adding after a moment, 'I'm running. I'm flying but I'm also running.'

'From whom?' asked Edie with interest.

'Anti-Semeets.'

'Anti-Semites?' they all said.

'That's from who.'

'What kind of anti-Semites bother a bird?' Edie asked.

'Any kind,' said the bird, 'also including eagles, vultures, and hawks. And once in a while some crows will take your eyes out.'

'But aren't you a crow?'

'Me? I'm a Jewbird.'

Cohen laughed heartily. 'What do you mean by that?'

The bird began dovening. He prayed without Book or tallith, but with passion. Edie bowed her head though not Cohen. And Maurie rocked back and forth with the prayer, looking up with one wide-open eye.

When the prayer was done Cohen remarked, 'No hat, no phylacteries?'

'I'm an old radical.'

'You're sure you're not some kind of a ghost or dybbuk?'

'Not a dybbuk,' answered the bird, 'though one of my relatives had such an experience once. It's all over now, thanks God. They freed her from a former lover, a crazy jealous man. She's now the mother of two wonderful children.'

'Birds?' Cohen asked slyly.

'Why not?'

'What kind of birds?'

'Like me. Jewbirds.'

Cohen tipped back in his chair and guffawed. 'That's a big laugh. I've heard of a Jewfish but not a Jewbird.'

'We're once removed.' The bird rested on one skinny leg, then on the other. 'Please, could you spare maybe a piece of herring with a small crust of bread?'

Edie got up from the table.

'What are you doing?' Cohen asked her.

'I'll clear the dishes.'

Cohen turned to the bird. 'So what's your name, if you don't mind saying?'

'Call me Schwartz.'

'He might be an old Jew changed into a bird by somebody,' said Edie, removing a plate.

'Are you?' asked Harry, lighting a cigar.

'Who knows?' answered Schwartz. 'Does God tell us everything?'

Maurie got up on his chair. 'What kind of herring?' he asked the bird in excitement.

'Get down, Maurie, or you'll fall,' ordered Cohen.

'If you haven't got matjes, I'll take schmaltz,' said Schwartz.

'All we have is marinated, with slices of onion—in a jar,' said Edie.

'If you'll open for me the jar I'll eat marinated. Do you have also, if you don't mind, a piece of rye bread—the spitz?'

Edie thought she had.

'Feed him out on the balcony,' Cohen said. He spoke to the bird. 'After that take off.'

Schwartz closed both bird eyes. 'I'm tired and it's a long way.'

'Which direction are you headed, north or south?'

Schwartz, barely lifting his wings, shrugged.

'You don't know where you're going?'

'Where there's charity I'll go.'

'Let him stay, papa,' said Maurie. 'He's only a bird.'

'So stay the night,' Cohen said, 'but no longer.'

In the morning Cohen ordered the bird out of the house but Maurie cried, so Schwartz stayed for a while. Maurie was still on vacation from school and his friends were away. He was lonely and Edie enjoyed the fun he had, playing with the bird.

'He's no trouble at all,' she told Cohen, 'and besides his appetite is very small.'

'What'll you do when he makes dirty?'

'He flies across the street in a tree when he makes dirty, and if nobody passes below, who notices?'

'So all right,' said Cohen, 'but I'm dead set against it. I warn you he ain't gonna stay here long.'

'What have you got against the poor bird?'

'Poor bird, my ass. He's a foxy bastard. He thinks he's a Jew.'

'What difference does it make what he thinks?'

'A Jewbird, what a chuzpah. One false move and he's out on his drumsticks.'

At Cohen's insistence Schwartz lived out on the balcony in a new wooden birdhouse Edie had bought him.

'With many thanks,' said Schwartz, 'though I would rather have a human roof over my head. You know how it is at my age. I like the warm, the windows, the smell of cooking. I would also be glad to see once in a while the *Jewish Morning Journal* and have now and then a schnapps because it helps my breathing, thanks God. But whatever you give me, you won't hear complaints.'

However, when Cohen brought home a bird feeder full of dried corn, Schwartz said, 'Impossible.'

Cohen was annoyed. 'What's the matter, crosseyes, is your life getting too good for you? Are you forgetting what it means to be migratory? I'll bet a helluva lot of crows you happen to be acquainted with, Jews or otherwise, would give their eyeteeth to eat this corn.'

Schwartz did not answer. What can you say to a grubber yung?

'Not for my digestion,' he later explained to Edie. 'Cramps. Herring is better even if it makes you thirsty. At least rainwater don't cost anything.' He laughed sadly in breathy caws.

And herring, thanks to Edie, who knew where to shop, was what Schwartz got, with an occasional piece of potato pancake, and even a bit of soupmeat when Cohen wasn't looking.

When school began in September, before Cohen would once again suggest giving the bird the boot, Edie prevailed on him to wait a little while until Maurie adjusted.

'To deprive him right now might hurt his school work, and you know what trouble we had last year.'

'So okay, but sooner or later the bird goes. That I promise you.'

Schwartz, though nobody had asked him, took on full responsibility for Maurie's performance in school. In return for favors granted, when he was let in for an hour or two at night, he spent most of his time overseeing the boy's lessons. He sat on top of the dresser near Maurie's desk as he laboriously wrote out his homework. Maurie was a restless type and Schwartz gently kept him to his studies. He also listened to him practice his screechy violin, taking a few minutes off now and then to rest his ears in the bathroom. And they afterwards played dominoes. The boy was an indifferent checker player and it was impossible to teach him chess. When he was sick, Schwartz read him comic books though he personally disliked them. But Maurie's work improved in school and even his violin teacher admitted his playing was better. Edie gave Schwartz credit for these improvements though the bird pooh-poohed them.

Yet he was proud there was nothing lower then C minuses on Maurie's report card, and on Edie's insistence celebrated with a little schnapps.

'If he keeps up like this,' Cohen said, 'I'll get him in an Ivy League college for sure.'

'Oh I hope so,' sighed Edie.

But Schwartz shook his head. 'He's a good boy—you don't have to worry. He won't be a shicker or a wifebeater, God forbid, but a scholar he'll never be, if you know what I mean, although maybe a good mechanic. It's no disgrace in these times.'

'If I were you,' Cohen said, angered, 'I'd keep my big snoot out of other people's private business.'

'Harry, please,' said Edie.

'My goddamn patience is wearing out. That crosseyes butts into everything.'

Though he wasn't exactly a welcome guest in the house, Schwartz gained a few ounces although he did not improve in appearance. He looked bedraggled as ever, his feathers unkempt, as though he had just flown out of a snowstorm. He spent, he admitted, little time taking care of himself. Too much to think about. 'Also outside plumbing,' he told Edie. Still there was more glow to his eyes so that though Cohen went on calling him crosseyes he said it less emphatically.

Liking his situation, Schwartz tried tactfully to stay out of Cohen's way, but one night when Edie was at the movies and Maurie was taking a hot shower, the frozen foods salesman began a quarrel with the bird.

'For Christ sake, why don't you wash yourself sometimes? Why must you always stink like a dead fish?'

'Mr Cohen, if you'll pardon me, if somebody eats garlic he will smell from garlic. I eat herring three times a day. Feed me flowers and I will smell like flowers.'

'Who's obligated to feed you anything at all? You're lucky to get herring.'

'Excuse me, I'm not complaining,' said the bird. 'You're complaining.'

'What's more,' said Cohen, 'even from out on the balcony I can hear you snoring away like a pig. It keeps me awake at night.'

'Snoring,' said Schwartz, 'isn't a crime, thanks God.'

'All in all you are a goddamn pest and free loader. Next thing you'll want to sleep in bed next to my wife.'

'Mr Cohen,' said Schwartz, 'on this rest assured. A bird is a bird.'

'So you say, but how do I know you're a bird and not some kind of a goddamn devil?'

'If I was a devil you would know already. And I don't mean because your son's good marks.'

'Shut up, you bastard bird,' shouted Cohen.

'Grubber yung,' cawed Schwartz, rising to the tips of his talons, his long wings outstretched.

Cohen was about to lunge for the bird's scrawny neck but Maurie came out of the bathroom, and for the rest of the evening until Schwartz's bedtime on the balcony, there was pretended peace.

But the quarrel had deeply disturbed Schwartz and he slept badly. His snoring woke him, and awake, he was fearful of what would become of him. Wanting to stay out of Cohen's way, he kept to the birdhouse as much as possible. Cramped by it, he paced back and forth on the balcony ledge, or sat on the birdhouse roof, staring into space. In the evenings, while overseeing Maurie's lessons, he often fell asleep. Awakening, he nervously hopped around exploring the four corners of the room. He spent much time in Maurie's closet, and carefully examined his bureau drawers when they were left open. And once when he found a large paper bag on the floor, Schwartz poked his way into it to investigate what possibilities were. The boy was amused to see the bird in the paper bag.

'He wants to build a nest,' he said to his mother.

Edie, sensing Schwartz's unhappiness, spoke to him quietly.

'Maybe if you did some of the things my husband wants you, you would get along better with him.'

'Give me a for instance,' Schwartz said.

'Like take a bath, for instance.'

'I'm too old for baths,' said the bird. 'My feathers fall out without baths.'

'He says you have a bad smell.'

'Everybody smells. Some people smell because of their thoughts or because who they are. My bad smell comes from the food I eat. What does his come from?'

'I better not ask him or it might make him mad,' said Edie.

In late November Schwartz froze on the balcony in the fog and cold, and especially on rainy days he woke with stiff joints and could barely move his wings. Already he felt twinges of rheumatism. He would have liked to spend more time in the warm house, particularly when Maurie was in school and Cohen at work. But though Edie was goodhearted and might have sneaked him in in the morning, just to thaw out, he was afraid to ask her. In the meantime Cohen, who had been reading articles about the migration of birds, came out on the balcony one night after work when Edie was in the kitchen preparing pot roast, and peeking into the birdhouse, warned Schwartz to be on his way soon if he knew what was good for him. 'Time to hit the flyways.'

'Mr Cohen, why do you hate me so much?' asked the bird. 'What did I do to you?'

'Because you're an A-number-one trouble maker, that's why. What's more, whoever heard of a Jewbird? Now scat or it's open war.'

But Schwartz stubbornly refused to depart so Cohen embarked on a campaign of harassing him, meanwhile hiding it from Edie and Maurie. Maurie hated violence and Cohen didn't want to leave a bad impression. He thought maybe if he played dirty tricks on the bird he would fly off without being physically kicked out. The vacation was over, let him make his easy living off the fat of somebody else's land. Cohen worried about the effect of the bird's departure on Maurie's schooling but decided to take the chance, first, because the boy now seemed to have the knack of studying—give the black bird-bastard credit—and second, because Schwartz was driving him bats by being there always, even in his dreams.

The frozen foods salesman began his campaign against the bird by mixing watery cat food with the herring slices in Schwartz's dish. He also blew up and popped numerous paper bags outside the birdhouse as the bird slept, and when he had got Schwartz good and nervous, though not enough to leave, he brought a full-grown cat into the house, supposedly a gift for little Maurie, who had always wanted a pussy. The cat never stopped springing up at Schwartz whenever he saw him, one day managing to claw out several of his tailfeathers. And even at lesson time, when the cat was usually excluded from Maurie's room, though somehow or other he quickly found his way in at the end of the lesson, Schwartz was desperately fearful of his life and flew from pinnacle to pinnacle—light fixture to clothes-tree to door-top—in order to elude the beast's wet jaws.

Once when the bird complained to Edie how hazardous his existence was, she said, 'Be patient, Mr Schwartz. When the cat gets to know you better he won't try to catch you any more.'

'When he stops trying we will both be in Paradise,' Schwartz answered. 'Do me a favor and get rid of him. He makes my whole life worry. I'm losing feathers like a tree loses leaves.'

'I'm awfully sorry but Maurie likes the pussy and sleeps with it.'

What could Schwartz do? He worried but came to no decision, being afraid to leave. So he ate the herring garnished with cat food, tried hard not to hear the paper bags bursting like fire crackers outside the birdhouse at night, and lived terror-stricken closer to the ceiling than the floor, as the cat, his tail flicking, endlessly watched him.

Weeks went by. Then on the day after Cohen's mother had died in her flat in the Bronx, when Maurie came home with a zero on an arithmetic test, Cohen, enraged, waited until Edie had taken the boy to his violin lesson, then openly attacked the bird. He chased him with a broom on the balcony and Schwartz frantically flew back and forth, finally escaping into his birdhouse. Cohen triumphantly reached in, and grabbing both skinny legs, dragged the bird out, cawing loudly, his wings wildly beating. He whirled the bird around and around his head. But Schwartz, as he moved in circles, managed to swoop down and catch Cohen's nose in his beak, and hung on for dear life. Cohen cried out in great pain, punched the bird with his fist, and tugging at its legs with all his might, pulled his nose free. Again he swung the yawking Schwartz around until the bird grew dizzy, then with a furious heave, flung him into the night. Schwartz sank like stone into the street. Cohen then tossed the birdhouse and feeder after him, listening at the ledge until they crashed on the sidewalk below. For a full hour, broom in hand, his heart palpitating and nose throbbing with pain, Cohen waited for Schwartz to return but the broken-hearted bird didn't.

That's the end of that dirty bastard, the salesman thought and went in. Edie and Maurie had come home.

'Look,' said Cohen, pointing to his bloody nose swollen three times its normal size, 'what that sonofabitchy bird did. It's a permanent scar.'

'Where is he now?' Edie asked, frightened.

'I threw him out and he flew away. Good riddance.'

Nobody said no, though Edie touched a handkerchief to her eyes and Maurie rapidly tried the nine times table and found he knew approximately half.

In the spring when the winter's snow had melted, the boy, moved by a memory, wandered in the neighborhood, looking for Schwartz. He found a dead black bird in a small lot near the river, his two wings broken, neck twisted, and both bird-eyes plucked clean.

'Who did it to you, Mr Schwartz?' Maurie wept.

'Anti-Semeets,' Edie said later.

1963

I. B. SINGER

Menaseh's Dream

MENASEH was an orphan. He lived with his uncle Mendel, who was a poor glazier and couldn't even manage to feed and clothe his own children. Menaseh had already completed his cheder studies and after the fall holidays was to be apprenticed to a bookbinder.

Menaseh had always been a curious child. He had begun to ask questions as soon as he could talk: 'How high is the sky?' 'How deep is the earth?' 'What is beyond the edge of the world?' 'Why are people born?' 'Why do they die?'

It was a hot and humid summer day. A golden haze hovered over the village. The sun was as small as a moon and yellow as brass. Dogs loped along with their tails between their legs. Pigeons rested in the middle of the marketplace. Goats sheltered themselves beneath the eaves of the huts, chewing their cuds and shaking their beards.

Menaseh quarreled with his aunt Dvosha and left the house without eating lunch. He was about twelve, with a longish face, black eyes, sunken cheeks. He wore a torn jacket and was barefoot. His only possession was a tattered storybook which he had read scores of times. It was called *Alone in the Wild Forest*. The village in which he lived stood in a forest that surrounded it like a sash and was said to stretch as far as Lublin. It was blueberry time and here and there one might also find wild strawberries. Menaseh made his way through pastures and wheat fields. He was hungry and he tore off a stalk of wheat to chew on the grain. In the meadows, cows were lying down, too hot even to whisk off the flies with their tails. Two horses stood, the head of one near the rump of the other, lost in their horse thoughts. In a field planted in buckwheat the boy was amazed to see a crow perched on the torn hat of a scarecrow.

Once Menaseh entered the forest, it was cooler. The pine trees stood straight as pillars and on their brownish bark hung golden necklaces, the light of the sun shining through the pine needles. The sounds of cuckoo and woodpecker were heard, and an unseen bird kept repeating the same eerie screech.

Menaseh stepped carefully over moss pillows. He crossed a shallow streamlet that purled joyfully over pebbles and stones. The forest was still, and yet full of voices and echoes.

He wandered deeper and deeper into the forest. As a rule, he left stone markers behind, but not today. He was lonely, his head ached and his knees felt weak. 'Am I getting sick?' he thought. 'Maybe I'm going to die. Then I will soon be with Daddy and Mama.' When he came to a blueberry patch, he sat down, picked one berry after another and popped them into his mouth. But they did not satisfy his hunger. Flowers with intoxicating odors grew among the blueberries. Without realizing it, Menaseh stretched full length on the forest floor. He fell asleep, but in his dream he continued walking.

The trees became even taller, the smells stronger, huge birds flew from branch to branch. The sun was setting. The forest grew thinner and he soon came out on a plain with a broad view of the evening sky. Suddenly a castle appeared in the twilight. Menaseh had never seen such a beautiful structure. Its roof was of silver and from it rose a crystal tower. Its many tall windows were as high as the building itself. Menaseh went up to one of the windows and looked in. On the wall opposite him, he saw his own portrait hanging. He was dressed in luxurious clothes such as he had never owned. The huge room was empty.

'Why is the castle empty?' he wondered. 'And why is my portrait hanging on the wall?' The boy in the picture seemed to be alive and waiting impatiently for someone to come. Then doors opened where there had been none before, and men and women came into the room. They were dressed in white satin and the women wore jewels and held holiday prayer books with gold-embossed covers. Menaseh gazed in astonishment. He recognized his father; his mother, his grandfathers and grandmothers, and other relatives. He wanted to rush over to them, hug and kiss them, but the window glass stood in his way. He began to cry. His paternal grandfather, Tobias the Scribe, separated himself from the group and came to the window. The old man's beard was as white as his long coat. He looked both ancient and young. 'Why are you crying?' he asked. Despite the glass that separated them, Menaseh heard him clearly.

'Are you my grandfather Tobias?'

'Yes, my child. I am your grandfather.'

'Who does this castle belong to?'

'To all of us.'

'To me too?'

'Of course, to the whole family.'

'Grandpa, let me in,' Menaseh called. 'I want to speak to my father and mother.'

His grandfather looked at him lovingly and said: 'One day you will live with us here, but the time has not yet come.'

'How long do I have to wait?'

'That is a secret. It will not be for many, many years.'

'Grandpa, I don't want to wait so long. I'm hungry and thirsty and tired. Please let me in. I miss my father and mother and you and Grandma. I don't want to be an orphan.'

'My dear child. We know everything. We think about you and we love you. We are all waiting for the time when we will be together, but you must be patient. You have a long journey to take before you come here to stay.'

'Please, just let me in for a few minutes.'

Grandfather Tobias left the window and took counsel with other members of the family. When he returned, he said: 'You may come in, but only for a little while. We will show you around the castle and let you see some of our treasures, but then you must leave.'

A door opened and Menaseh stepped inside. He was no sooner over the threshold than his hunger and weariness left him. He embraced his parents and they kissed and hugged him. But they didn't utter a word. He felt strangely light. He floated along and his family floated with him. His grandfather opened door after door and each time Menaseh's astonishment grew.

One room was filled with racks of boys' clothing—pants, jackets, shirts, coats. Menaseh realized that these were the clothes he had worn as far back as he could remember. He also recognized his shoes, socks, caps, and nightshirts.

A second door opened and he saw all the toys he had ever owned: the tin soldiers his father had bought him; the jumping clown his mother had brought back from the fair at Lublin; the whistles and harmonicas; the teddy bear Grandfather had given him one Purim and the wooden horse that was the gift of Grandmother Sprintze on his sixth birthday. The notebooks in which he had practiced writing, his pencils and Bible lay on a table. The Bible was open at the title page, with its familiar engraving of Moses holding the holy tablets and Aaron in his priestly robes, both framed by a border of six-winged angels. He noticed his name in the space allowed for it.

Menaseh could hardly overcome his wonder when a third door opened. This room was filled with soap bubbles. They did not burst as soap bubbles do, but floated serenely about, reflecting all the colors of the rainbow. Some of them mirrored castles, gardens, rivers, windmills, and many other sights. Menaseh knew that these were the bubbles he used to blow from his favorite bubble pipe. Now they seemed to have a life of their own.

A fourth door opened. Menaseh entered a room with no one in it, yet it was full of the sounds of happy talk, song, and laughter. Menaseh heard his own voice and the songs he used to sing when he lived at home with his parents. He also heard the voices of his former playmates, some of whom he had long since forgotten.

The fifth door led to a large hall. It was filled with the characters in the stories his parents had told him at bedtime and with the heroes and heroines of *Alone in the Wild Forest*. They were all there: David the Warrior and the Ethiopian princess, whom David saved from captivity; the highwayman Bandurek, who robbed the rich and fed the poor; Velikan the giant, who had one eye in the center of his forehead and who carried a fir tree as a staff in his right hand and a snake in his left; the midget Pitzeles, whose beard dragged on the ground and who was jester to the fearsome King Merodach; and the two-headed wizard Malkizedek, who by witchcraft spirited innocent girls into the desert of Sodom and Gomorrah.

Menaseh barely had time to take them all in when a sixth door opened. Here everything was changing constantly. The walls of the room turned like a carousel. Events flashed by. A golden horse became a blue butterfly; a rose as bright as the sun became a goblet out of which flew fiery grasshoppers, purple fauns, and silver bats. On a glittering throne with seven steps leading up to it sat King Solomon, who somehow resembled Menaseh. He wore a crown and at his feet knelt the Queen of Sheba. A peacock spread his tail and addressed King Solomon in Hebrew. The priestly Levites played their lyres. Giants waved their swords in the air and Ethiopian slaves riding lions served goblets of wine and trays filled with pomegranates. For a moment Menaseh did not understand what it all meant. Then he realized that he was seeing his dreams.

Behind the seventh door, Menaseh glimpsed men and women, animals, and many things that were completely strange to him. The images were not as vivid as they had been in the other rooms. The figures were transparent and surrounded by mist. On the threshold there stood a girl Menaseh's own age. She had long golden braids. Although Menaseh could not see her clearly, he liked her at once. For the first time he turned to his grandfather. 'What is all this?' he asked. And his grandfather replied: 'These are the people and events of your future.'

'Where am I?' Menaseh asked.

'You are in a castle that has many names. We like to call it the place where nothing is lost. There are many more wonders here, but now it is time for you to leave.'

Menaseh wanted to remain in this strange place forever, together with his parents and grandparents. He looked questioningly at his grandfather, who shook his head. Menaseh's parents seemed to want him both to remain and to leave as quickly as possible. They still did not speak, but signaled to him, and Menaseh understood that he was in grave danger. This must be a forbidden place. His parents silently bade him farewell and his face became wet and hot from their kisses. At that moment everything disappeared—the castle, his parents, his grandparents, the girl.

Menaseh shivered and awoke. It was night in the forest. Dew was falling. High above the crowns of the pine trees, the full moon shone and the stars twinkled. Menaseh looked into the face of a girl who was bending over him. She was barefoot and wore a patched skirt; her long braided hair shone golden in the moonlight. She was shaking him and saying: 'Get up, get up. It is late and you can't remain here in the forest.'

Menaseh sat up. 'Who are you?'

'I was looking for berries and I found you here. I've been trying to wake you.'

'What is your name?'

'Channeleh. We moved into the village last week.'

She looked familiar, but he could not remember meeting her before. Suddenly he knew. She was the girl he had seen in the seventh room, before he woke up.

'You lay there like dead. I was frightened when I saw you. Were you dreaming? Your face was so pale and your lips were moving.'

'Yes, I did have a dream.'

'What about?'

'A castle.'

'What kind of castle?'

Menaseh did not reply and the girl did not repeat her question. She stretched out her hand to him and helped him get up. Together they started toward home. The moon had never seemed so light or the stars so close. They walked with their shadows behind them. Myriads of crickets chirped. Frogs croaked with human voices.

Menaseh knew that his uncle would be angry at him for coming home late. His aunt would scold him for leaving without his lunch. But these things no longer mattered. In his dream he had visited a mysterious world. He had found a friend. Channeleh and he had already decided to go berry picking the next day.

Among the undergrowth and wild mushrooms, little people in red jackets, gold caps, and green boots emerged. They danced in a circle and sang a song which is heard only by those who know that everything lives and nothing in time is ever lost.

1968

DONALD BARTHELME

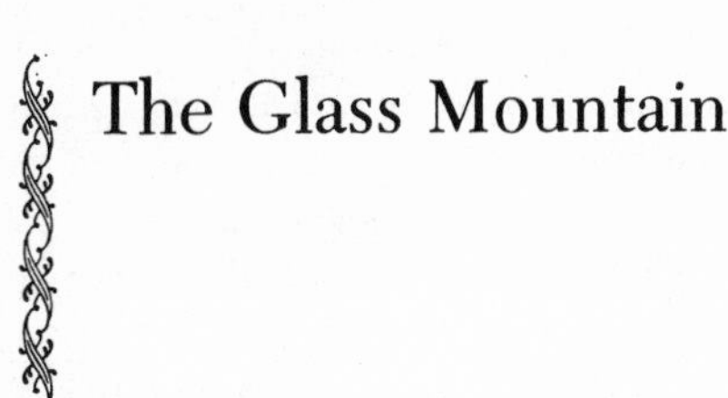

The Glass Mountain

1. I was trying to climb the glass mountain.
2. The glass mountain stands at the corner of Thirteenth Street and Eighth Avenue.
3. I had attained the lower slope.
4. People were looking up at me.
5. I was new in the neighborhood.
6. Nevertheless I had acquaintances.
7. I had strapped climbing irons to my feet and each hand grasped a sturdy plumber's friend.
8. I was 200 feet up.
9. The wind was bitter.
10. My acquaintances had gathered at the bottom of the mountain to offer encouragement.
11. 'Shithead.'
12. 'Asshole.'
13. Everyone in the city knows about the glass mountain.
14. People who live here tell stories about it.
15. It is pointed out to visitors.
16. Touching the side of the mountain, one feels coolness.
17. Peering into the mountain, one sees sparkling blue-white depths.
18. The mountain towers over that part of Eighth Avenue like some splendid, immense office building.
19. The top of the mountain vanishes into the clouds, or on cloudless days, into the sun.
20. I unstuck the righthand plumber's friend leaving the lefthand one in place.
21. Then I stretched out and reattached the righthand one a little higher up, after which I inched my legs into new positions.
22. The gain was minimal, not an arm's length.

23. My acquaintances continued to comment.
24. 'Dumb motherfucker.'
25. I was new in the neighborhood.
26. In the streets were many people with disturbed eyes.
27. Look for yourself.
28. In the streets were hundreds of young people shooting up in doorways, behind parked cars.
29. Older people walked dogs.
30. The sidewalks were full of dogshit in brilliant colors: ocher, umber, Mars yellow, sienna, viridian, ivory black, rose madder.
31. And someone had been apprehended cutting down trees, a row of elms broken-backed among the VWs and Valiants.
32. Done with a power saw, beyond a doubt.
33. I was new in the neighborhood yet I had accumulated acquaintances.
34. My acquaintances passed a brown bottle from hand to hand.
35. 'Better than a kick in the crotch.'
36. 'Better than a poke in the eye with a sharp stick.'
37. 'Better than a slap in the belly with a wet fish.'
38. 'Better than a thump on the back with a stone.'
39. 'Won't he make a splash when he falls, now?'
40. 'I hope to be here to see it. Dip my handkerchief in the blood.'
41. 'Fart-faced fool.'
42. I unstuck the lefthand plumber's friend leaving the righthand one in place.
43. And reached out.
44. To climb the glass mountain, one first requires a good reason.
45. No one has ever climbed the mountain on behalf of science, or in search of celebrity, or because the mountain was a challenge.
46. Those are not good reasons.
47. But good reasons exist.
48. At the top of the mountain there is a castle of pure gold, and in a room in the castle tower sits . . .
49. My acquaintances were shouting at me.
50. 'Ten bucks you bust your ass in the next four minutes!'
51. . . . a beautiful enchanted symbol.

52. I unstuck the righthand plumber's friend leaving the lefthand one in place.
53. And reached out.
54. It was cold there at 206 feet and when I looked down I was not encouraged.
55. A heap of corpses both of horses and riders ringed the bottom of the mountain, many dying men groaning there.
56. 'A weakening of the libidinous interest in reality has recently come to a close.' (Anton Ehrenzweig)
57. A few questions thronged into my mind.
58. Does one climb a glass mountain, at considerable personal discomfort, simply to disenchant a symbol?
59. Do today's stronger egos still *need* symbols?
60. I decided that the answer to these questions was 'yes.'
61. Otherwise what was I doing there, 206 feet above the power-sawed elms, whose white meat I could see from my height?
62. The best way to fail to climb the mountain is to be a knight in full armor—one whose horse's hoofs strike fiery sparks from the sides of the mountain.
63. The following-named knights had failed to climb the mountain and were groaning in the heap: Sir Giles Guilford, Sir Henry Lovell, Sir Albert Denny, Sir Nicholas Vaux, Sir Patrick Grifford, Sir Gisbourne Gower, Sir Thomas Grey, Sir Peter Coleville, Sir John Blunt, Sir Richard Vernon, Sir Walter Willoughby, Sir Stephen Spear, Sir Roger Faulconbridge, Sir Clarence Vaughan, Sir Hubert Ratcliffe, Sir James Tyrrel, Sir Walter Herbert, Sir Robert Brakenbury, Sir Lionel Beaufort, and many others.
64. My acquaintances moved among the fallen knights.
65. My acquaintances moved among the fallen knights, collecting rings, wallets, pocket watches, ladies' favors.
66. 'Calm reigns in the country, thanks to the confident wisdom of everyone.' (M. Pompidou)
67. The golden castle is guarded by a lean-headed eagle with blazing rubies for eyes.
68. I unstuck the lefthand plumber's friend, wondering if—
69. My acquaintances were prising out the gold teeth of not-yet-dead knights.
70. In the streets were people concealing their calm behind a façade of vague dread.

71. 'The conventional symbol (such as the nightingale, often associated with melancholy), even though it is recognized only through agreement, is not a sign (like the traffic light) because, again, it presumably arouses deep feelings and is regarded as possessing properties beyond what the eye alone sees.' (*A Dictionary of Literary Terms*)
72. A number of nightingales with traffic lights tied to their legs flew past me.
73. A knight in pale pink armor appeared above me.
74. He sank, his armor making tiny shrieking sounds against the glass.
75. He gave me a sideways glance as he passed me.
76. He uttered the word '*Muerte*' as he passed me.
77. I unstuck the righthand plumber's friend.
78. My acquaintances were debating the question, which of them would get my apartment?
79. I reviewed the conventional means of attaining the castle.
80. The conventional means of attaining the castle are as follows: 'The eagle dug its sharp claws into the tender flesh of the youth, but he bore the pain without a sound, and seized the bird's two feet with his hands. The creature in terror lifted him high up into the air and began to circle the castle. The youth held on bravely. He saw the glittering palace, which by the pale rays of the moon looked like a dim lamp; and he saw the windows and balconies of the castle tower. Drawing a small knife from his belt, he cut off both the eagle's feet. The bird rose up in the air with a yelp, and the youth dropped lightly onto a broad balcony. At the same moment a door opened, and he saw a courtyard filled with flowers and trees, and there, the beautiful enchanted princess.' (*The Yellow Fairy Book*)
81. I was afraid.
82. I had forgotten the Bandaids.
83. When the eagle dug its sharp claws into my tender flesh—
84. Should I go back for the Bandaids?
85. But if I went back for the Bandaids I would have to endure the contempt of my acquaintances.
86. I resolved to proceed without the Bandaids.
87. 'In some centuries, his [man's] imagination has made life an intense practice of all the lovelier energies.' (John Masefield)

88. The eagle dug its sharp claws into my tender flesh.
89. But I bore the pain without a sound, and seized the bird's two feet with my hands.
90. The plumber's friends remained in place, standing at right angles to the side of the mountain.
91. The creature in terror lifted me high in the air and began to circle the castle.
92. I held on bravely.
93. I saw the glittering palace, which by the pale rays of the moon looked like a dim lamp; and I saw the windows and balconies of the castle tower.
94. Drawing a small knife from my belt, I cut off both the eagle's feet.
95. The bird rose up in the air with a yelp, and I dropped lightly onto a broad balcony.
96. At the same moment a door opened, and I saw a courtyard filled with flowers and trees, and there, the beautiful enchanted symbol.
97. I approached the symbol, with its layers of meaning, but when I touched it, it changed into only a beautiful princess.
98. I threw the beautiful princess headfirst down the mountain to my acquaintances.
99. Who could be relied upon to deal with her.
100. Nor are eagles plausible, not at all, not for a moment.

1970

TANITH LEE

Prince Amilec

In a palace by the sea lived a beautiful princess. She had eyes as green as apples, long red hair, and a very nasty temper indeed.

One day, her father said that she should think about getting married. Lots of princes had thought they would like to marry her in the past, but once she had flown into a rage with them a few times, they changed their minds and went off to find someone a bit quieter.

'I don't want to get married,' said the princess. 'And I *won't*!' And she picked up a china dog to throw at her father.

He dodged out of the room, just in time.

'I'm getting too old for all this running about,' he said to his secretary. 'I really must get her married off, and then her husband can deal with her.'

'Don't worry,' said the secretary. 'I'll get some messengers to ride around the other kingdoms with the princess's portrait. A lot of people will be interested, and they won't know till they get here what she's like.'

Now, in one of the kingdoms that the messenger visited there lived a handsome young prince named Amilec. When the portrait was brought in, the prince immediately fell in love with the princess in the picture. No sooner were the messengers out of the room than Amilec grabbed his cloak and was riding full speed up the road toward the sea.

When he reached the palace, he found a good many suitors there already. They were waiting for a glimpse of the princess, who was supposed to appear at one of the palace windows and graciously wave to them.

Suddenly a window shot up and a red head appeared.

'Go *away*!' bawled the princess, and went in again.

'Ha, ha!' cried the king, who was leaning anxiously out of another window. 'She will have her little joke!'

The suitors laughed uneasily, and the butler came and showed them to their rooms.

'My dear,' said the king cautiously, 'couldn't you just try? All those nice young men have come such a long way.'

'Don't worry,' said the princess. 'I'll soon get rid of them. I'm going to set them such impossible tasks that they'll give up and go home inside a week.'

Sure enough, when all the suitors had gathered in the dining room, the princess's page came down, asked for silence, and took out a scroll.

'The princess,' read the page, 'wishes to inform you that she has thrown her ruby bracelet out the tower window into the sea. Before any of you think of asking for her hand, you must dive down and get it back for her.'

None of the suitors felt like eating their dinner after that. Some went out immediately and began to pack their bags; others flew into a rage and banged their fists on the table. Prince Amilec put his head in his hands and groaned. The page felt rather sorry for Amilec, who was the only suitor who had not been rude to him.

'If I were you, I should just go home and forget about the princess,' whispered the page. 'She's frightful!'

'I can't,' said Amilec sorrowfully. 'I shall have to look for her bracelet, even if I drown in the attempt.'

'I'll tell you what, then,' said the page. 'Farther up the beach lives a witch in a cave. Nobody's ever seen her, but she's supposed to be very clever. She might help you, if you asked her nicely.'

Prince Amilec wasn't too keen on visiting a witch, but he thanked the page, and went and thought about it. Eventually he decided that it was a lot safer than diving into the sea on his own.

When the palace clock struck twelve, he went quietly out and down the path to the beach. At last he came to a cave with a big front door set in it, covered over by a lot of seaweed. Prince Amilec knocked, feeling rather nervous.

'The witch will probably be horribly ugly, with three eyes and a wart on the end of her nose,' Amilec said to himself. 'But I mustn't let her see that I'm not completely used to people with three eyes and warts, or she may be offended and not help me.'

Just then the door opened, and there stood a very pretty girl holding a lantern.

'Can I help you?' asked the girl.

'Oh—er, yes. I was looking for the witch,' said Amilec, brushing off the seaweed that had fallen on him when the door opened.

'I *am* the witch,' said the girl. 'Do come in.'

She took him down a long cave-corridor, and at the end was another door. The witch hung the lantern on a hook in the wall, opened the door, and led the prince into a small cozy room, where a fire was burning on the hearth.

'I thought,' said Amilec, 'that witches were old and ugly, and lived in ruined castles full of bats.'

'Well,' said the girl, 'I do have a bat.' She pointed and Amilec saw a furry shape with folded wings hanging upside down in an armchair on the other side of the fire.

'Do sit down,' said the witch. 'I'll just hang Basil on the mantelpiece.' Which she did. 'Now tell me what the trouble is.'

So Amilec told her about the red-haired princess, and how he wanted very much to marry her, and about the ruby bracelet that he had to try to find.

'I don't mind helping you,' said the witch when he had finished, 'but the princess will never marry you, you know. And she has a dreadful temper. I can hear her down here sometimes when she starts shouting.'

Prince Amilec sighed and said he thought that might be the case but he just couldn't help being in love with the princess, however awful she was.

'All right,' said the witch. 'Leave everything to me. You just stay here and look after Basil, and see that the fire doesn't go out.'

So saying, she left the room, and a minute later Amilec heard the cave door shut. He couldn't help being curious, and as there was a small round window in the cave wall, he looked out of it to see where she had gone. The witch was walking along some rocks that ran into the sea. Suddenly she changed into a dolphin, leaped forward, and vanished in the water.

Amilec tried not to be too surprised. He put some more wood on the fire and hung Basil a bit farther along the mantelpiece, so the smoke wouldn't get into his fur.

Not long after, the cave door opened and the witch, no longer looking a bit like a dolphin, came back into the room. She held a ruby bracelet in her hand.

'Sorry I was so long,' said the witch, 'but a sea serpent had got his tail stuck in it, and I had to pull him out.'

'How can I ever thank you?' gasped Amilec.

'Don't give it a second thought,' said the witch. 'I haven't had an excuse to change into a dolphin for ages.'

Next morning, the princess called all the suitors into a big room and asked them if they had had any luck. Of course, none of them had, although some of them had been swimming about since dawn, and most of them had caught bad colds.

Just then Amilec came in. He walked up to the princess, bowed, and handed her the ruby bracelet.

'Oh!' screamed the princess. 'This can't be the right one,' she added hysterically. However, the king, who had given it to her in the first place, came up and had a look, and declared that it was.

'Well done, my boy,' he added to Amilec.

'Yes, well done,' said the princess quickly. 'Now that you've succeeded with the first task, you can go on and do the second. Everybody else is disqualified.'

So the other suitors snuffled and coughed and complained their way out, and went home.

'The second task,' said the princess, smiling a nasty smile, 'is to find my golden girdle. I have tied it to the arrow of one of the bowmen, and when he fires, goodness only knows where it will end up.'

'Perhaps I could tell him where to aim,' whispered the king's secretary.

'Certainly not!' cried the princess, who had overheard, and she went to the window and gave a signal to the bowman.

When Prince Amilec went out to look for the golden girdle, he found that there was a forest growing at the back of the palace, full of thick fern. He searched till dusk, and then he sat down on a stone, because he was worn out.

'I'll never find it,' he said. 'I might just as well go home right now.'

Just then a figure came through the forest toward him, carrying a lantern and a bat.

'Hello,' said the witch. 'I've just been taking Basil for a fly. You look unhappy. Is it that princess again?'

'I'm afraid it is,' said Amilec.

'I thought so,' said the witch. 'What does she want now?'

Amilec told her about the golden girdle which had been fired into the forest.

'Oh, that's easy,' said the witch, 'Hold Basil, and I'll see what I can do.'

So Amilec sat on the stone with the lantern and Basil. One by one the stars came out, and the sea sounded very drowsy, as if it were going to sleep. Prince Amilec closed his eyes and dreamed that he had won the hand of the red-haired princess and she had just thrown her crown at him. He woke up with a start and found an owl sitting on the ground, a golden girdle in its beak.

'Here we are,' said the owl, changing back into the witch. 'Sorry I was so long, but some doves had got it tangled up in their nest and I had to get it out and put the nest back for them afterward.'

'How can I ever repay you?' implored Amilec.

'Come and have a cup of tea with me,' said the witch. So he did.

The next morning the princess called him into the big room and sneered at him.

'There's no need to say anything,' said the princess. 'Just pack your things and go home. Of course, if you'd care to send me a golden girdle to replace the old one, I might give you a kiss.'

Prince Amilec took out the girdle and handed it to her.

The princess screamed, 'I don't believe it! It's not mine—it isn't!'

'Oh, yes, it is,' said the king. 'Well done, my boy.'

'Yes,' said the princess. 'Well done. You can now go on to the third task.'

Amilec paled.

'You see this pearl necklace,' said the princess with a ghastly smile. 'On it are one hundred and fifty pearls.'

The king realized what was coming and tried to stop her, but it was too late. The princess tugged on the silver chain and it broke, and the pearls flew everywhere.

'I want you to find them all, and return them to me in one hour's time.' And she glided out.

'We can help,' cried the king and the secretary.

'No, you can't!' cried the princess, rushing back. The king and the secretary hurried away.

'Now what am I to do?' Amilec wondered. He looked out the window, and who should be walking along the seashore below the palace, gathering seaweed in a basket, but the witch. The prince leaped downstairs and out through the garden gate and along the beach.

'Hello!' said Amilec. 'How's Basil?'

'Basil's very well, thank you,' replied the witch. 'How are you getting on with the princess?'

Amilec sighed deeply and told her about the pearl necklace.

'Wouldn't it be simpler to forget all about it and go home?' asked the witch.

'I'm afraid I love the princess much too much to do that,' murmured Amilec.

'All right,' said the witch. 'You collect me some seaweed in the basket, and I'll go and do what I can.' And she changed into a mouse and ran through the palace gate.

About half an hour later, the witch came back and handed Amilec a velvet bag.

'You said one hundred and fifty pearls, but I found one hundred and fifty-one, so your princess can't count.'

'Here's your seaweed. I can never thank you enough,' said Amilec.

'That's all right.' said the witch. 'Come down and tell me what happens.'

When Prince Amilec went back into the palace, the princess was already sitting waiting. Amilec went forward and tipped all the pearls into her lap.

'Count them!' shouted the princess.

The secretary hurriedly obeyed. 'One hundred and fifty-one,' he declared at last. The princess shrieked and fainted with fury.

'This is the happiest day of my life,' beamed the king. 'At last—you can marry her.'

Just then the princess revived.

'Very well,' she snapped. 'I'm yours, you pest, but before I marry you I want a splendid wedding dress, and if I don't like it, I shall change my mind.'

By this time Amilec was getting a bit fed up with her tantrums, but he thought that, of all her demands, this was the most reasonable. So he bowed and said that he'd do his best.

'See that you do!' yelled the princess, and flounced out.

That night the prince made his way down the path to the witch's cave and knocked. The witch let him in, hung Basil on the mantelpiece, and sat him in the armchair.

'She wants a wedding dress now,' said Amilec, as the witch put on the kettle.

'Oh, does she?' said the witch. 'Well, how would you like me to make her a special magic one?'

'You're marvelous!' said Amilec.

The witch smiled and lifted Basil down for a moment while she got out the teapot. 'I'll bring it up to the castle first thing tomorrow morning.'

'How can I ever thank you enough?' asked Amilec.

'I'll think of something,' said the witch.

The next morning the whole court gathered worriedly.

When Amilec came in, the princess jumped up and demanded, 'Where is it?'

'A friend of mine is bringing it,' said the prince. 'She'll be here any moment.'

Just then the doors opened and in came the most beautiful girl the prince had ever seen. The court sighed with wonder, the secretary dropped his pen, and even the princess forgot to be rude.

The girl came up to the prince, curtsied, and said, 'Here is the dress. I thought it would look better with someone wearing it. It's made of moonlight and star-glow, and the glitter on a mermaid's tail. I hope you like it.'

'But who are you?' gasped Amilec.

'I am the witch.'

Then, in front of the court, the king, and the princess, Amilec bowed to the witch and said, 'How can I have been so blind! You are the most beautiful girl I have ever met. You are also the kindest. May I humbly ask you to be my wife? I promise to look after Basil, and I'll live in the cave, if it will make things easier.'

'Dear Amilec,' said the witch. 'Basil and I both love you very much, and will be delighted to accept.'

So Amilec and the witch got married and lived happily ever after with Basil, although Basil was asleep most of the time.

As for the red-haired princess, when she had finished shouting and screaming, she told the king that what she really wanted to do was travel. The king was only too glad to see her go, and packed her off as soon as possible.

One day, however, the princess came to a kingdom where there was a very handsome prince, and she thought he was just the kind of young man who would be good enough for her. So she went and knocked on the door of his palace.

'Come in, my dear,' said the queen. 'How pretty you are! I should like my son to marry someone like you. The only trouble is, he always makes princesses complete dreadful tasks before he'll even look at them. The latest thing he wants done is for somebody to make an apple tree grow upside down from his bedroom ceiling. However, I've heard that there's a wizard who lives in the wood, and he might help you, if you asked him nicely.'

'Oh, well,' thought the princess, setting off for the wood. 'The wizard will probably be horribly ugly, with three eyes, and a wart on the end of his nose, but it can't be helped.'

1972

JAY WILLIAMS

Petronella

IN the kingdom of Skyclear Mountain, three princes were always born to the king and queen. The oldest prince was always called Michael, the middle prince was always called George, and the youngest was always called Peter. When they were grown, they always went out to seek their fortunes. What happened to the oldest prince and the middle prince no one ever knew. But the youngest prince always rescued a princess, brought her home, and in time ruled over the kingdom. That was the way it had always been. And so far as anyone knew, that was the way it would always be.

Until now.

Now was the time of King Peter the twenty-sixth and Queen Blossom. An oldest prince was born, and a middle prince. But the youngest prince turned out to be a girl.

'Well,' said the king gloomily, 'we can't call her Peter. We'll have to call her Petronella. And what's to be done about it, I'm sure I don't know.'

There was nothing to be done. The years passed, and the time came for the princes to go out and seek their fortunes. Michael and George said good-bye to the king and queen and mounted their horses. Then out came Petronella. She was dressed in traveling clothes, with her bag packed and a sword by her side.

'If you think,' she said, 'that I'm going to sit at home, you are mistaken. I'm going to seek my fortune, too.'

'Impossible' said the king.

'What will people say?' cried the queen.

'Look,' said Prince Michael, 'be reasonable, Pet. Stay home. Sooner or later a prince will turn up here.'

Petronella smiled. She was a tall, handsome girl with flaming red hair and when she smiled in that particular way it meant she was trying to keep her temper.

'I'm going with you,' she said. 'I'll find a prince if I have to rescue one from something myself. And that's that.'

The grooms brought out her horse, she said good-bye to her parents, and away she went behind her two brothers.

They traveled into the flatlands below Skyclear Mountain. After many days, they entered a great dark forest. They came to a place where the road divided into three, and there at the fork sat a little, wrinkled old man covered with dust and spiderwebs.

Prince Michael said haughtily, 'Where do these roads go, old man?'

'The road on the right goes to the city of Gratz,' the man replied. 'The road in the center goes to the castle of Blitz. The road on the left goes to the house of Albion the enchanter. And that's one.'

'What do you mean by "And that's one"?' asked Prince George.

'I mean,' said the old man, 'that I am forced to sit on this spot without stirring, and that I must answer one question from each person who passes by. And that's two.'

Petronella's kind heart was touched. 'Is there anything I can do to help you?' she asked.

The old man sprang to his feet. The dust fell from him in clouds.

'You have already done so,' he said. 'For that question is the one which releases me. I have sat here for sixty-two years waiting for someone to ask me that.' He snapped his fingers with joy. 'In return, I will tell you anything you wish to know.'

'Where can I find a prince?' Petronella said promptly.

'There is one in the house of Albion the enchanter,' the old man answered.

'Ah,' said Petronella, 'then that is where I am going.'

'In that case I will leave you,' said her oldest brother. 'For I am going to the castle of Blitz to see if I can find my fortune there.'

'Good luck,' said Prince George. 'For I am going to the city of Gratz. I have a feeling my fortune is there.'

They embraced her and rode away.

Petronella looked thoughtfully at the old man, who was combing spiderwebs and dust out of his beard. 'May I ask you something else?' she said.

'Of course. Anything.'

'Suppose I wanted to rescue that prince from the enchanter. How would I go about it? I haven't any experience in such things, you see.'

The old man chewed a piece of his beard. 'I do not know everything,' he said, after a moment. 'I know that there are three magical secrets which, if you can get them from him, will help you.'

'How can I get them?' asked Petronella.

'Offer to work for him. He will set you three tasks, and if you can do them you may demand a reward for each. You must ask him for a comb for your hair, a mirror to look into, and a ring for your finger.'

'And then?'

'I do not know. I only know that when you rescue the prince, you can use these things to escape from the enchanter.'

'It doesn't sound easy,' sighed Petronella.

'Nothing we really want is easy,' said the old man. 'Look at me—I have wanted my freedom, and I've had to wait sixty-two years for it.'

Petronella said good-bye to him. She mounted her horse and galloped along the third road.

It ended at a low, rambling house with a red roof. It was a comfortable-looking house, surrounded by gardens and stables and trees heavy with fruit.

On the lawn, in an armchair, sat a handsome young man with his eyes closed and his face turned to the sky.

Petronella tied her horse to the gate and walked across the lawn.

'Is this the house of Albion the enchanter?' she said.

The young man blinked up at her in surprise.

'I think so,' he said. 'Yes, I'm sure it is.'

'And who are you?'

The young man yawned and stretched. 'I am Prince Ferdinand of Firebright,' he replied. 'Would you mind stepping aside? I'm trying to get a suntan and you're standing in the way.'

Petronella snorted. 'You don't sound like much of a prince,' she said.

'That's funny,' said the young man, closing his eyes. 'That's what my father always says.'

At that moment the door of the house opened. Out came a man dressed all in black and silver. He was tall and thin, and his eyes were as black as a cloud full of thunder. Petronella knew at once that he must be the enchanter.

He bowed to her politely. 'What can I do for you?'

'I wish to work for you,' said Petronella boldly.

Albion nodded. 'I cannot refuse you,' he said. 'But I warn you, it will be dangerous. Tonight I will give you a task. If you do it, I will reward you. If you fail, you must die.'

Petronella glanced at the prince and sighed. 'If I must, I must,' she said. 'Very well.'

That evening they all had dinner together in the enchanter's cozy kitchen. Then Albion took Petronella out to a stone building and unbolted its door. Inside were seven huge black dogs.

'You must watch my hounds all night,' said he.

Petronella went in, and Albion closed and locked the door.

At once the hounds began to snarl and bark. They bared their teeth at her. But Petronella was a real princess. She plucked up her courage. Instead of backing away, she went toward the dogs. She began to speak to them in a quiet voice. They stopped snarling and sniffed at her. She patted their heads.

'I see what it is,' she said. 'You are lonely here. I will keep you company.'

And so all night long, she sat on the floor and talked to the hounds and stroked them. They lay close to her, panting.

In the morning Albion came and let her out. 'Ah,' said he, 'I see that you are brave. If you had run from the dogs, they would have torn you to pieces. Now you may ask for what you want.'

'I want a comb for my hair,' said Petronella.

The enchanter gave her a comb carved from a piece of black wood.

Prince Ferdinand was sunning himself and working at a crossword puzzle. Petronella said to him in a low voice, 'I am doing this for you.'

'That's nice,' said the prince. 'What's "selfish" in nine letters?'

'You are,' snapped Petronella. She went to the enchanter. 'I will work for you once more,' she said.

That night Albion led her to a stable. Inside were seven huge horses.

'Tonight,' he said, 'you must watch my steeds.'

He went out and locked the door. At once the horses began to rear and neigh. They pawed at her with their iron hoofs.

But Petronella was a real princess. She looked closely at them and saw that their coats were rough and their manes and tails full of burrs.

'I see what it is,' she said. 'You are hungry and dirty.'

She brought them as much hay as they could eat, and began to brush them. All night long she fed them and groomed them, and they stood quietly in their stalls.

In the morning Albion let her out. 'You are as kind as you are brave,' said he. 'If you had run from them they would have trampled you under their hoofs. What will you have as a reward?'

'I want a mirror to look into,' said Petronella.

The enchanter gave her a mirror made of silver.

She looked across the lawn at Prince Ferdinand. He was doing exercises leisurely. He was certainly handsome. She said to the enchanter, 'I will work for you once more.'

That night Albion led her to a loft above the stables. There, on perches, were seven great hawks.

'Tonight,' said he, 'you must watch my falcons.'

As soon as Petronella was locked in, the hawks began to beat their wings and scream at her.

Petronella laughed. 'That is not how birds sing,' she said. 'Listen.'

She began to sing in a sweet voice. The hawks fell silent. All night long she sang to them, and they sat like feathered statues on their perches, listening.

In the morning Albion said, 'You are as talented as you are kind and brave. If you had run from them, they would have pecked and clawed you without mercy. What do you want now?'

'I want a ring for my finger,' said Petronella.

The enchanter gave her a ring made from a single diamond.

All that day and all that night Petronella slept, for she was very tired. But early the next morning, she crept into Prince Ferdinand's room. He was sound asleep, wearing purple pajamas.

'Wake up,' whispered Petronella. 'I am going to rescue you.'

Ferdinand awoke and stared sleepily at her. 'What time is it?'

'Never mind that,' said Petronella. 'Come on!'

'But I'm sleepy,' Ferdinand objected. 'And it's so pleasant here.'

Petronella shook her head. 'You're not much of a prince,' she said grimly. 'But you're the best I can do.'

She grabbed him by the wrist and dragged him out of bed. She hauled him down the stairs. His horse and hers were in a separate stable, and she saddled them quickly. She gave the prince a shove, and he mounted. She jumped on her own horse, seized the prince's reins, and away they went like the wind.

They had not gone far when they heard a tremendous thumping. Petronella looked back. A dark cloud rose behind them, and beneath it she saw the enchanter. He was running with great strides, faster than the horses could go.

'What shall we do?' she cried.

'Don't ask me,' said Prince Ferdinand grumpily. 'I'm all shaken to bits by this fast riding.'

Petronella desperately pulled out the comb. 'The old man said this would help me!' she said. And because she didn't know what else to do with it, she threw the comb on the ground. At once a forest rose up. The trees were so thick that no one could get between them.

Away went Petronella and the prince. But the enchanter turned himself into an ax and began to chop. Right and left he chopped, slashing, and the trees fell before him.

Soon he was through the wood, and once again Petronella heard his footsteps thumping behind.

She reined in the horses. She took out the mirror and threw it on the ground. At once a wide lake spread out behind them, gray and glittering.

Off they went again. But the enchanter sprang into the water, turning himself into a salmon as he did so. He swam across the lake and leaped out of the water on to the other bank. Petronella heard him coming—*thump! thump!*—behind them again.

This time she threw down the ring. It didn't turn into anything, but lay shining on the ground.

The enchanter came running up. And as he jumped over the ring, it opened wide and then snapped up around him. It held his arms tight to his body, in a magical grip from which he could not escape.

'Well,' said Prince Ferdinand, 'that's the end of him.'

Petronella looked at him in annoyance. Then she looked at the enchanter, held fast in the ring.

'Bother!' she said. 'I can't just leave him here. He'll starve to death.'

She got off her horse and went up to him. 'If I release you,' she said, 'will you promise to let the prince go free?'

Albion stared at her in astonishment. 'Let him go free?' he said. 'What are you talking about? I'm glad to get rid of him.'

It was Petronella's turn to look surprised. 'I don't understand,' she said. 'Weren't you holding him prisoner?'

'Certainly not,' said Albion. 'He came to visit me for a weekend. At the end of it, he said, "It's so pleasant here, do you mind if I stay on for another day or two?" I'm very polite and I said, "Of course." He stayed on, and on, and on. I didn't like to be rude to a guest and I couldn't just kick him out. I don't know what I'd have done if you hadn't dragged him away.'

'But then—'said Petronella, 'but then—why did you come running after him this way?'

'I wasn't chasing him,' said the enchanter, 'I was chasing *you*. You are just the girl I've been looking for. You are brave and kind and talented, and beautiful as well.'

'Oh,' said Petronella. 'I see.'

'Hmm,' said she. 'How do I get this ring off you?'

'Give me a kiss.'

She did so. The ring vanished from around Albion and reappeared on Petronella's finger.

'I don't know what my parents will say when I come home with you instead of a prince,' she said.

'Let's go and find out, shall we?' said the enchanter cheerfully.

He mounted one horse and Petronella the other. And off they trotted, side by side, leaving Prince Ferdinand of Firebright to walk home as best he could.

1973

JOAN AIKEN

The Man Who Had Seen the Rope Trick

'MISS Drake,' said Mrs Minser, 'when ye've finished with the salt and pepper will ye please put them *together*?'

'Sorry, I'm sorry,' mumbled Miss Drake. 'I can't see very well as you know, I can't see very well.' Her tremulous hands worked out like tendrils across the table and succeeded in knocking the mustard on to its side. An ochre blob defiled the snowy stiffness of the tablecloth. Mrs Minser let a slight hiss escape her.

'That's the *third* tablecloth ye've dirtied in a week, Miss Drake. Do ye know I had to get up at four o'clock this morning to do all the washing? I shan't be able to keep ye if ye go on like this, ye know.'

Without waiting for the whispered apologies she turned towards the dining-room door, pushing the trolley with the meat plates before her. Her straw-grey hair was swept to a knot on the top of her head, her grey eyes were as opaque as bottle-tops, her mouth was screwed tight shut against the culpabilities of other people.

'Stoopid business, gettin' up at four in the mornin',' muttered old Mr Hill, but he muttered it quietly to himself. 'Who cares about a blob of mustard on the tablecloth, anyway? Who cares about a tablecloth, or a separate table, if the food's good? If she's got to get up at four, why don't she make us some decent porridge instead of the slime she gives us?'

He bowed his head prayerfully over his bread plate as Mrs Minser returned, weaving her way with the neatness of long practice between the white-covered tables, each with its silent, elderly, ruminating diner.

The food was *not* good. 'Rice shape or banana, Mr Hill?' Mrs Minser asked, pausing beside him.

'Banana, thank'ee.' He repressed a shudder as he looked at the colourless, glutinous pudding. The bananas were unripe, and bad for his indigestion, but at least they were palatable.

'Mr Wakefield! Ye've spotted yer shirt with gravy! That means more washing, and I've got a new guest coming tomorrow. I cann't think how you old people can be so inconsiderate.'

'I'll wash it, I'll wash it myself, Mrs Minser.' The old man put an anxious, protective hand over the spot.

'Ye'll do no such thing!'

'Who is the new guest then, Mrs Minser?' Mr Hill asked, more to distract her attention from his neighbour's misfortune than because he wanted to know.

'A Mr Ollendod. Retired from India. I only hope,' said Mrs Minser, forebodingly, 'that he won't have a great deal of luggage, else where we shall put it all I cann't imagine.'

'India,' murmured Mr Hill to himself. 'From India, eh? He'll certainly find it different here.' And he looked round the dining-room of the Balmoral Guest House. The name Balmoral, and Mrs Minser's lowland accent, constituted the only Scottish elements in the guest house, which was otherwise pure Westcliff. The sea, half a mile away, invisible from the house, was implicit in the bracingness of the air and the presence of so many elderly residents pottering out twice a day to listen to the municipal orchestra. Nobody actually swam in the sea, or even looked at it much, but there it was anyway, a guarantee of ozone and fresh fish on the tables of the residential hotels.

Mr Ollendod arrived punctually next day, and he did have a lot of luggage.

Mrs Minser's expression became more and more ominous as trunks and cases—some of them very foreign-looking and made of straw—boxes and rolls and bundles were unloaded.

'Where does he think all that is going?' she said, incautiously loudly, to her husband who was helping to carry in the cases.

Mr Ollendod was an elderly, very brown, shrivelled little man, but he evidently had all his faculties intact, for he looked up from paying the cab-driver to say, 'In my room, I trust, naturally. It is a double room, is it not? Did I not stipulate for a double room?'

Mrs Minser's idea of a double room was one into which a double bed could be squeezed. She eyed Mr Ollendod measuringly, her lips pursed together. Was he going to be the sort who gave trouble? If so, she'd soon find a reason for giving him his notice.

Summer was coming, when prices and the demand for rooms went up; one could afford to be choosy. Still, ten guineas a week was ten guineas; it would do no harm to wait and see.

The Minser children, Martin and Jenny, came home from school and halted, fascinated, amongst Mr Ollendod's possessions.

'Look, a screen, all covered with pictures!'

'He's got spears!'

'A tigerskin!'

'An elephant's foot!'

'What's this, a shield?'

'No, it's a fan, made of peacocks' feathers.' Mr Ollendod smiled at them benevolently. Jenny thought that his face looked like the skin on top of cocoa, wrinkling when you stir it.

'Is he an Indian, Mother?' she asked when they were in the kitchen.

'No, of course he's not. He's just brown because he's lived in a hot climate,' Mrs Minser said sharply. 'Run and do yer homework and stay out from under my feet.'

The residents, also, were discussing Mr Ollendod.

'Do you think he can be—*foreign*?' whispered Mrs Pursey. 'He is such an odd-looking man. His eyes are so bright—just like diamonds. What do you think, Miss Drake?'

'How should I know?' snapped Miss Drake. 'You seem to forget I haven't been able to see across the room for the last five years.'

The children soon found their way to Mr Ollendod's room. They were strictly forbidden to speak to or mix with the guests in any way, but there was an irresistible attraction about the little bright-eyed man and his belongings.

'Tell us about India,' Jenny said, stroking the snarling tiger's head with its great yellow glass eyes.

'India? The hills are blue and wooded, they look as innocent as Essex but they're full of tigers and snakes and swinging, chattering monkeys. In the villages you can smell dust and dung-smoke and incense; there are no brown or grey clothes, but flashing pinks and blood-reds, turquoises and saffrons; the cows have horns three yards wide.'

'Shall you ever go back there?' Martin asked, wondering how anybody could bear to exchange such a place for the worn grey, black and fawn carpeting, the veneer wardrobe and plate-glass, the limp yellow sateen coverlid of a Balmoral bedroom.

'No,' said Mr Ollendod, sighing. 'I fell ill. And no one wants me there now. Still,' he added more cheerfully, 'I have brought back plenty of reminders with me; enough to keep India alive in my mind. Look at this—and this—and this.'

Everything was wonderful—the curved leather slippers, the richly patterned silk of Mr Ollendod's dressing-gown and scarves, the screen with its exotic pictures ('I'm not letting *that* stay there long,' said Mrs Minser), the huge pink shells with a sheen of pearl, the gnarled and grinning images, the hard, scented sweets covered with coloured sugar.

'You are *not* to go up there. And if he offers you anything to eat you are to throw it straight away,' Mrs Minser said, but she might as well have spoken to the wind. The instant the children had done their homework they were up in Mr Ollendod's room, demanding stories of snakes and were-wolves, of crocodiles who lived for a hundred years, of mysterious ceremonies in temples, ghosts who walked with their feet swivelled backwards on their ankles, and women with the evil eye who could turn milk sour and rot the unripe fruit on a neighbour's vine.

'You've really seen it? You've seen them? You've seen a snake-charmer and a snake standing on its tail? And a lizard break in half and each half run away separately? And an eagle fly away with a live sheep?'

'All those things,' he said. 'I'll play you a snake-charmer's tune if you like.'

He fished a little bamboo flageolet out of a cedarwood box and began to play a tune that consisted of no more than a few trickling, monotonous notes, repeated over and over again. Tuffy, the aged, moth-eaten black cat who followed the children everywhere when they were at home and dozed in Mr Ollendod's armchair when they were at school, woke up, and pricked up his ears; downstairs Jip, the bad-tempered Airedale, growled gently in his throat; and Mrs Minser, sprinkling water on her starched ironing, paused and angrily rubbed her ear as if a mosquito had tickled it.

'And I've seen another thing: a rope that stands on its tail when the man says a secret word to it, stands straight up on end! And a boy climbs up it, right up! Higher and higher, till he finally disappears out of sight.'

'Where does he go to?' the children asked, huge-eyed.

'A country where the grass grows soft and patterned like a

carpet, where the deer wear gold necklets and come to your hand for pieces of bread, where the plums are red and sweet and as big as oranges, and the girls have voices like singing birds.'

'Does he never come back?'

'Sometimes he jumps down out of the sky with his hands full of wonderful grass and fruits. But sometimes he never comes back.'

'Do *you* know the word they say to the rope?'

'I've heard it, yes.'

'If I were the boy I wouldn't come back,' said Jenny.

'Tell us some more. About the witch woman who fans herself.'

'She fans herself with a peacock-feather fan,' Mr Ollendod said. 'And when she does that she becomes a snake and slips away into the forest. And when she is tired of being a snake and wants to turn into a woman again she taps her husband's foot with her cold head till he waves the fan over her.'

'Is the fan just like yours on the wall?'

'Just like it.'

'Oh, may we fan ourselves with it, may we?'

'And turn yourselves into little snakes? What would your mother say?' asked Mr Ollendod, laughing heartily.

Mrs Minser had plenty to say as it was. When the children told her a garbled mixture of the snakes and the deer and the live rope and girls with birds' voices and plums as big as oranges she pursed her lips together tight.

'A pack of moonshine and rubbish! I've a good mind to forbid him to speak to them.'

'Oh, come, Hannah,' her husband said mildly. 'He keeps them out of mischief for hours on end. You know you can't stand it if they come into the kitchen or make a noise in the garden. And he's only telling them Indian fairy tales.'

'Well anyway ye're not to believe a word he says,' Mrs Minser ordered the children. 'Not a *single* word.'

She might as well have spoken to the wind . . .

Tuffy the cat fell ill and lay with faintly heaving sides in the middle of the hallway. Mrs Minser exclaimed angrily when she found Mr Ollendod bending over him.

'That dirty old cat! It's high time he was put away.'

'It is a cold he has, nothing more,' Mr Ollendod said mildly. 'If you will allow me, I shall take him to my room and treat him. I have some Indian gum which is very good for inhaling.'

But Mrs Minser refused to consider the idea. She rang up the vet, and when the children came home from school, Tuffy was gone.

They found their way up to Mr Ollendod's room, speechless with grief.

He looked at them thoughtfully for a while and then said, 'Shall I tell you a secret?'

'Yes, what? What?' Martin said, and Jenny cried, 'You've got Tuffy hidden here, is that it?'

'Not exactly,' said Mr Ollendod, 'but you see that mirror on the wall?'

'The big one covered with a fringy shawl, yes?'

'Once upon a time that mirror belonged to a queen, in India. She was very beautiful, so beautiful that it was said sick people could be cured of their illnesses just by looking at her. In course of time she grew old, and lost her beauty. But the mirror remembered how beautiful she had been and showed her still the lovely face she had lost. And one day she walked right into the mirror and was never seen again. So if you look into it you do not see things as they are now, but beautiful as they were in their youth.'

'May we look?'

'Just for a short time you may. Climb on that chair,' Mr Ollendod said, smiling, and they climbed up and peered into the mirror, while he steadied them with a hand on each of their necks.

'Oh!' cried Jenny, 'I can see him! I can see Tuffy! He's a kitten again, chasing grasshoppers.'

'I can see him too!' shouted Martin, jumping up and down. The chair overbalanced and tipped them on to the floor.

'Let us look again, please let us!'

'Not today,' said Mr Ollendod. 'If you look too long into that mirror you, like the queen, might vanish into it for good. That is why I keep it covered with a shawl.'

The children went away comforted, thinking of Tuffy young and frolicsome once more, chasing butterflies in the sun. Mr Ollendod gave them a little ivory chess set, to distract them from missing their cat, but Mrs Minser, saying it was too good for the children and that they would only spoil it, sold it and put the money in the Post Office 'for later on'.

It was July now. The weather grew daily warmer and closer.

Mrs Minser told Mr Ollendod that she was obliged to raise his rent by three guineas 'for the summer prices'. She rather hoped this would make him leave, but he paid up.

'I'm old and tired,' he said. 'I don't want to move again, for I may not be here very long. One of these days my heart will carry me off.'

And in fact one oppressive, thundery day he had a bad heart attack and had to stay in bed for a week.

'I certainly don't want him if he's going to be ill all the time,' Mrs Minser said to her husband. 'I shall tell him that we want his room as soon as he's better.' In the meantime she put away as many as possible of the Indian things, saying that they were a dust-collecting nuisance in the sick-room. She left the swords and the fan and the mirror, because they hung on the wall, out of harm's way.

As she had promised, the minute Mr Ollendod was up and walking around again, she told him his room was wanted and he must go.

'But where?' he said, standing so still, leaning on his stick, that Mrs Minser had the uneasy notion for a moment that the clock on the wall had stopped ticking to listen for her answer.

'That's no concern of mine,' she said coldly. 'Go where you please, wherever anyone can be found who will take you with all this rubbish.'

'I must think it over,' said Mr Ollendod. He put on his panama hat and walked slowly down to the beach. The tide was out, revealing a mile of flat, pallid mud studded with baked-bean tins. Jenny and Martin were there, listlessly trying to fly a home-made kite. Not a breath of wind stirred and the kite kept flopping down in the mud, but they knew that if they went home before six their mother would send them out again.

'There's Mr Ollendod,' said Jenny.

'Perhaps he could fly the kite,' said Martin.

They ran to him, leaving two black parallel trails in the shining goo.

'Mr Ollendod, can you fly our kite?'

'It needs someone to run with it *very* fast.'

He smiled at them kindly. Even the slowest stroll now made his heart begin to race and stumble.

'Let's see,' he said. He held the string for a moment in his hands and was silent; then he said, 'I can't run with it, but perhaps I can persuade it to go up of its own accord.'

The children watched, silent and attentive, while he murmured something to the rope in a low voice that they could not quite catch.

'Look, it's moving,' whispered Martin.

The kite, which had been hanging limp, suddenly twitched and jerked like a fish at the end of a line, then, by slow degrees, drew itself up and, as if invisibly pulled from above, began to climb higher and higher into the warm grey sky. Mr Ollendod kept his eyes fixed on it; Jenny noticed that his hands were clenched and the sweat was rolling off his forehead.

'It's like the story!' exclaimed Martin. 'The man with the rope and the magic word and the boy who climbs it—may we climb it? We've learnt how to at school.'

Mr Ollendod couldn't speak, but they took his silence as consent. They flung themselves at the rope and swarmed up it. Mr Ollendod, still holding on to the end of the rope, gradually lowered himself to the ground and sat with his head bowed over his knees; then with a slow subsiding motion fell over on to his side. His hands relaxed on the rope which swung softly upwards and disappeared; after a while the tide came in and washed away three sets of foot-prints.

'Those children are very late,' said Mrs Minser at six o'clock. 'Are they up in Mr Ollendod's room?'

She went up to see. The room was empty.

'I shall let it to a couple, next time,' reflected Mrs Minser, picking up the peacock-feather fan and fanning herself, for the heat was oppressive. 'A couple will pay twice the rent and they are most likely to eat meals out. I wonder where those children can have got to . . . ?'

An hour later old Mr Hill, on his way down to supper, looked through Mr Ollendod's open door and saw a snake wriggling about on the carpet. He called out excitedly. By the time Mr Minser had come up, the snake had slid under the bed and Mrs Pursey was screaming vigorously. Mr Minser rattled a stick and the snake shot out towards his foot, but he was ready with a sharp scimitar snatched from the wall, and cut off its head. The old people, clustered in a dithering group outside the door, applauded his quickness.

'Fancy Mr Ollendod's keeping a pet snake all this time and we never knew!' shuddered Mrs Pursey. 'I hope he hasn't anything else of the kind in here.' Inquisitively she ventured in. 'Why, what a beautiful mirror!' she cried. The others followed, pushing and chattering, looking about greedily.

Mr Minser brushed through the group, irritably, and went downstairs with the decapitated snake. 'I shall sound the gong for supper in five minutes,' he called. 'Hannah, Hannah! Where are you? Nothing's going as it should in this house today.'

But Hannah, needless to say, did not reply, and when he banged the gong in five minutes, nobody came down but blind old Miss Drake who said, rather peevishly, that all the others had slipped away and left her behind in Mr Ollendod's room.

'Slipped away! And left me! Among all his horrid things! Without saying a word, so inconsiderate! Anything might have happened to me.'

And she started quickly eating up Mrs Pursey's buttered toast.

1976

ANGELA CARTER

The Courtship of Mr Lyon

Outside her kitchen window, the hedgerow glistened as if the snow possessed a light of its own; when the sky darkened towards evening, an unearthly, reflected pallor remained behind upon the winter's landscape, while still the soft flakes floated down. This lovely girl, whose skin possesses that same, inner light so you would have thought she, too, was made all of snow, pauses in her chores in the mean kitchen to look out at the country road. Nothing has passed that way all day; the road is white and unmarked as a spilled bolt of bridal satin.

Father said he would be home before nightfall.

The snow brought down all the telephone wires; he couldn't have called, even with the best of news.

The roads are bad. I hope he'll be safe.

But the old car stuck fast in a rut, wouldn't budge an inch; the engine whirred, coughed and died and he was far from home. Ruined, once; then ruined again, as he had learnt from his lawyers that very morning; at the conclusion of the lengthy, slow attempt to restore his fortunes, he had turned out his pockets to find the cash for petrol to take him home. And not even enough money left over to buy his Beauty, his girl-child, his pet, the one white rose she said she wanted; the only gift she wanted, no matter how the case went, how rich he might once again be. She had asked for so little and he had not been able to give it to her. He cursed the useless car, the last straw that broke his spirit; then, nothing for it but to fasten his old sheepskin coat around him, abandon the heap of metal and set off down the snow-filled lane to look for help.

Behind wrought iron gates, a short, snowy drive performed a reticent flourish before a miniature, perfect, Palladian house that seemed to hide itself shyly behind snow-laden skirts of an antique cypress. It was almost night; that house, with its sweet, retiring melancholy grace, would have seemed deserted but for a light that flickered in an upstairs window, so vague it might have been the reflection of a star, if any stars could have penetrated the snow that whirled yet more thickly. Chilled through, he pressed the latch of the gate and saw, with a pang, how, on the withered ghost of a tangle of thorns, there clung, still, the faded rag of a white rose.

The gate clanged loudly shut behind him; too loudly. For an instant, that reverberating clang seemed final, emphatic, ominous as if the gate, now closed, barred all within it from the world outside the walled, wintry garden. And, from a distance, though from what distance he could not tell, he heard the most singular sound in the world: a great roaring, as of a beast of prey.

In too much need to allow himself to be intimidated, he squared up to the mahogany door. This door was equipped with a knocker in the shape of a lion's head, with a ring through the nose; as he raised his hand towards it, it came to him this lion's head was not, as he had thought at first, made of brass, but, instead, of solid gold. Before, however, he could announce his presence, the door swung silently inward on well-oiled hinges and he saw a white hall where the candles of a great chandelier cast their benign light upon so many, many flowers in great, free-standing jars of crystal that it seemed the whole of spring drew him into its warmth with a profound intake of perfumed breath. Yet there was no living person in the hall.

The door behind him closed as silently as it had opened, yet, this time, he felt no fear although he knew by the pervasive atmosphere of a suspension of reality that he had entered a place of privilege where all the laws of the world he knew need not necessarily apply, for the very rich are often very eccentric and the house was plainly that of an exceedingly wealthy man. As it was, when nobody came to help him with his coat, he took it off himself. At that, the crystals of the chandelier tinkled a little, as if emitting a pleased chuckle, and the door of a cloakroom opened of its own accord. There were, however, no clothes at all in this cloakroom, not even the statutory country-house garden mackintosh to greet his own squirearchal sheepskin, but, when he emerged again into the hall, he found a greeting waiting for him at last—there was, of all things, a liver and white King Charles spaniel crouched, with head intelligently cocked, on the Kelim runner. It gave him further, comforting proof of his unseen host's wealth and eccentricity to see the dog wore, in place of a collar, a diamond necklace.

The dog sprang to its feet in welcome and busily shepherded him (how amusing!) to a snug little leather-panelled study on the first floor, where a low table was drawn up to a roaring log fire. On the table, a silver tray; round the neck of the whisky decanter, a silver tag with the legend: *Drink me*, while the cover of the silver dish was engraved with the exhortation: *Eat me*, in a flowing hand. This dish contained sandwiches of thick-cut roast beef, still bloody. He drank the one with soda and ate the other with some excellent mustard thoughtfully provided in a stoneware pot, and, when the spaniel saw to it he had served himself, she trotted off about her own business.

All that remained to make Beauty's father entirely comfortable was to find, in a curtained recess, not only a telephone but the card of a garage that advertised a twenty-four-hour rescue service; a couple of calls later and he had confirmed, thank God, there was no serious trouble, only the car's age and the cold weather . . . could he pick it up from the village in an hour? And directions to the village, but half a mile away, were supplied, in a new tone of deference, as soon as he described the house from where he was calling.

And he was disconcerted but, in his impecunious circumstances, relieved to hear the bill would go on his hospitable if absent host's account; no question, assured the mechanic. It was the master's custom.

Time for another whisky as he tried, unsuccessfully, to call Beauty and tell her he would be late; but the lines were still down, although, miraculously, the storm had cleared as the moon rose and now a glance between the velvet curtains revealed a landscape as of ivory with an inlay of silver. Then the spaniel appeared again, with his hat in her careful mouth, prettily wagging her tail, as if to tell him it was time to be gone, that this magical hospitality was over.

As the door swung to behind him, he saw the lion's eyes were made of agate.

Great wreaths of snow now precariously curded the rose trees and, when he brushed against a stem on his way to the gate, a chill armful softly thudded to the ground to reveal, as if miraculously preserved beneath it, one last, single, perfect rose that might have been the last rose left living in all the white winter, and of so intense and yet delicate a fragrance it seemed to ring like a dulcimer on the frozen air.

How could his host, so mysterious, so kind, deny Beauty her present?

Not now distant but close at hand, close as that mahogany front door, rose a mighty, furious roaring; the garden seemed to hold its breath in apprehension. But still, because he loved his daughter, Beauty's father stole the rose.

At that, every window of the house blazed with furious light and a fugal baying, as of a pride of lions, introduced his host.

There is always a dignity about great bulk, an assertiveness, a quality of being more *there* than most of us are. The being who now confronted Beauty's father seemed to him, in his confusion, vaster than the house he owned, ponderous yet swift, and the moonlight glittered on his great, mazy head of hair, on the eyes green as agate, on the golden hairs of the great paws that grasped his shoulders so that their claws pierced the sheepskin as he shook him like an angry child shakes a doll.

This leonine apparition shook Beauty's father until his teeth rattled and then dropped him sprawling on his knees while the spaniel, darting from the open door, danced round them, yapping distractedly, like a lady at whose dinner party blows have been exchanged.

'My good fellow—' stammered Beauty's father; but the only response was a renewed roar.

'Good fellow? I am no good fellow! I am the Beast, and you must call me Beast, while I call you, Thief!'

'Forgive me for robbing your garden, Beast!'

Head of a lion; mane and mighty paws of a lion; he reared on his hind legs like an angry lion yet wore a smoking jacket of dull red brocade and was the owner of that lovely house and the low hills that cupped it.

'It was for my daughter,' said Beauty's father. 'All she wanted, in the whole world, was one white, perfect rose.'

The Beast rudely snatched the photograph her father drew from his wallet and inspected it, first brusquely, then with a strange kind of wonder, almost the dawning of surmise. The camera had captured a certain look she had, sometimes, of absolute sweetness and absolute gravity, as if her eyes might pierce appearances and see your soul. When he handed the picture back, the Beast took good care not to scratch the surface with his claws.

'Take her the rose, then, but bring her to dinner,' he growled; and what else was there to be done?

Although her father had told her of the nature of the one who waited for her, she could not control an instinctual shudder of fear when she saw him, for a lion is a lion and a man is a man and, though lions are more beautiful by far than we are, yet they belong to a different order of beauty and, besides, they have no respect for us: why should they? Yet wild things have a far more rational fear of us than is ours of them, and some kind of sadness in his agate eyes, that looked almost blind, as if sick of sight, moved her heart.

He sat, impassive as a figurehead, at the top of the table; the dining room was Queen Anne, tapestried, a gem. Apart from an aromatic soup kept hot over a spirit lamp, the food, though exquisite, was cold—a cold bird, a cold soufflé, cheese. He asked her father to serve them from a buffet and, himself, ate nothing. He grudgingly admitted what she had already guessed, that he disliked the presence of servants because, she thought, a constant human presence would remind him too bitterly of his otherness, but the spaniel sat at his feet throughout the meal, jumping up from time to time to see that everything was in order.

How strange he was. She found his bewildering difference from herself almost intolerable; its presence choked her. There seemed a heavy, soundless pressure upon her in his house, as if it lay under water, and when she saw the great paws lying on the arm of his chair, she thought: they are the death of any tender herbivore. And such a one she felt herself to be, Miss Lamb, spotless, sacrificial.

Yet she stayed, and smiled, because her father wanted her to do so; and when the Beast told her how he would aid her father's appeal against the judgement, she smiled with both her mouth and her eyes. But when, as they sipped their brandy, the Beast, in the diffuse, rumbling purr with which he conversed, suggested, with a hint of shyness, of fear of refusal, that she should stay here, with him, in comfort, while her father returned to London to take up the legal cudgels again, she forced a smile. For she knew with a pang of dread, as soon as he spoke, that it would be so and her visit to the Beast must be, on some magically reciprocal scale, the price of her father's good fortune.

Do not think she had no will of her own; only, she was possessed by a sense of obligation to an unusual degree and, besides, she would gladly have gone to the ends of the earth for her father, whom she loved dearly.

Her bedrom contained a marvellous glass bed; she had a bathroom, with towels thick as fleece and vials of suave unguents; and a little parlour of her own, the walls of which were covered with an antique paper of birds of paradise and Chinamen, where there were precious books and pictures and the flowers grown by invisible gardeners in the Beast's hothouses. Next morning, her father kissed her and drove away with a renewed hope about him that made her glad, but, all the same, she longed for the shabby home of their poverty. The unaccustomed luxury about her she found poignant, because it gave no pleasure to its possessor and himself she did not see all day as if, curious reversal, she frightened him, although the spaniel came and sat with her, to keep her company. Today, the spaniel wore a neat choker of turquoises.

Who prepared her meals? Loneliness of the Beast; all the time she stayed there, she saw no evidence of another human presence but the trays of food that arrived on a dumb waiter inside a mahogany cupboard in her parlour. Dinner was eggs Benedict and grilled veal; she ate it as she browsed in a book she had found in the rosewood revolving bookcase, a collection of courtly and elegant French fairy tales about white cats who were transformed princesses and fairies who were birds. Then she pulled a sprig of muscat grapes from a fat bunch for her dessert and found herself yawning; she discovered she was bored. At that, the spaniel took hold of her skirt with its velvet mouth and gave it a firm but gentle tug. She allowed the dog to trot before her to the study in which her father had been entertained and there, to her well-disguised dismay, she found her host, seated beside the fire with a tray of coffee at his elbow from which she must pour.

The voice that seemed to issue from a cave full of echoes, his dark, soft rumbling growl; after her day of pastel-coloured idleness, how could she converse with the possessor of a voice that seemed an instrument created to inspire the terror that the chords of great organs bring? Fascinated, almost awed, she watched the firelight play on the gold fringes of his mane; he was irradiated, as if with a kind of halo, and she thought of the first great beast of the Apocalypse, the winged lion with his paw upon the Gospel, Saint Mark. Small talk turned to dust in her mouth; small talk had never, at the best of times, been Beauty's forte, and she had little practice at it.

But he, hesitantly, as if he himself were in awe of a young girl who looked as if she had been carved out of a single pearl, asked after her father's law case; and her dead mother; and how they, who had been so rich, had come to be so poor. He forced himself to master his shyness, which was that of a wild creature, and so she contrived to master her own—to such effect that soon she was chattering away to him as if she had known him all her life. When the little cupid in the gilt clock on the mantelpiece struck its miniature tambourine, she was astonished to discover it did so twelve times.

'So late! You will want to sleep,' he said.

At that, they both fell silent, as if these strange companions were suddenly overcome with embarrassment to find themselves together, alone, in that room in the depths of the winter's night. As she was about to rise, he flung himself at her feet and buried his head in her lap. She stayed stock-still, transfixed; she felt his hot breath on her fingers, the stiff bristles of his muzzle grazing her skin, the rough lapping of his tongue and then, with a flood of compassion, understood: all he is doing is kissing my hands.

He drew back his head and gazed at her with his green, inscrutable eyes, in which she saw her face repeated twice, as small as if it were in bud. Then, without another word, he sprang from the room and she saw, with an indescribable shock, he went on all fours.

Next day, all day, the hills on which the snow still settled echoed with the Beast's rumbling roar: has master gone a-hunting? Beauty asked the spaniel. But the spaniel growled, almost bad-temperedly, as if to say, that she would not have answered, even if she could have.

Beauty would pass the day in her suite reading or, perhaps, doing a little embroidery; a box of coloured silks and a frame had been provided for her. Or, well wrapped up, she wandered in the walled garden, among the leafless roses, with the spaniel at her heels, and did a little raking and rearranging. An idle, restful time; a holiday. The enchantment of that bright, sad, pretty place enveloped her and she found that, against all her expectations, she was happy there. She no longer felt the slightest apprehension at her nightly interviews with the Beast. All the natural laws of the world were held in suspension, here, where an army of invisibles tenderly waited on her, and she would talk with the lion, under the patient chaperonage of the brown-eyed dog, on the nature of the moon and its borrowed light, about the stars and the substances of which they were made, about the variable transformations of the weather. Yet still his strangeness made her shiver; and when he helplessly fell before her to kiss her hands, as he did every night when they parted, she would retreat nervously into her skin, flinching at his touch.

The telephone shrilled; for her. Her father. Such news!

The Beast sunk his great head on to his paws. You will come back to me? It will be lonely here, without you.

She was moved almost to tears that he should care for her so. It was in her heart to drop a kiss upon his shaggy mane but, though she stretched out her hand towards him, she could not bring herself to touch him of her own free will, he was so different from herself. But, yes, she said; I will come back. Soon, before the winter is over. Then the taxi came and took her away.

You are never at the mercy of the elements in London, where the huddled warmth of humanity melts the snow before it has time to settle; and her father was as good as rich again, since his hirsute friend's lawyers had the business so well in hand that his credit brought them nothing but the best. A resplendent hotel; the opera, theatres; a whole new wardrobe for his darling, so she could step out on his arm to parties, to receptions, to restaurants, and life was as she had never known it, for her father had ruined himself before her birth killed her mother.

Although the Beast was the source of this new-found prosperity and they talked of him often, now that they were so far away from the timeless spell of his house it seemed to possess the radiant and finite quality of dream and the Beast himself, so monstrous, so benign, some kind of spirit of good fortune who had smiled on them and let them go. She sent him flowers, white roses in return for the ones he had given her; and when she left the florist, she experienced a sudden sense of perfect freedom, as if she had just escaped from an unknown danger, had been grazed by the possibility of some change but, finally, left intact. Yet, with this exhilaration, a desolating emptiness. But her father was waiting for her at the hotel; they had planned a delicious expedition to buy her furs and she was as eager for the treat as any girl might be.

Since the flowers in the shop were the same all the year round, nothing in the window could tell her that winter had almost gone.

Returning late from supper after the theatre, she took off her earrings in front of the mirror; Beauty. She smiled at herself with satisfaction. She was learning, at the end of her adolescence, how to be a spoiled child and that pearly skin of hers was plumping out, a little, with high living and compliments. A certain inwardness was beginning to transform the lines around her mouth, those signatures of the personality, and her sweetness and her gravity could sometimes turn a mite petulant when things went not quite as she wanted them to go. You could not have said that her freshness was fading but she smiled at herself in mirrors a little too often, these days, and the face that smiled back was not quite the one she had seen contained in the Beast's agate eyes. Her face was acquiring, instead of beauty, a lacquer of the invincible prettiness that characterizes certain pampered, exquisite, expensive cats.

The soft wind of spring breathed in from the near-by park through the open windows; she did not know why it made her want to cry.

There was a sudden, urgent, scrabbling sound, as of claws, at her door.

Her trance before the mirror broke; all at once, she remembered everything perfectly. Spring was here and she had broken her promise. Now the Beast himself had come in pursuit of her! First, she was frightened of his anger; then, mysteriously joyful, she ran to open the door. But it was his liver and white spotted spaniel who hurled herself into the girl's arms in a flurry of little barks and gruff murmurings, of whimpering and relief.

Yet where was the well-brushed, jewelled dog who had sat beside her embroidery frame in the parlour with birds of paradise nodding on the walls? This one's fringed ears were matted with mud, her coat was dusty and snarled, she was thin as a dog that has walked a long way and, if she had not been a dog, she would have been in tears.

After that first, rapturous greeting, she did not wait for Beauty to order her food and water; she seized the chiffon hem of her evening dress, whimpered and tugged. Threw back her head, howled, then tugged and whimpered again.

There was a slow, late train that would take her to the station where she had left for London three months ago. Beauty scribbled a note for her father, threw a coat round her shoulders. Quickly, quickly, urged the spaniel soundlessly; and Beauty knew the Beast was dying.

In the thick dark before dawn, the station master roused a sleepy driver for her. Fast as you can.

It seemed December still possessed his garden. The ground was hard as iron, the skirts of the dark cypress moved on the chill wind with a mournful rustle and there were no green shoots on the roses as if, this year, they would not bloom. And not one light in any of the windows, only, in the topmost attic, the faintest smear of radiance on a pane, the thin ghost of a light on the verge of extinction.

The spaniel had slept a little, in her arms, for the poor thing was exhausted. But now her grieving agitation fed Beauty's urgency and, as the girl pushed open the front door, she saw, with a thrust of conscience, how the golden door knocker was thickly muffled in black crêpe.

The door did not open silently, as before, but with a doleful groaning of the hinges and, this time, on to perfect darkness. Beauty clicked her gold cigarette lighter; the tapers in the chandelier had drowned in their own wax and the prisms were wreathed with drifting arabesques of cobwebs. The flowers in the glass jars were dead, as if nobody had had the heart to replace them after she was gone. Dust, everywhere; and it was cold. There was an air of exhaustion, of despair in the house and, worse, a kind of physical disillusion, as if its glamour had been sustained by a cheap conjuring trick and now the conjurer, having failed to pull the crowds, had departed to try his luck elsewhere.

Beauty found a candle to light her way and followed the faithful spaniel up the staircase, past the study, past her suite, through a house echoing with desertion up a little back staircase dedicated to mice and spiders, stumbling, ripping the hem of her dress in her haste.

What a modest bedroom! An attic, with a sloping roof, they might have given the chambermaid if the Beast had employed staff. A night light on the mantelpiece, no curtains at the windows, no carpet on the floor and a narrow, iron bedstead on which he lay, sadly diminished, his bulk scarcely disturbing the faded patchwork quilt, his mane a greyish rat's nest and his eyes closed. On the stick-backed chair where his clothes had been thrown, the roses she had sent him were thrust into the jug from the washstand but they were all dead.

The spaniel jumped up on the bed and burrowed her way under the scanty covers, softly keening.

'Oh, Beast,' said Beauty. 'I have come home.'

His eyelids flickered. How was it she had never noticed before that his agate eyes were equipped with lids, like those of a man? Was it because she had only looked at her own face, reflected there?

'I'm dying, Beauty,' he said in a cracked whisper of his former purr. 'Since you left me, I have been sick. I could not go hunting, I found I had not the stomach to kill the gentle beasts, I could not eat. I am sick and I must die; but I shall die happy because you have come to say good-bye to me.'

She flung herself upon him, so that the iron bedstead groaned, and covered his poor paws with her kisses.

'Don't die, Beast! If you'll have me, I'll never leave you.'

When her lips touched the meat-hook claws, they drew back into their pads and she saw how he had always kept his fists clenched but now, painfully, tentatively, at last began to stretch his fingers. Her tears fell on his face like snow and, under their soft transformation, the bones showed through the pelt, the flesh through the wide, tawny brow. And then it was no longer a lion in her arms but a man, a man with an unkempt mane of hair and, how strange, a broken nose, such as the noses of retired boxers, that gave him a distant, heroic resemblance to the handsomest of all the beasts.

'Do you know,' said Mr Lyon, 'I think I might be able to manage a little breakfast today, Beauty, if you would eat something with me.'

Mr and Mrs Lyon walk in the garden; the old spaniel drowses on the grass, in a drift of fallen petals.

1979

JEANNE
DESY

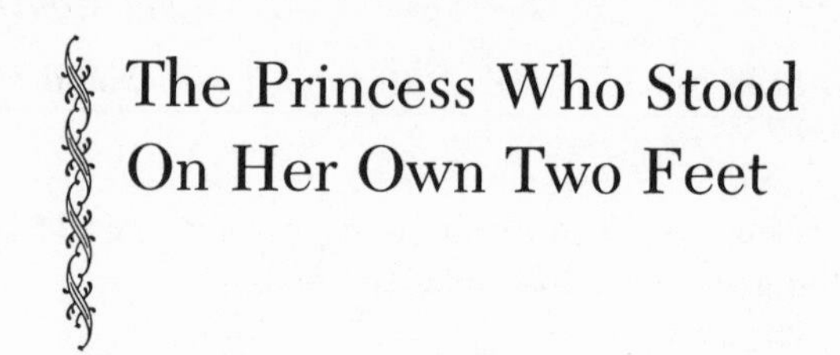

The Princess Who Stood On Her Own Two Feet

A LONG time ago in a kingdom by the sea there lived a Princess tall and bright as a sunflower. Whatever the royal tutors taught her, she mastered with ease. She could tally the royal treasure on her gold and silver abacus, and charm even the Wizard with her enchantments. In short, she had every gift but love, for in all the kingdom there was no suitable match for her.

So she played the zither and designed great tapestries and trained her finches to eat from her hand, for she had a way with animals.

Yet she was bored and lonely, as princesses often are, being a breed apart. Seeing her situation, the Wizard came to see her one day, a strange and elegant creature trotting along at his heels. The Princess clapped her hands in delight, for she loved anything odd.

'What is it?' she cried. The Wizard grimaced.

'Who knows?' he said. 'It's supposed to be something enchanted. I got it through the mail.' The Royal Wizard looked a little shamefaced. It was not the first time he had been taken in by mail-order promises.

'It won't turn into anything else,' he explained. 'It just is what it is.'

'But what is it?'

'They call it a dog,' the Wizard said. 'An Afghan hound.'

Since in this kingdom dogs had never been seen, the Princess was quite delighted. When she brushed the silky, golden dog, she secretly thought it looked rather like her, with its thin aristocratic features and delicate nose. Actually, the Wizard had thought so too, but you can never be sure what a Princess will take as an insult. In any case, the Princess and the dog became constant companions. It followed her on her morning rides and slept at the foot of her bed every night. When she talked, it watched her so attentively that she often thought it understood.

Still, a dog is a dog and not a Prince, and the Princess longed to marry. Often she sat at her window in the high tower, her embroidery idle in her aristocratic hands, and gazed down the road, dreaming of a handsome prince in flashing armor.

One summer day word came that the Prince of a neighboring kingdom wished to discuss an alliance. The royal maids confided that he was dashing and princely, and the Princess's heart leaped with joy. Eagerly she awaited the betrothal feast.

When the Prince entered the great banquet hall and cast his dark, romantic gaze upon her, the Princess nearly swooned in her chair. She sat shyly while everyone toasted the Prince and the golden Princess and peace forever between the two kingdoms. The dog watched quietly from its accustomed place at her feet.

After many leisurely courses, the great feast ended, and the troubadors began to play. The Prince and Princess listened to the lyrical songs honoring their love, and she let him hold her hand under the table—an act noted with triumphant approval by the King and Queen. The Princess was filled with happiness that such a man would love her.

At last the troubadors swung into a waltz, and it was time for the Prince and Princess to lead the dance. Her heart bursting with joy, the Princess rose to take his arm. But as she rose to her feet, a great shadow darkened the Prince's face, and he stared at her as if stricken.

'What is it?' she cried. But the Prince would not speak, and dashed from the hall.

For a long time the Princess studied her mirror that night, wondering what the Prince had seen.

'If you could talk,' she said to the dog, 'you could tell me, I know it,' for the animal's eyes were bright and intelligent. 'What did I do wrong?'

The dog, in fact, *could* talk; it's just that nobody had ever asked him anything before.

'You didn't do anything,' he said. 'It's your height.'

'My height?' The Princess was more astonished by what the dog said than the fact that he said it. As an amateur wizard, she had heard of talking animals.

'But I am a Princess!' she wailed. 'I'm supposed to be tall.' For in her kingdom, all the royal family was tall, and the Princess the tallest of all, and she had thought that was the way things were supposed to be.

The dog privately marveled at her naiveté, and explained that in the world outside this kingdom, men liked to be taller than their wives.

'But why?' asked the Princess.

The dog struggled to explain. 'They think if they're not, they can't . . . train falcons as well. Or something.' Now that he thought for a moment, he didn't know either.

'It's my legs,' she muttered. 'When we were sitting down, everything was fine. It's these darn long legs.' The dog cocked his head. He thought she had nice legs, and he was in a position to know. The Princess strode to the bell pull and summoned the Wizard.

'Okay,' she said when he arrived. 'I know the truth.'

'Who told you?' the Wizard asked. Somebody was in for a bit of a stay in irons.

'The dog.' The Wizard sighed. In fact, he had *known* the creature was enchanted.

'It's my height,' she continued bitterly. The Wizard nodded. 'I want you to make me shorter,' she said. 'A foot shorter, at least. Now.'

Using all his persuasive powers, which were considerable, the Wizard explained to her that he could not possibly do that. 'Fatter,' he said, 'yes. Thinner, yes. Turn you into a raven, maybe. But shorter, no. I cannot make you even an inch shorter, my dear.'

The Princess was inconsolable.

Seeing her sorrow, the King sent his emissary to the neighboring kingdom with some very attractive offers. Finally the neighboring King and Queen agreed to persuade the Prince to give the match another chance. The Queen spoke to him grandly of chivalry and honor, and the King spoke to him privately of certain gambling debts.

In due course he arrived at the castle, where the Princess had taken to her canopied bed. They had a lovely romantic talk, with him at the bedside holding her hand, and the nobility, of course, standing respectfully at the foot of the bed, as such things are done. In truth, he found the Princess quite lovely when she was sitting or lying down.

'Come on,' he said, 'let's get some fresh air. We'll go riding.'

He had in mind a certain dragon in these parts, against whom he might display his talents. And so the Prince strode and the Princess slouched to the stables.

On a horse, as in a chair, the Princess was no taller than he, so they cantered along happily. Seeing an attractive hedge ahead, the Prince urged his mount into a gallop and sailed the hedge proudly. He turned to see her appreciation, only to find the Princess doing the same, and holding her seat quite gracefully. Truthfully, he felt like leaving again.

'Didn't anyone ever tell you,' he said coldly, 'that ladies ride sidesaddle?' Well, of course they had, but the Princess always thought that that was a silly, unbalanced position that took all the fun out of riding. Now she apologized prettily and swung her legs around.

At length the Prince hurdled another fence, even more dashingly than before, and turned to see the Princess attempting to do the same thing. But riding sidesaddle, she did not have a sure seat, and tumbled to the ground.

'Girls shouldn't jump,' the Prince told the air, as he helped her up.

But on her feet, she was again a head taller than he. She saw the dim displeasure in his eyes. Then, with truly royal impulsiveness, she made a decision to sacrifice for love. She crumpled to the ground.

'My legs,' she said. 'I can't stand.' The Prince swelled with pride, picked her up, and carried her back to the castle.

There the Royal Physician, the Wizard, and even the Witch examined her legs, with the nobility in attendance.

She was given infusions and teas and herbs and packs, but nothing worked. She simply could not stand.

'When there is nothing wrong but foolishness,' the Witch muttered, 'you can't fix it.' And she left. She had no patience with lovesickness.

The Prince lingered on day after day, as a guest of the King, while the Princess grew well and happy, although she did not stand. Carried to the window seat, she would sit happily and watch him stride around the room, describing his chivalric exploits, and she would sigh with contentment. The loss of the use of her legs seemed a small price to pay for such a man. The dog observed her without comment.

Since she was often idle now, the Princess practiced witty and amusing sayings. She meant only to please the Prince, but he turned on her after one particularly subtle and clever remark and said sharply, 'Haven't you ever heard that women should be seen and not heard?'

The Princess sank into thought. She didn't quite understand the saying, but she sensed that it was somehow like her tallness. For just as he preferred her sitting, not standing, he seemed more pleased when she listened, and more remote when she talked.

The next day when the Prince came to her chambers he found the royal entourage gathered around her bed.

'What's the matter?' he asked. They told him the Princess could not speak, not for herbs or infusions or magic spells. And the Prince sat by the bed and held her hand and spoke to her gently, and she was given a slate to write her desires. All went well for several days. But the Prince was not a great reader, so she put the slate aside, and made conversation with only her eyes and her smile. The Prince told her daily how lovely she was, and then he occupied himself with princely pastimes. Much of the time her only companion was the dog.

One morning the Prince came to see her before he went hunting. His eyes fixed with disgust on the dog, who lay comfortably over her feet.

'Really,' the Prince said, 'sometimes you surprise me.' He went to strike the dog from the bed, but the Princess stayed his hand. He looked at her in amazement.

That night the Princess lay sleepless in the moonlight, and at last, hearing the castle fall silent, and knowing that nobody would catch her talking, she whispered to the dog, 'I don't know what I would do without you.'

'You'd better get used to the idea,' said the dog. 'The Prince doesn't like me.'

'He will never take you away.' The Princess hugged the dog fiercely. The dog looked at her skeptically and gave a little doggy cough.

'He took everything else away,' he said.

'No,' she said. 'I did that. I made myself . . . someone he could love.'

'I love you, too,' the dog said.

'Of course you do.' She scratched his ears.

'And,' said the dog, 'I loved you *then*.' The Princess lay a long time thinking before she finally slept.

The next morning the Prince strode in more handsome and dashing than ever, although, oddly enough, the Princess could have sworn he was getting shorter.

As he leaned down to kiss her, his smile disappeared. She frowned a question at him: What's the matter?

'You've still *got* that thing,' he said, pointing to the dog. The Princess grabbed her slate.

'He is all I have,' she wrote hastily. The lady-in-waiting read it to the Prince.

'You have *me*,' the Prince said, his chin high. 'I believe you love that smelly thing more than you love me.' He strode (he never walked any other way) to the door.

'I *was* going to talk to you about the wedding feast,' he said, as he left. 'But now, never mind!'

The Princess wept softly and copiously, and the dog licked a tear from her trembling hand.

'What does he *want*?' she asked the dog.

'Roast dog for the wedding feast, I'd imagine,' he said. The Princess cried out in horror.

'Oh, not literally,' the dog said. 'But it follows.' And he would say no more.

At last the Princess called the Wizard and wrote on her slate what the dog had said. The Wizard sighed. How awkward. Talking animals were always so frank. He hemmed and hawed until the Princess glared to remind him that Wizards are paid by royalty to advise and interpret—not to sigh.

'All right,' he said at last. 'Things always come in threes. Everything.'

The Princess looked at him blankly.

'Wishes always come in threes,' the Wizard said. 'And sacrifices, too. So far, you've given up walking. You've given up speech. One more to go.'

'Why does he want me to give up the dog?' she wrote.

The Wizard looked sorrowfully at her from under his bushy brows.

'Because you love it,' he said.

'But that takes nothing from him!' she scribbled. The Wizard smiled, thinking that the same thing could be said of her height and her speech.

'If you could convince him of that, my dear,' he said, 'you would be more skilled in magic than I.'

When he was gone, the Princess reached for her cards and cast her own fortune, muttering to herself. The dog watched bright-eyed as the wands of growth were covered by the swords of discord. When the ace of swords fell, the Princess gasped. The dog put a delicate paw on the card.

'You poor dumb thing,' she said, for it is hard to think of a dog any other way, whether it talks or not. 'You don't understand. That is death on a horse. Death to my love.'

'His banner is the white rose,' said the dog, looking at the card intently. 'He is also rebirth.' They heard the Prince's striding step outside the door.

'Quick,' the Princess said. 'Under the bed.' The dog's large brown eyes spoke volumes, but he flattened and slid under the bed. And the Prince's visit was surprisingly jolly.

After some time the Prince looked around with imitation surprise. 'Something's missing,' he said. 'I know. It's that creature of yours. You know, I think I was allergic to it. I feel much better now that it's gone.' He thumped his chest to show how clear it was. The Princess grabbed her slate, wrote furiously, and thrust it at the Royal Physician.

'"He loved me,"' the Royal Physician read aloud.

'Not as I love you,' the Prince said earnestly. The Princess gestured impatiently for the reading to continue.

'That's not all she wrote,' the Royal Physician said. 'It says, "The dog loved me *then*."'

When everyone was gone, the dog crept out to find the Princess installed at her window seat thinking furiously.

'If I am to keep you,' she said to him. 'we shall have to disenchant you with the spells book.' The dog smiled, or seemed to. She cast dice, she drew pentagrams, she crossed rowan twigs and chanted every incantation in the index. Nothing worked. The dog was still a dog, silken, elegant, and seeming to grin in the heat. Finally the Princess clapped shut the last book and sank back.

'Nothing works,' she said. 'I don't know what we shall do. Meanwhile, when you hear anyone coming, hide in the cupboard or beneath the bed.'

'You're putting off the inevitable,' the dog told her sadly.

'I'll think of something,' she said. But she couldn't.

At last it was the eve of her wedding day. While the rest of the castle buzzed with excitement, the Princess sat mute in her despair.

'I can't give you up and I can't take you!' she wailed. And the dog saw that she was feeling grave pain.

'Sometimes,' the dog said, looking beyond her shoulder, 'sometimes one must give up everything for love.' The Princess's lip trembled and she looked away.

'What will I *do*?' she cried again. The dog did not answer. She turned toward him and then fell to her knees in shock, for the dog lay motionless on the floor. For hours she sat weeping at his side, holding his lifeless paw.

At last she went to her cupboard and took out her wedding dress, which was of the softest whitest velvet. She wrapped the dog in its folds and picked him up gently.

Through the halls of the castle the Princess walked, and the nobility and chambermaids and royal bishops stopped in their busy preparations to watch her, for the Princess had not walked now for many months. To their astonished faces she said, 'I am going to bury the one who really loved me.'

On the steps of the castle she met the Prince, who was just dismounting and calling out jovial hearty things to his companions. So surprised was he to see her walking that he lost his footing and tumbled to the ground. She paused briefly to look down at him, held the dog closer to her body, and walked on. The Prince got up and went after her.

'What's going on here?' he asked. 'What are you doing? Isn't that your wedding dress?' She turned so he could see the dog's head where it nestled in her left arm.

'I thought you got rid of that thing weeks ago,' the Prince said. It was difficult for him to find an emotion suitable to this complex situation. He tried feeling hurt.

'What you call "this thing,"' the Princess said, 'died to spare me pain. And I intend to bury him with honor.' The Prince only half-heard her, for he was struck by another realization.

'You're talking!'

'Yes.' She smiled.

Looking down at him, she said, 'I'm talking. The better to tell you good-bye. So good-bye.' And off she went. She could stride too, when she wanted to.

'Well, my dear,' the Queen said that night, when the Princess appeared in the throne room. 'You've made a proper mess of things. We have alliances to think of. I'm sure you're aware of the very complex negotiations you have quite ruined. Your duty as a Princess . . .'

'It is not necessarily my duty to sacrifice everything,' the Princess interrupted. 'And I have other duties: a Princess says what she thinks. A Princess stands on her own two feet. A Princess stands tall. And she does not betray those who love her.' Her royal parents did not reply. But they seemed to ponder her words.

The Princess lay awake that night for many hours. She was tired from the day's exertions, for she let no other hand dig the dog's grave or fill it, but she could not sleep without the warm weight of the dog across her feet, and the sound of his gentle breathing. At last she put on her cloak and slippers and stole through the silent castle out to the gravesite. There she mused upon love, and what she had given for love, and what the dog had given.

'How foolish we are,' she said aloud. 'For a stupid Prince I let my wise companion die.'

At last the Princess dried her tears on her hem and stirred herself to examine the white rose she had planted on the dog's grave. She watered it again with her little silver watering can. It looked as though it would live.

As she slipped to the castle through the ornamental gardens, she heard a quiet jingling near the gate. On the bridge there was silhouetted a horseman. The delicate silver bridles of his horse sparkled in the moonlight. She could see by his crested shield that he must be nobility, perhaps a Prince. Well, there was many an empty room in the castle tonight, with the wedding feast canceled and all the guests gone home. She approached the rider.

He was quite an attractive fellow, thin with silky golden hair. She smiled up at him, admiring his lean and elegant hand on the reins.

'Where have you come from?' she asked.

He looked puzzled. 'Truthfully,' he replied, 'I can't remember. I know I have traveled a long dark road, but that is all I know.' He gave an odd little cough.

The Princess looked past him, where the road was bright in the moonlight.

'I see,' she said slowly. 'And what is your banner?' For she could not quite decipher it waving above him. He moved it down. A white rose on a black background.

'Death,' she breathed.

'No, no,' he said, smiling. 'Rebirth. And for that, a death is sometimes necessary.' He dismounted and bent to kiss the Princess's hand. She breathed a tiny prayer as he straightened up, but it was not answered. Indeed, he was several inches shorter than she was. The Princess straightened her spine.

'It is a pleasure to look up to a proud and beautiful lady,' the young Prince said, and his large brown eyes spoke volumes. The Princess blushed.

'We're still holding hands,' she said foolishly. The elegant Prince smiled, and kept hold of her hand, and they went toward the castle.

In the shadows the Wizard watched them benignly until they were out of sight. Then he turned to the fluffy black cat at his feet.

'Well, Mirabelle,' he said. 'One never knows the ways of enchantments.' The cat left off from licking one shoulder for a moment and regarded him, but said nothing. Mirabelle never had been much of a conversationalist.

'Ah, well,' the Wizard said. 'I gather from all this—I shall make a note—that sometimes one must sacrifice for love.'

Mirabelle looked intently at the Wizard. 'On the other hand,' the cat said at last, 'sometimes one must *refuse* to sacrifice.'

'Worth saying,' said the Wizard approvingly. 'And true. True.' And then, because he had a weakness for talking animals, he took Mirabelle home for an extra dish of cream.

1982

URSULA
LE GUIN

The Wife's Story

HE was a good husband, a good father. I don't understand it. I don't believe in it. I don't believe that it happened. I saw it happen but it isn't true. It can't be. He was always gentle. If you'd have seen him playing with the children, anybody who saw him with the children would have known that there wasn't any bad in him, not one mean bone. When I first met him he was still living with his mother, over near Spring Lake, and I used to see them together, the mother and the sons, and think that any young fellow that was that nice with his family must be one worth knowing. Then one time when I was walking in the woods I met him by himself coming back from a hunting trip. He hadn't got any game at all, not so much as a field mouse, but he wasn't cast down about it. He was just larking along enjoying the morning air. That's one of the things I first loved about him. He didn't take things hard, he didn't grouch and whine when things didn't go his way. So we got to talking that day. And I guess things moved right along after that, because pretty soon he was over here pretty near all the time. And my sister said—see, my parents had moved out the year before and gone south, leaving us the place—my sister said, kind of teasing but serious, 'Well! If he's going to be here every day and half the night, I guess there isn't room for me!' And she moved out—just down the way. We've always been real close, her and me. That's the sort of thing doesn't ever change. I couldn't ever have got through this bad time without my sis.

Well, so he come to live here. And all I can say is, it was the happy year of my life. He was just purely good to me. A hard worker and never lazy, and so big and fine-looking. Everybody looked up to him, you know, young as he was. Lodge Meeting nights, more and more often they had him to lead the singing. He had such a beautiful voice, and he'd lead off strong, and the others following and joining in, high voices and low. It brings the shivers on me now to think of it, hearing it, nights when I'd stayed home from meeting when the children was babies—the singing coming up through the trees there, and the moonlight, summer nights, the full moon shining. I'll never hear anything so beautiful. I'll never know a joy like that again.

It was the moon, that's what they say. It's the moon's fault, and the blood. It was in his father's blood. I never knew his father, and now I wonder what become of him. He was from up Whitewater way, and had no kin around here. I always thought he went back there, but now I don't know. There was some talk about him, tales, that come out after what happened to my husband. It's something runs in the blood, they say, and it may never come out, but if it does, it's the change of the moon that does it. Always it happens in the dark of the moon. When everybody's home and asleep. Something comes over the one that's got the curse in his blood, they say, and he gets up because he can't sleep, and goes out into the glaring sun, and goes off all alone—drawn to find those like him.

And it may be so, because my husband would do that. I'd half rouse and say, 'Where you going to?' and he'd say, 'Oh, hunting, be back this evening,' and it wasn't like him, even his voice was different. But I'd be so sleepy, and not wanting to wake the kids, and he was so good and responsible, it was no call of mine to go asking 'Why?' and 'Where?' and all like that.

So it happened that way maybe three times or four. He'd come back late, and worn out, and pretty near cross for one so sweet-tempered—not wanting to talk about it. I figured everybody got to bust out now and then, and nagging never helped anything. But it did begin to worry me. Not so much that he went, but that he come back so tired and strange. Even, he smelled strange. It made my hair stand up on end. I could not endure it and I said, 'What is that—those smells on you? All over you!' And he said, 'I don't know,' real short, and made like he was sleeping. But he went down when he thought I wasn't noticing, and washed and washed himself. But those smells stayed in his hair, and in our bed, for days.

And then the awful thing. I don't find it easy to tell about this. I want to cry when I have to bring it to my mind. Our youngest, the little one, my baby, she turned from her father. Just overnight. He come in and she got scared-looking, stiff, with her eyes wide, and then she begun to cry and try to hide behind me. She didn't yet talk plain but she was saying over and over, 'Make it go away! Make it go away!'

The look in his eyes, just for one moment, when he heard that. That's what I don't want ever to remember. That's what I can't forget. The look in his eyes looking at his own child.

I said to the child, 'Shame on you, what's got into you!'—scolding, but keeping her right up close to me at the same time, because I was frightened too. Frightened to shaking.

He looked away then and said something like, 'Guess she just waked up dreaming,' and passed it off that way. Or tried to. And so did I. And I got real mad with my baby when she kept on acting crazy scared of her own dad. But she couldn't help it and I couldn't change it.

He kept away that whole day. Because he knew, I guess. It was just beginning dark of the moon.

It was hot and close inside, and dark, and we'd all been asleep some while, when something woke me up. He wasn't there beside me. I heard a little stir in the passage, when I listened. So I got up, because I could bear it no longer. I went out into the passage, and it was light there, hard sunlight coming in from the door. And I saw him standing just outside, in the tall grass by the entrance. His head was hanging. Presently he sat down, like he felt weary, and looked down at his feet. I held still, inside, and watched—I didn't know what for.

And I saw what he saw. I saw the changing. In his feet, it was, first. They got long, each foot got longer, stretching out, the toes stretching out and the foot getting long, and fleshy, and white. And no hair on them.

The hair begun to come away all over his body. It was like his hair fried away in the sunlight and was gone. He was white all over, then, like a worm's skin. And he turned his face. It was changing while I looked. It got flatter and flatter, the mouth flat and wide, and the teeth grinning flat and dull, and the nose just a knob of flesh with nostril holes, and the ears gone, and the eyes gone blue—blue, with white rims around the blue—staring at me out of that flat, soft, white face.

He stood up then on two legs.

I saw him, I had to see him, my own dear love, turned into the hateful one.

I couldn't move, but as I crouched there in the passage staring out into the day I was trembling and shaking with a growl that burst out into a crazy, awful howling. A grief howl and a terror howl and a calling howl. And the others heard it, even sleeping, and woke up.

It stared and peered, that thing my husband had turned into, and shoved its face up to the entrance of our house. I was still bound by mortal fear, but behind me the children had waked up, and the baby was whimpering. The mother anger come into me then, and I snarled and crept forward.

The man thing looked around. It had no gun, like the ones from the man places do. But it picked up a heavy fallen tree branch in its long white foot, and shoved the end of that down into our house, at me. I snapped the end of it in my teeth and started to force my way out, because I knew the man would kill our children if it could. But my sister was already coming. I saw her running at the man with her head low and her mane high and her eyes yellow as the winter sun. It turned on her and raised up that branch to hit her. But I come out of the doorway, mad with the mother anger, and the others all were coming answering my call, the whole pack gathering, there in that blind glare and heat of the sun at noon.

The man looked round at us and yelled out loud, and brandished the branch it held. Then it broke and ran, heading for the cleared fields and plowlands, down the mountainside. It ran, on two legs, leaping and weaving, and we followed it.

I was last, because love still bound the anger and the fear in me. I was running when I saw them pull it down. My sister's teeth were in its throat. I got there and it was dead. The others were drawing back from the kill, because of the taste of the blood, and the smell. The younger ones were cowering and some crying, and my sister rubbed her mouth against her forelegs over and over to get rid of the taste. I went up close because I thought if the thing was dead the spell, the curse must be done, and my husband could come back—alive, or even dead, if I could only see him, my true love, in his true form, beautiful. But only the dead man lay there white and bloody. We drew back and back from it, and turned and ran, back up into the hills, back to the woods of the shadows and the twilight and the blessed dark.

1982

JANE YOLEN 

The River Maid

THERE was once a rich farmer named Jan who decided to expand his holdings. He longed for the green meadow that abutted his farm with a passion that amazed him. But a swift river ran between the two. It was far too wide and far too deep for his cows to cross.

He stood on the riverbank and watched the water hurtle over its rocky course.

'I could build a bridge,' he said aloud. 'But, then, any fool could do that. And I am no fool.'

At his words the river growled, but Jan did not heed it.

'No!' Jan said with a laugh, 'I shall build no bridge across this water. I shall make the river move aside for me.' And so he planned how he would dam it up, digging a canal along the outer edge of the meadow, and thus allow his cows the fresh green grass.

As if guessing Jan's thought, the river roared out, tumbling stones in its rush to be heard. But Jan did not understand it. Instead, he left at once to go to the town where he purchased the land and supplies.

The men Jan hired dug and dug for weeks until a deep ditch and a large dam had been built. Then they watched as the river slowly filled up behind the dam. And when, at Jan's signal, the gate to the canal was opened, the river was forced to move into its new course and leave its comfortable old bed behind.

At that, Jan was triumphant. He laughed and turned to the waiting men. 'See!' he called out loudly, 'I am not just Jan the Farmer. I am Jan the River Tamer. A wave of my hand, and the water must change its way.'

His words troubled the other men. They spat between their fingers and made other signs against the evil eye. But Jan paid them no mind. He was the last to leave the river's side that evening and went home well after dark.

The next morning Jan's feeling of triumph had not faded and he went down again to the path of the old river, which was now no more than mud and mire. He wanted to look at the desolation and dance over the newly dried stones.

But when he got to the river's old bed, he saw someone lying face up in the center of the waterless course. It was a girl clothed only in a white shift that clung to her body like a skin.

Fearing her dead, Jan ran through the mud and knelt by her side. He put out his hand but could not touch her. He had never seen anyone so beautiful.

Fanned out about her head, her hair was a fleece of gold, each separate strand distinguishable. Fine gold hairs lay molded on her forearms and like wet down upon her legs. On each of her closed eyelids a drop of river water glistened and reflected back to him his own staring face.

At last Jan reached over and touched her cheek, and at his touch, her eyes opened wide. He nearly drowned in the blue of them.

He lifted the girl up in his arms, never noticing how cold her skin or how the mud stuck nowhere to her body or her shift, and he carried her up onto the bank. She gestured once toward the old riverbank and let out a single mewling cry. Then she curled in toward his body, nestling, and seemed to sleep.

Not daring to wake her again, Jan carried her home and put her down by the hearth. He lit the fire, though it was late spring and the house already quite warm. Then he sat by the sleeping girl and stared.

She lay in a curled position for some time. Only the slow pulsing of her back told him that she breathed. But, as dusk settled about the house, bringing with it a half-light, the girl gave a sudden sigh and stretched. Then she sat up and stared. Her arms went out before her as if she were swimming in the air. Jan wondered for a moment if she were blind.

Then the girl leaped up in one fluid movement and began to sway, to dance upon the hearthstones. Her feet beat swiftly and she turned round and round in dizzying circles. She stopped so suddenly that Jan's head still spun. He saw that she was now perfectly dry except for one side of her shift; the left hem and skirt were still damp and remained molded against her.

'Turn again,' Jan whispered hoarsely, suddenly afraid.

The girl looked at him and did not move.

When he saw that she did not understand his tongue, Jan walked over to her and led her back to the fire. Her hand was quite cold in his. But she smiled shyly up at him. She was small, only chest high, and Jan himself was not a large man. Her skin, even in the darkening house, was so white it glowed with a fierce light. Jan could see the rivulets of her veins where they ran close to the surface, at her wrists and temples.

He stayed with her by the fire until the heat made him sweat. But though she stood silently, letting the fire warm her first one side and then the other, her skin remained cold, and the left side of her shift would not dry.

Jan knelt down before her and touched the damp hem. He put his cheek against it.

'Huttah!' he cried at last. 'I know you now. You are a river maid. A water spirit. I have heard of such. I believed in them when I was a child.'

The water girl smiled steadily down at him and touched his hair with her fingers, twining the strands round and about as if weaving a spell.

Jan felt the touch, cold and hot, burn its way down the back of his head and along his spine. He remembered with dread all the old tales. To hold such a one against her will meant death. To love such a one meant despair.

He shook his head violently and her hand fell away. 'How foolish,' Jan thought. 'Old wives and children believe such things. I do not love her, beautiful as she is. And as for the other, how am I to know what is her will? If we cannot talk the same tongue, I can only guess her wants.' He rose and went to the cupboard and took out bread and cheese and a bit of salt fish which he put before her.

The water maid are nothing. Not then or later. She had only a few drops of water before the night settled in.

When the moon rose, the river maid began to pace restlessly about the house. From one wall to the other, she walked. She went to the window and put her hand against the glass. She stood by the closed door and put her shoulder to the wood, but she would not touch the metal latch.

It was then that Jan was sure of her. 'Cold iron will keep her in.' He was determined she would stay at least until morning.

The river maid cried all the night, a high keening that rose and fell like waves. But in the morning she seemed accommodated to the house and settled quietly to sleep by the fire. Once in a while, she would stretch and stand, the damp left side of her shift clinging to her thigh. In the half-light of the hearth she seemed even more beautiful than before.

Jan left a bowl of fresh water near the fire, with some cress by it, before he went to feed the cows. But he checked the latch on the windows and set a heavy iron bar across the outside of the door.

'I will let you go tonight,' he promised slowly. '*Tonight*,' he said, as if speaking to a child. But she did not know his language and could not hold him to his vow.

By the next morning, he had forgotten making it.

For a year Jan kept her. He grew to like the wavering sounds she made as she cried each night. He loved the way her eyes turned a deep green when he touched her. He was fascinated by the blue veins that meandered at her throat, along the backs of her knees, and laced each small breast. Her mouth was always cold under his.

Fearing the girl might guess the working of window or gate, Jan fashioned iron chains for the glass and an ornate grillwork for the door. In that way, he could open them to let in air and let her look out at the sun and moon and season's changes. But he did not let her go. And as she never learned to speak with him in his tongue and thereby beg for release, Jan convinced himself that she was content.

Then it was spring again. Down from the mountains came the swollen streams, made big with melted snow. The river maid drank whole glasses of water now, and put on weight. Jan guessed that she carried his child, for her belly grew, she moved slowly and no longer tried to dance. She sat by the window at night with her arms raised and sang strange, wordless tunes, sometimes loud and sometimes soft as a cradle song. Her voice was as steady as the patter of the rain, and underneath Jan fancied he heard a growing strength. His nights became as restless as hers, his sleep full of watery dreams.

The night of the full moon, the rain beat angrily against the glass as if insisting on admission. The river maid put her head to one side, listening. Then she rose and left her window place. She stretched and put her hands to her back, then traced them slowly around her sides to the front. She moved heavily to the hearth and sat. Bracing both hands on the stones behind her, she spread her legs, crooked at the knees.

Jan watched as her belly rolled in great waves under the tight white shift.

She threw her head back, gasped at the air, and then, with a great cry of triumph, expelled the child. It rode a gush of water between her legs and came to rest at Jan's feet. It was small and fishlike, with a translucent tail. It looked up at him with blue eyes that were covered with a veil of skin. The skin lifted once, twice, then closed again as the child slept.

Jan cried because it was a beast.

At that very moment, the river outside gave a shout of release. With the added waters from the rain and snow, it had the strength to push through the earth dam. In a single wave, that gathered force as it rolled, it rushed across the meadow, through the farmyard and barn, and overwhelmed the house. It broke the iron gates and grilles as if they were brittle sticks, washing them away in its flood. Then it settled back into its old course, tumbling over familiar rocks and rounding the curves it had cut in its youth.

When the neighbors came the next day to assess the damage, they found no trace of the house or of Jan.

'Gone,' said one.

'A bad end,' said another.

'Never change a river,' said a third.

They spat through their fingers and made other signs against evil. Then they went home to their own fires and gave it no more thought.

But a year later, in a pocket of the river, in a quiet place said to house a great fish with a translucent tail, an inquisitive boy found a jumble of white bones.

His father and the other men guessed the bones to be Jan's, and they left them to the river instead of burying them.

When the boy asked why, his father said, 'Huttah! Hush, boy, and listen.'

The boy listened and heard the river playing merrily over the bones. It was a high, sweet, bubbling song. And anyone with half an ear could hear that the song, though wordless, or at least in a language unknown to men, was full of freedom and a conquering joy.

1982

RICHARD KENNEDY

The Porcelain Man

ONCE upon a time at the edge of town lived a harsh man with a timid daughter who had grown pale and dreamy from too much obedience. The man kept the girl busy and hardly ever let her go out of doors. 'You're lucky to be inside where it's safe and sound,' he would say to her. 'It's dog-eat-dog out there. The world is full of bottle-snatchers, ragmongers, and ratrobbers. Believe you me!'

The girl believed him.

Each morning the man left the house in his rickety wagon pulled by his rackety horse. All day long he, would go up and down the streets of the town, into the countryside and to neighboring villages to find what he could find. He would bring home old broken wheels, tables and chairs with the legs gone from them, pots and pans with holes in them, scraps of this and pieces of that. His daughter would then mend and repair the junk during her long and lonely days inside the house, and the man would take the things away and sell them as secondhand goods. This is the way they lived.

One morning the man left the house and gave his daughter his usual instruction and warning. 'If someone passes on the road, stay away from the windows. If someone knocks, don't answer. I could tell you terrible stories.' Then he left, and the girl began work on a broken lantern.

Now this morning some good luck happened to the man. As he was passing a rich man's house, a clumsy kitchen maid chased two cats out the front door with a broom and knocked over a large porcelain vase. The vase rolled out the door and down the steps and path, and shattered to pieces against a marble pillar near the roadway. The man stopped, and watched. The maid closed the door. The man waited there for ten minutes. No one came out.

Then he leaped down from his cart and gathered up the broken porcelain. He set the pieces gently in his cart and hurried off toward home.

His daughter, as usual, was safe and sound inside. 'This is fine porcelain,' the men said. 'Drop whatever you're doing and patch it up. We'll get a good price for it.'

It was early in the day yet, so the man left again to see what else he might find. He remembered to pause at the door and say, 'Stay inside. Terrible things are going on out there. Dog-eat-dog, the devil take the hindmost, and so forth.' Then he left.

The girl turned a piece of porcelain in her fingers, admiring its beauty. She carefully laid the pieces on a blanket and got out the glue. Then, humming to herself and musing on fanciful thoughts in the way she had acquired from being so much alone, she began to put the pieces of porcelain together. She worked quickly and neatly even though her thoughts were completely elsewhere, and at the end of a couple of hours she was amazed to see that she had just set the last piece in place on a full-sized porcelain man. And at that moment the porcelain man spoke.

'I love you,' he said, taking a step toward the girl.

'Gracious!' gasped the girl, snatching up the blanket and throwing it about the man. 'Gracious!' she gasped again as the porcelain man encircled her in his arms and kissed her.

While this was happening, the girl's father returned to the house. And right at this moment he opened the door to the room.

'Whoa!' he bellowed.

He grabbed a chair, raised it above his head and brought it down squarely on top of the head of the porcelain man with a blow that shattered him from head to toe, and the porcelain pieces scattered over the floor.

'Godalmighty!' the man cried, 'I've fractured his skull!' The girl let out a wail, and the man dropped to his knees, stunned with the catastrophe. But the girl explained that it had not been a real man, but only one made of the porcelain.

'A porcelain man who could move!?'

'And he could talk as well,' said the girl.

'Fantastic!' said the man. 'Quick, put him together again before you forget how you did it. I'll make a cage for him and take him to the county fair. I'll charge a dollar to see him. He can learn to dance. I'll make a big sign saying, "See the dancing pot," or something like that. I'll make thousands! Quick, put him together again!'

So the girl collected the pieces on the blanket and slowly began gluing them together again. Her father sat down and watched her for a while, but he found it to be boring and he dropped off to sleep.

The girl worked on, very sad that the porcelain man would be taken away in a cage. She was so distressed by her thoughts that she did not notice until putting the last piece in place that she had built a small porcelain horse. And the horse neighed.

The man woke up.

'What's that!' he said. 'That's no man, that's a horse. Now you'll just have to do it all over again. And this time, *concentrate*!' Saying this, the man took up the chair over his head so as to smash the horse.

But the horse said to the girl, 'Quickly, jump on my back!' She did, and in a second the horse leaped out the window with the girl, and they galloped across the countryside as the man stood waving the chair at them through the window and shouting words they could not hear.

After running for several miles, the horse stopped in a small meadow, in the center of which stood a tree.

'Get down,' said the horse. The girl did. 'Now,' said the horse, 'I will run into the tree and break myself to pieces, and then you are to put me back together as a man again.' Then the horse added, 'Remember—I love you,' and before the girl could say a word, the horse dashed toward the tree and crashed into it at full gallop and broke into hundreds of pieces.

The girl cried out, and then sat down under the tree and wept, for she had no glue.

Now on a path nearby came along a young man pushing a wheelbarrow. He stopped when he saw the girl by the tree, and went to comfort her.

'Don't cry,' he said. 'Here, let us gather up the pieces and put them in the wheelbarrow. Come along with me and we'll fix everything almost as good as new.'

So they loaded up the broken porcelain, and they went to the man's cottage and spread the pieces out on a blanket.

'It must have been a beautiful set of dishes,' said the man, and he began to glue some pieces together. They talked as they worked and told each other all about themselves. The girl admired how well and quickly the young man worked with his hands, and in a short while they had put together a dozen dishes, eight saucers and teacups, six bowls, two large serving platters, a milk pitcher and two small vases.

They cooked supper then. Their eyes met often as they moved about. Now and again their hands touched, and they brushed against each other going to and fro.

They set the table with the porcelain ware, and when they were eating, the girl's plate whispered up at her, 'I still love you.'

'Hush!' she said.

'I beg your pardon!' the young man said.

'Oh, nothing,' said the girl.

And they lived happily ever after.

1987

LOUISE ERDRICH

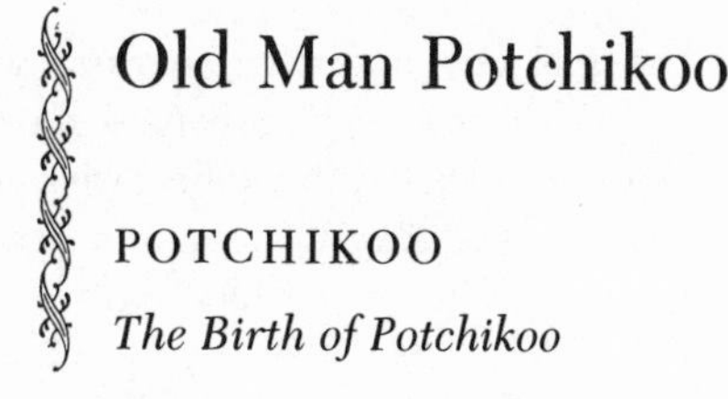

Old Man Potchikoo

POTCHIKOO

The Birth of Potchikoo

You don't have to believe this, I'm not asking you to.

But Potchikoo claims that his father is the sun in heaven that shines down on us all.

There was once a very pretty Chippewa girl working in a field. She was digging potatoes for a farmer some place around Pembina when suddenly the wind blew her dress up around her face and wrapped her apron so tightly around her arms that she couldn't move. She lay helplessly in the dust with her potato sack, this poor girl, and as she lay there she felt the sun shining down very steadily upon her.

Then she felt something else. You know what. I don't have to say it. She cried out for her mother.

This girl's mother came running and untangled her daughter's clothes. When she freed the girl, she saw that there were tears in her daughter's eyes. Bit by bit, the mother coaxed out the story. After the girl told what had happened to her, the mother just shook her head sadly.

'I don't know what we can expect now,' she said.

Well, nine months passed and he was born looking just like a potato, with tough warty skin and a puckered round shape. All the ladies came to visit the girl and left saying things behind their hands.

'That's what she gets for playing loose in the potato fields,' they said.

But the girl didn't care what they said after a while, because she used to go and stand alone in a secret clearing in the woods and let the sun shine steadily upon her. Sometimes she took her little potato boy. She noticed when the sun shone on him he grew and became a little more human-looking.

One day the girl fell asleep in the sun with her potato boy next to her. The sun beat down so hard on him that he had an enormous spurt of growth. When the girl woke up, her son was fully grown. He said goodbye to his mother then, and went out to see what was going on in the world.

Potchikoo Marries

After he had several adventures, the potato boy took the name Potchikoo and decided to try married life.

I'll just see what it's like for a while, he thought, and then I'll start wandering again.

How very inexperienced he was!

He took the train to Minneapolis to find a wife, and as soon as he got off he saw her. She was a beautiful Indian girl standing at the door to a little shop where they sold cigarettes and pipe tobacco. How proud she looked! How peaceful. She was so lovely that she made Potchikoo shy. He could hardly look at her.

Potchikoo walked into the store and bought some cigarettes. He lit one up and stuck it between the beautiful woman's lips. Then he stood next to her, still too shy to look at her, until he smelled smoke. He saw that she had somehow caught fire.

'Oh, I'll save you!' cried Potchikoo.

He grabbed his lady love and ran with her to the lake, which was, handily, across the street. He threw her in. At first he was afraid she would drown, but soon she floated to the surface and kept floating away from Potchikoo. This made him angry.

'Trying to run away already!' he shouted.

He leaped in to catch her. But he had forgotten that he couldn't swim. So Potchikoo had to hang on to his wooden sweetheart while she drifted slowly all the way across the lake. When they got to the other side of the lake, across from Minneapolis, they were in wilderness. As soon as the wooden girl touched the shore she became alive and jumped up and dragged Potchikoo out of the water.

'I'll teach you to shove a cigarette between my lips like that,' she said, beating him with her fists, which were still hard as wood. 'Now that you're my husband, you'll do things my way!'

That was how Potchikoo met and married Josette. He was

married to her all his life. After she had made it clear what she expected of her husband, Josette made a little toboggan of cut saplings and tied him upon it. Then she decided she never wanted to see Minneapolis again. She wanted to live in the hills. That is why she dragged Potchikoo all the way back across Minnesota to the Turtle Mountains, where they spent all the years of their wedded bliss.

How Potchikoo Got Old

As a young man, Potchikoo sometimes embarrassed his wife by breaking wind during Holy Mass. It was for this reason that Josette whittled him a little plug out of ash wood and told him to put it in that place before he entered Saint Ann's church.

Potchikoo did as she asked, and even said a certain charm over the plug so that it would not be forced out, no matter what. Then the two of them entered the church to say their prayers.

That Sunday, Father Belcourt was giving a special sermon on the ascension of the Lord Christ to heaven. It happened in the twinkling of an eye, he said, with no warning, because Christ was more pure than air. How surprised everyone was to see, as Father Belcourt said this, the evil scoundrel Potchikoo rising from his pew!

His hands were folded, and his closed eyes and meek face wore a look of utter piety. He didn't even seem to realize he was rising, he prayed so hard.

Up and up he floated, still in the kneeling position, until he reached the dark blue vault of the church. He seemed to inflate, too, until he looked larger than life to the people. They were on the verge of believing it a miracle when all of a sudden it happened. Bang! Even with the charm, the little ash-wood plug could not contain the wind of Potchikoo. Out it popped, and Potchikoo went buzzing and sputtering around the church the way balloons do when children let go of the ends.

Holy Mass was cancelled for a week so the church could be aired, but to this day a faint scent still lingers and Potchikoo, sadly enough, was shrivelled by his suddden collapse and flight through the air. For when Josette picked him up to bring him home, she found that he was now wrinkled and dry like an old man.

Louise Erdrich

The Death of Potchikoo

Once there were three stones sitting in a patch of soft slough mud. Each of these stones had the smooth round shape of a woman's breast, but no one had ever noticed this—that is, not until Old Man Potchikoo walked through the woods. He was the type who always noticed this kind of thing. As soon as he saw the three stones, Potchikoo sat down on a small bank of grass to enjoy what he saw.

He was not really much of a connoisseur, the old man. He just knew what he liked when he saw it. The three stones were light brown in colour, delicately veined, and so smooth that they almost looked slippery. Old Man Potchikoo began to wonder if they really were slippery, and then he thought of touching them.

They were in the middle of the soft slough mud, so the old man took his boots and socks off. Then he thought of his wife Josette and what she would say if he came home with mud on his clothes. He took off his shirt and pants. He never wore any undershorts. Wading towards those stones, he was as naked as them.

He had to kneel in the mud to touch the stones, and when he did this he sank to his thighs. But oh, when he touched the stones, he found that they were bigger than they looked from the shore and so shiny, so slippery. His hands polished them, and polished them some more, and before he knew it Potchikoo was making love to the slough.

Years passed by. The Potchikoos got older and more frail. One day, Josette went into town, and as he always did as soon as she was out of sight, Potchikoo sat down on his front steps to do nothing.

As he sat there, he saw three women walk very slowly out of the woods. They came across the field and then walked slowly towards him. As they drew near, Potchikoo saw that they were just his kind of women. They were large, their hair was black and very long, and because they wore low-cut blouses, he could see that their breasts were beautiful—light brown, delicately veined, and so smooth they looked slippery.

'We are your daughters,' they said, standing before him. 'We are from the slough.'

A faint memory stirred in Potchikoo as he looked at their breasts, and he smiled.

'Oh my daughters,' he said to them. 'Yes, I remember you. Come sit on your daddy's lap and get acquainted.'

The daughters moved slowly towards Potchikoo. As he saw their skin up close, he marvelled at how fine it was, smooth as polished stone. The first daughter sank upon his knee and clasped her arms around him. She was so heavy the old man couldn't move. Then the others sank upon him, blocking away the sun with their massive bodies. The old man's head began to swim and yellow stars turned in his skull. He hardly knew it when all three daughters laid their heads dreamily against his chest. They were cold, and so heavy that his ribs snapped apart like little dry twigs.

POTCHIKOO'S LIFE AFTER DEATH

How They Don't Let Potchikoo into Heaven

After Old Man Potchikoo died, the people had a funeral for his poor crushed body, and everyone felt sorry for the things they had said while he was alive. Josette went home and set some bread by the door for him to take on his journey to the next world. Then she began to can a box of plums she'd bought cheap, because they were overripe.

As she canned, she thought how it was. Now she'd have to give away these sweet plums since they had been her husband's favourites. She didn't like plums. Her tastes ran sour. Everything about her did. As she worked, she cried vinegar tears into the jars before she sealed them. People would later remark on her ingenuity. No one else on the reservation pickled plums.

Now, as night fell, Potchikoo got out of his body, and climbed up through the dirt. He took the frybread Josette had left in a towel, his provisions. He looked in the window, saw she was sleeping alone, and he was satisfied. Of course, since he could never hold himself back, he immediately ate the bread as he walked the long road, a mistake. Two days later, he was terribly hungry, and there was no end in sight. He came to the luscious berry he knew he shouldn't eat if he wanted to enter the heaven all the priests and nuns described. He took a little bite, and told himself he'd not touch the rest. But it tasted so good tears came in his eyes. It took a minute, hardly that, for him to stuff the whole berry by handfuls into his mouth.

He didn't know what would happen now, but the road was still there. He kept walking, but he'd become so fat from his greed that when he came to the log bridge, a test for good souls, he couldn't balance to cross it, fell in repeatedly, and went on cold and shivering. But he was dry again, and warmer, by the time he reached the pearly gates.

Saint Peter was standing there, dressed in a long brown robe, just as the nuns and priests had always said. He examined Potchikoo back and front for berry stains, but they had luckily washed away when Potchikoo fell off the bridge.

'What's your name?' Saint Peter asked.

Potchikoo told him, and then Saint Peter pulled a huge book out from his robe. As the saint's finger travelled down the lists, Potchikoo became frightened to think how many awful deeds would be recorded after his name. But as it happened, there was only one word there. The word *Indian*.

'Too bad,' Saint Peter said. 'You'll have to keep walking.'

Where Potchikoo Goes Next

So he kept on. As he walked, the road, which had been nicely paved and lit when it got near heaven, narrowed and dipped. Soon it was only gravel, then dirt, then mud, then just a path beaten in the grass. The land around it got poor too, dry and rocky. And when Potchikoo got to the entrance of the Indian heaven it was no gate of pearl, just a simple pasture-gate of weathered wood. There was no one standing there to guard it, either, so he went right in.

On the other side of the gate there were no tracks, so Potchikoo walked aimlessly. All along the way, there were choke-cherry bushes, not quite ripe. But Potchikoo was so hungry again that he raked them off the stems by the handful and gobbled them down, not even spitting out the pits.

Soon it was worse than hunger, the dreadful stomach aches he got, and every few steps poor Potchikoo had to relieve himself. On and on he went, day after day, eating berries to keep his strength up and staggering from the pain and shitting until he felt so weak and famished that he had to sit down. Some time went by, and then people came to sit around him. They got to talking. Someone built a fire, and soon they were roasting venison.

The taste of it made Potchikoo lonesome. Josette always fried her meat with onions.

'Well,' he said, standing up when he was full, 'it's time to go home now.'

The people didn't say goodbye though, just laughed. There were no markers in this land, nothing but extreme and gentle emptiness. It was made to be confusing. There were no landmarks, no look-outs. The wind was strong, and the bushes grew quickly, so that every path made was instantly obscured.

But not Potchikoo's path. At regular intervals new choke-cherry bushes had sprung up from the seeds that had passed through his body. So he had no trouble finding his way to the gate, out through it and back on to the road.

Potchikoo's Detour

Along the way back, he got curious and wondered what the hell for white people could be like.

As he passed the pearly gates, Saint Peter was busy checking in a busload of Mormons, and so he didn't even look up and see Potchikoo take the dark fork in the road.

Walking along, Potchikoo began to think twice about what he was doing. The air felt warm and humid, and he expected it to get worse, much worse. Soon the screams of the damned would ring out and the sky would turn pitch-black. But his curiosity was, as always, stronger than his fear. He kept walking until he came to what looked like a giant warehouse.

It was a warehouse, and it was hell.

There was a little sign above the metal door marked *Entrance, Hell.* Potchikoo got a thrill of terror in his stomach. He carefully laid his ear against the side of the building, expecting his blood to curdle. But all he heard was the sound of rustling pages. And so, gathering his courage, he bent to the keyhole and looked in to see what it was the white race suffered.

He started back, shook his head, then bent to the keyhole again.

It was worse than flames.

They were all chained, hand and foot and even by the neck, to old Sears, Roebuck catalogues. Around and around the huge warehouse they dragged the heavy paper books, mumbling, collapsing from time to time to flip through the pages. Each person was bound to five or six, bent low beneath the weight. Potchikoo had always wondered where old Sears catalogues went, and now he knew the devil gathered them, that they were instruments of torment.

The words of the damned, thin and drained, rang in his ears all the way home.

Look at that wall unit. What about this here recliner? We could put up that home gym in the basement.

Potchikoo Greets Josette

On his journey through hell and heaven, Potchikoo had been a long time without sex. It was night when he finally got back home, and he could hardly wait to hold Josette in his arms. Therefore, after he had entered the house and crept up to her bed, the first words he uttered to his wife in greeting were, 'Let's Pitch Whoopee.'

Josette yelled and grabbed the swatter she kept next to her bed to kill mosquitoes in the dark. She began to lambast Potchikoo until she realized who it was, and that this was no awful dream.

Then they lay down in bed and had no more thoughts.

Afterwards, lying there happily, Potchikoo was surprised to find that he was still passionate. They began to make love again, and still again, and over and over. At first Josette returned as good as Potchikoo gave her, but after a while it seemed that the more he made love, the more need he felt, the more heat he gave off. He was unquenchable fire.

Finally, Josette fell asleep, and let him go on and on. He was so glad to be alive again that he could never remember, afterwards, how many times he had sex that night. Even he lost count. But when he woke up late the next day, Potchikoo felt a little strange, as though there was something missing. And sure enough, there was.

When Potchikoo looked under the covers, he found that his favourite part of himself was charred black, and thin as a burnt twig.

Potchikoo Restored

It was terrible to have burnt his pride and joy down to nothing. It was terrible to have to face the world, especially Josette, without it. Potchikoo put his pants on and sat in the shade to think. But not until Josette left for daily Mass and he was alone, did Potchikoo have a good idea.

He went inside and found a block of the paraffin wax that Josette used to seal her jars of plum pickles. He stirred the coals in Josette's stove and melted the wax in an old coffee can. Then he dipped in his penis. It hurt the first time, but after that not so much and then not at all. He kept dipping and dipping. It got back to the normal size, and he should have been pleased with that. But Potchikoo got grandiose ideas.

He kept dipping and dipping. He melted more wax, more and more, and kept dipping, until he was so large he could hardly stagger out the door. Luckily, the wheelbarrow was sitting in the path. He grabbed the handles and wheeled it before him into town.

There was only one road in the village then. Potchikoo went there with his wheelbarrow, calling for women. He crossed the village twice. Mothers came out in wonder, saw what was in the wheelbarrow, and whisked their daughters inside. Everybody was disgusted and scolding and indignant, except for one woman. She lived at the end of the road. Her door was always open, and she was large.

Even now, we can't use her name, this Mrs B. No man satisfied her. But that day Potchikoo wheeled his barrow in and then, for once, her door was shut.

Potchikoo and Mrs B went rolling through the house. The walls shuddered and people standing around outside thought the whole place might collapse. Potchikoo was shaken from side to side, powerfully, as if he were on a ride at the carnival. But eventually, of course, the heat of their union softened and wilted Potchikoo back to nothing. Mrs B was disgusted and threw him out back, into the weeds. From there he crept home to Josette, and on the crooked path he took to avoid others, he tried to think of new ways he might please her.

Potchikoo's Mean Twin

To his relief, nature returned manhood to Potchikoo in several weeks. But his troubles weren't over. One day, there was the police. They said Potchikoo had been seen stealing fence-posts down the road. But they found no stolen fence-posts on his property, so they did not arrest him.

More accusations were heard, none true.

Potchikoo threw rocks at a nun, howled like a dog and barked until she chased him off. He got drunk and tossed a pool cue out the window of the Stumble Inn. The pool cue hit the Tribal Chairman on the shoulder and caused a bruise. Potchikoo ran down the street laughing, flung off his clothes, ran naked through the trading store. He ripped antennas from twenty cars. He broke a portable radio that belonged to a widow, her only comfort. If a friendly dog came up to this bad Potchikoo, he lashed out with his foot. He screamed at children until tears came into their eyes, and then he knocked down the one road sign the government had seen fit to place on the reservation.

It was red, in the very middle of town, and it said STOP. People were naturally proud of the sign. So there was finally a decision to lock Potchikoo in jail. When the police came to get him, he went quite willingly because he was so confused.

But here's what happened.

While Potchikoo was locked up, under the eyes of the Tribal Sheriff, his mean twin went out and caused some mischief near the school by starting a grass-fire. So now the people knew the trouble wasn't caused by Old Man Potchikoo. And next time the bad twin was seen, Josette followed him. He ran very fast, until he reached the chain-link fence around the graveyard. Josette saw him jump over the fence and dodge among the stones. Then the twin got to the place where Potchikoo had been buried, lifted the ground like a lid, and wiggled under.

How Josette Takes Care Of It

So the trouble was that Potchikoo had left his body in the ground, empty, and something bad found a place to live.

The people said the only thing to do was trap the mean twin and then get rid of him. But no one could agree on how to do it. People just talked and planned, no one acted. Finally Josette had to take the matter into her hands.

One day she made a big pot of stew and into it she put a bird. Into the roasted bird, she put a bit of blue plaster that had fallen off the Blessed Virgin's robe while Josette cleaned the altar. She took the stew and left the whole pot just outside the cemetery fence. From her hiding-place deep in a lilac bush, she saw the mean twin creep forth. He took the pot in his hands and gulped down every morsel, then munched the bird up bones and all. Stuffed full, he lay down to sleep. He snored. After a while, he woke and looked around himself, very quiet. That was when Josette came out of the bush.

'In the name of the Holy Mother of God!' she cried. 'Depart!'

So the thing stepped out of Potchikoo's old body, all hairless and smooth and wet and grey. But Josette had no pity. She pointed sternly at the dark stand of pines, where no one went, and slowly, with many a sigh and backward look, the thing walked over there.

Potchikoo's old body lay, crumpled like a worn suit of clothes, where the thing had stepped out. Right there, Josette made a fire, a little fire. When it was very hot, she threw in the empty skin and it crackled in the flames, shed sparks and was finally reduced to a crisp of ashes, which Josette brushed carefully into a little sack, and saved in her purse.

1989

BIOGRAPHICAL NOTES

JOAN AIKEN (1924–) was born in Sussex into a literary family: her father was the American poet Conrad Aiken. She began to publish poems and stories in her teens, and was soon a full-time writer. Aiken has now written over fifty books for children and over twenty for adults, several plays, and a guide called *The Way to Write for Children* (1983). Several of her most famous books, *The Wolves of Willoughby Chase* (1962), *Black Hearts in Battersea* (1965), and their many sequels, take place in an imaginary nineteenth-century England still ruled by Stuart kings. Aiken has also published many collections of ghost stories and magical and humorous tales, including *A Small Pinch of Weather* (1969) and *A Bundle of Nerves* (1976), which contains 'The Man Who Had Seen the Rope Trick'.

DONALD BARTHELME (1931–89) was one of the most interesting and experimental writers to appear in the 1960s. He was born in Texas, the son of an avant-garde architect; when he was a child, people used to park their cars on the prairie and stare at his parents' house. His work is full of fantastic humour, absurd situations, and skilled parody, often inspired by characters or themes from folklore. It includes the novels *Snow White* (1967) and *The Dead Father* (1975), and seven collections of stories, of which perhaps the best are *Unspeakable Practices, Unnatural Acts* (1968) and *City Life* (1970), from which 'The Glass Mountain' is taken. Barthelme also published a book for children, *The Slightly Irregular Fire Engine* (1971).

L. F. BAUM (1856–1919) was born and raised in up-state New York. As a young man he worked with varying success as an actor, journalist, and travelling salesman, and founded a magazine for shop-window dressers. After the birth of his four sons he turned to writing for children. *The Wonderful Wizard of Oz* was published in 1900; it was followed by many popular sequels. *American Fairy Stories*, which includes 'The Queen of Quok', appeared the following year. Baum saw movie possibilities in *The Wonderful Wizard of Oz*, and made a series of short hand-coloured films based on the book, using trick photography; there were also several early stage versions. In 1911 he and his family moved to Hollywood, where he tried without success to establish an 'Oz land' along the lines of Disneyland.

Biographical Notes

FRANCES BROWNE (1816–79) was the seventh of twelve children of a poor Irish postmaster. She was born and grew up in a remote mountain village in Donegal, and became blind as an infant. Yet in spite of these handicaps she managed to get an education and then to support herself (and one of her sisters, who was her amanuensis) as a writer in Edinburgh and London. She became a popular poet and the author of many books for children, of which the best known is *Granny's Wonderful Chair* (1857). Though some of its tales draw on Irish folk tradition, they also tend to have an improving moral, like most fairy stories of the period. But Frances Browne's lessons are leavened with wit and sometimes, as in 'The Story of Fairyfoot', with an almost Swiftian social satire.

ANGELA CARTER (1940–92), English author of fantasy and magic-realist fiction, was born in Sussex and educated at Bristol University. Among her best-known works are *The Magic Toyshop* (1967) and two collections of tales, many of them based on traditional fairy-tale motifs: *The Bloody Chamber* (1979), which contains 'The Courtship of Mr Lyon', and *Fireworks* (1984). Her non-fiction includes *The Sadean Woman* (1979) and *Nothing Sacred* (1982). Carter had a lifelong interest in folklore and spent several years collecting and singing English folk songs; she also translated *The Fairy Tales of Charles Perrault* and edited *The Virago Book of Fairy Tales* (1990).

LUCY LANE CLIFFORD (1853–1929) was born on the island of Barbados and came to England as a young woman to study art. In 1875 she married William Clifford, a professor of mathematics. Four years later he died, leaving her with two young daughters and very little money. To support them and herself she began writing: romantic novels, verse, plays, and short stories—so successfully that she was able to become the hostess of a London literary salon frequented by novelists such as Henry James and Rudyard Kipling. Her strange and haunting tale 'The New Mother' was first published in *Anyhow Stories, Moral and Otherwise* (1882); it has some interesting similarities to her friend Henry James's *The Turn of the Screw* (1898).

JOHN COLLIER (1901–80) was born in England and during the 1920s and 1930s was poetry editor of *Time and Tide*. In 1935, however, he moved to America and eventually became a Hollywood screen-writer. He is remembered today for the satirical novel *His Monkey Wife* (1930), about an explorer who marries a chimpanzee, and for his short stories, which combine fantasy and black humour. The most famous of these, 'Green Thoughts', which features a plant that eats people, inspired the hit film *Little Shop of Horrors*. 'The Chaser' was first published in *Presenting Moonshine* (1941).

Biographical Notes

Walter de la Mare (1873–1956), English poet, short-story writer, and anthologist, began his career as a 16-year-old clerk in the London offices of the Anglo-American Oil Company. Though he soon started to publish poems and stories, he could not afford to leave the job he hated for eighteen years. Today he is best known for his volumes of poetry for children, among them *Peacock Pie* (1913), *Down-Adown-Derry* (1928), and *Bells and Grass* (1941), and for his fantastic and original novels and stories, such as *The Three Mulla-Mulgars* (1910), *Memoirs of a Midget* (1921), and *Broomsticks* (1925), which includes 'The Lovely Myfanwy'. De la Mare also published one of the best anthologies of poetry for children, *Come Hither* (1923), and *Early One Morning* (1935), a study of childhood as revealed in memoirs and writings.

Mary de Morgan (1850–1907) was the daughter of a famous London mathematician and the sister of the novelist and illustrator William De Morgan. For many years she and her brother shared a house in Chelsea which became a centre for writers and artists of the time. Her three collections of original fairy tales began as stories told to the children of her friends—one of her young listeners was Rudyard Kipling. 'A Toy Princess' comes from the first of these books, *On a Pincushion and Other Tales* (1877). It was followed by *The Necklace of Princess Fiorimonde* (1880) and *The Windfairies* (1900). The books were illustrated by her brother in the Pre-Raphaelite style, and the stories in them often, unlike traditional fairy tales, question the value of wealth and power.

Jeanne Desy (1942–   ) is a lifelong resident of Columbus, Ohio. She credits lack of television as a child with her interest in literature, especially fairy tales. She has held a number of editorial jobs and has taught college courses, and is currently enrolled as a Ph.D. student at Ohio State University, concentrating on narrative theory of creative writing. Although she has had many essays and humorous pieces in magazines, 'The Princess Who Stood On Her Own Two Feet' is her only published story for children.

Philip K. Dick (1928–82) was born in Chicago and attended the University of California at Berkeley. Today he is regarded as one of the finest science-fiction writers of his time. Dick said that he created worlds of his own because 'the world we actually have is not up to my standards'. A major subject of his fiction is the difficulty of distinguishing between the real and the unreal. Recently there has been a revival of enthusiasm for Dick's work, largely caused by the film *Blade Runner*, which was based on Dick's novel, *Do Androids Dream of Electric Sheep?* (1968). 'The King of the Elves' was first published in the magazine *Beyond Fantasy Fiction* in 1953.

Biographical Notes

CHARLES DICKENS (1812–70), the most famous novelist of the Victorian age, wrote a number of books for children, of which the best known is *A Christmas Carol* (1843). 'The Magic Fishbone' was first published in 1868 in the American magazine *Our Young Folks*, as Part II of 'A Holiday Romance'. In relating this story Dickens adopted the persona of 'Miss Alice Rainbird, Aged Seven'. Though its length and vocabulary are well beyond the capacities of most 7-year-olds, it also shows a remarkable sense of how children's minds work. 'The Magic Fishbone' can also be read as an allegory of Dickens's own situation in earlier years, when he was the harassed father of ten children, with an ailing wife, and in constant need of more money.

LORD DUNSANY (1878–1957). Edward John Moreton Drax Plunkett, eighteenth Baron Dunsany, was born in London of Anglo-Irish parents and educated in England. After serving in the Boer War, he moved to Dublin, where he became associated with William Butler Yeats and the Irish revival movement. The first of his many plays, *The Glittering Gate* (1909), was performed at the Abbey Theatre. He also wrote novels, essays, poetry, and autobiography; but he is best known today for his fantastic tales, collected in many volumes, including *Time and the Gods* (1906), *The Sword of Welleran* (1908), and *A Dreamer's Tales* (1910), which contains 'The Kith of the Elf-Folk'.

LOUISE ERDRICH (1954–) was born in Little Falls, Minnesota, the eldest of seven part-Chippewa children, and educated at Indian boarding schools in North Dakota. She graduated from Dartmouth College and received a graduate degree in writing from Johns Hopkins University. Her unusual, poetic novels about Native American life—*Love Medicine* (1984), *The Beet Queen* (1986), and *Tracks* (1988)—have won many awards and established her as one of the most gifted of young American writers. Erdrich often draws on Native American traditions and beliefs in her fiction. In 'Old Man Potchikoo' she creates, or re-creates, an Indian trickster legend. Louise Erdrich now lives in Montana with her husband, the writer Michael Dorris, who is a member of the Modoc tribe, and their six children.

JULIANA HORATIA EWING (1841–85) was born in Yorkshire, the daughter of Margaret Gatty, a prolific author of serious moral tales and fables drawn from natural history. From the start Juliana was a born story-teller, and by the time she was in her twenties she was one of the best-known children's authors in England. She published realistic tales like *Lob Lie-by-the-Fire* (1874) and *Jackanapes* (1879), based on her experiences as an Army wife. Today she is best remembered for her lively and amusing fairy stories, collected in *The Brownies* (1870) and *Old-Fashioned Fairy Tales* (1882), which contains 'Good Luck Is Better Than Gold'. In spite of the title, these 'old-fashioned' tales are surprisingly modern for their time, and their morals are well disguised with humour.

KENNETH GRAHAME (1859–1932) was born in Edinburgh, but when he was 4 his mother died and his alcoholic father sent him and his sister and brothers to live with their grandmother in a village near the Thames in Berkshire. Since she was elderly and uninterested in children, the four Grahames were free to roam the countryside and invent their own world. Kenneth enjoyed school and hoped to go on to Oxford, but his well-to-do relatives were unwilling to spend the money, and insisted instead that he become a clerk in the Bank of England. While there he began to write the reminiscences of childhood collected in *The Golden Age* (1895) and *Dream Days* (1898), which contains 'The Reluctant Dragon'. Although now a successful writer, Grahame continued to work at the Bank. It was not until 1908 that he published his next and most famous work, *The Wind in the Willows*, based on bedtime stories invented for his son Alastair.

NATHANIEL HAWTHORNE (1804–64) was born in Salem, Massachusetts; and his Puritan ancestors included a judge at the Salem witch trials who appears under another name in his first famous novel, *The House of the Seven Gables* (1851). Hawthorne was also the author of *The Scarlet Letter* (1850) and *The Blithedale Romance* (1852). He wrote several books for children, including *A Wonder Book* (1852) and *Tanglewood Tales* (1853), sensitive and poetic retellings of the Greek myths. Many of Hawthorne's short stories contain supernatural characters or events, and are set in a New England which is at once down to earth and fantastic. Mother Rigby, the witch in 'Feathertop', for instance, smokes an old, smelly pipe and transforms a scarecrow into a fine gentleman. The tale first appeared in *Mosses from an Old Manse* (1846).

LAURENCE HOUSMAN (1865–1959) was the younger brother of the poet A. E. Housman, and a well-known dramatist, poet, and graphic artist of his time. As a young man studying art in London he became attracted to socialist ideas and, with George Bernard Shaw and H. G. Wells, was one of the founders of the Fabian Society. He was the author of four volumes of fairy tales, all of them intended for adults as well as children, and illustrated with his own strange and characteristically art nouveau designs. 'The Rooted Lover' comes from Housman's first collection of tales, *A Farm in Fairyland* (1894). It was followed by *The House of Joy* (1895), *The Field of Clover* (1898), and *The Blue Moon* (1904).

RICHARD HUGHES (1900–74), novelist, playwright, and author of stories for children, was born in Surrey and educated at Oxford. He became famous in 1929 with his first novel, *A High Wind in Jamaica*, the poetic but remarkably unsentimental tale of a family of children captured by pirates in the West Indies. He also published two collections of unusual and idiosyncratic stories for children, *The Spider's Palace* (1931) and *Don't Blame Me* (1940), the latter of which contains 'Gertrude's Child'. After a long silence, he again attracted public attention with two ambitious political novels, *The Fox in the Attic* (1961) and *The Wooden Shepherdess* (1973).

RICHARD KENNEDY (1932–) has spent most of his adult life in Oregon, except for a stint in the US Air Force from 1951 to 1954. He has held a variety of jobs, including elementary school teacher and woodcutter, and since 1974 has been the custodian at Oregan State University Marine Center. Kennedy's recent works include a young-adult novel, *Amy's Eyes* (1985) and *Collected Stories* (1987), which includes 'The Porcelain Man'. Many of his tales are based on traditional folk motifs, but there is usually a twist to the happily-ever-after ending. Is there any stranger ever-after than the fate of the Porcelain Man, who becomes a set of dishes for his true love and her new husband to dine upon?

TANITH LEE (1947–) was born in London and educated at Croydon Art College. After working at a variety of clerical jobs, she became a full-time writer in 1974. She now lives in Sidcup, Kent, where her interests include ancient civilizations and psychic powers. Lee has written many fantasies for both children and adults, including *Red as Blood* (1983), a collection of feminist fairy tales, *Dreams of Dark and Light* (1986), and *Princess Hynchatti* (1972), which contains 'Prince Amilec'. Many of her tales put an ironic twist on a classic plot line. In one, a prince mistakenly believes a real swan is an enchanted princess, and in 'Prince Amilec' it is the witch who lives happily ever after.

URSULA LE GUIN (1929–) was born in California and educated at Radcliffe College and Columbia University. She is one of the most distinguished contemporary writers of fantasy and science fiction, and also a poet and essayist. Her adult novels include *The Left Hand of Darkness* (1969) and *The Dispossessed* (1974); but she is even better known for the remarkable Earthsea series of fantasies for children: *A Wizard of Earthsea* (1968), *The Tombs of Atuan* (1972), *The Farthest Shore* (1973), and *Tehanu* (1990). In the first book of the series the hero battles an evil shadow which turns out to bear his name. This is characteristic of Le Guin: as in 'The Wife's Story', which was first published in 1982 in *The Compass Rose*, she deftly points out that the other may be you.

GEORGE MACDONALD (1824–1905) was born in Aberdeenshire, Scotland, the son of a farmer. He won a scholarship to King's College in Aberdeen, and later studied for the ministry in London. But after serving as a Congregational minister for two years, he was dismissed for his heretical views: he believed, for example, that even heathens would one day be saved. For the rest of his life he supported himself and his wife and family as a novelist, essayist, and author of imaginative and mystical children's stories, notably *At the Back of the North Wind* (1871), *The Princess and the Goblin* (1872), and *The Princess and Curdie* (1877). 'The Light Princess' was first published in 1864 as part of a novel for adults, *Adela Cathcart*.

BERNARD MALAMUD (1914–86), American novelist, was born in Brooklyn, the son of Russian Jewish immigrants, and attended City College and Columbia University, supporting himself by teaching high school. It was not until he was almost 40 that he began to publish the stories and novels that made his reputation, beginning with *The Natural* (1952) and *The Assistant* (1957); in 1957 his fourth novel, *The Fixer*, won both the National Book Award and the Pulitzer Prize. His stories, many of them based on Jewish folk tradition, were collected in *The Magic Barrel* (1958) and *Idiots First* (1963), which contains 'The Jewbird'.

NAOMI MITCHISON (1897–), the British novelist, was born in Edinburgh but brought up in Oxford. Like her husband, the Labour politician G. R. Mitchison, she was devoted to progressive causes. She has published many successful historical novels, including *The Conquered* (1923), *The Corn King and the Spring Queen* (1931), and *The Blood of the Martyrs* (1939). She has also written a number of fantasy novels and stories and three volumes of autobiography. 'In the Family' comes from a collection of tales based on Celtic folklore and entitled *Three Men and a Swan* (1957).

E. (EDITH) NESBIT (1858–1924) began writing in her teens and published her first poem at 15. With her husband, Hubert Bland, she was one of the first members of the Fabian Society, and though he was well known as a journalist it was she who provided most of the income. For years Nesbit produced only hack-work, composed as rapidly as possible to support their growing family. In 1898 she published the first of her remarkable books for children, *The Treasure Seekers*. It was followed by many others, some realistic and others containing her characteristic blend of comedy and magic. Among the most popular today are *Five Children and It* (1902), *The Phoenix and the Carpet* (1904), *The Story of the Amulet* (1905), *The Railway Children* (1906), and *The Enchanted Castle* (1907). 'The Book of Beasts' first appeared in *The Book of Dragons* (1900).

HOWARD PYLE (1853–1911), the late nineteenth-century American artist and illustrator, was born and raised as a Quaker in Delaware. Though trained as a painter, he soon began to both write and illustrate his own books for children. His first collection of fairy tales based on traditional European motifs, *Pepper and Salt or Seasoning for Young Folk* (1886), included 'The Apple of Contentment'; it was followed by *The Wonder Clock* (1888) and in 1895 by a novel-length fantasy, *The Garden Behind the Moon*. Pyle wrote and illustrated historical novels for older children, including *Otto of the Silver Hand* (1888) and *Men of Iron* (1892); he also published new versions of the tales of Robin Hood and King Arthur and his knights. Though the principal influence on his art was Dürer, his books also recall those of Walter Crane and William Morris.

JOHN RUSKIN (1819–1900), Victorian art historian, social critic, and man of letters, wrote *The King of the Golden River* in 1841 for a little girl named Effie Gray. Seven years later, when he was 29 and she was 19, they were married, and the story was finally published in 1850. Its romantic descriptions of wild mountain landscapes recall the paintings of Turner, whose work Ruskin defended and popularized in his series of essays on contemporary British art, later collected in the four volumes of *Modern Painters* (1834–56). Among Ruskin's other works are *The Seven Lamps of Architecture* (1849), *The Stones of Venice* (1851–3), and *Sesame and Lilies* (1865), a collection of lectures on social issues.

CARL SANDBURG (1878–1967), American poet and biographer of Abraham Lincoln, was born in rural Illinois of Swedish immigrant parents, and spent most of his early years as an itinerant labourer and journalist. He was in his thirties when his free verse, strongly influenced by Walt Whitman, first attracted public attention with the publication of *Chicago Poems* (1916). Sandburg was also an important early collector—and gifted performer—of American folksong; his *The American Songbag* was a landmark in the field. The bedtime tales Sandburg told his twin daughters were the basis of all three of his books for children: *Rootabaga Stories* (1922), which includes 'The Story of Blixie Bimber', *Rootabaga Pigeons* (1923), and *Potato Face* (1930). A rootabaga (more commonly today, *rutabaga*) is a large coarse yellow root vegetable, also known at the time as a 'Swedish turnip' or more simply—like Sandburg and most of his relatives—as a 'swede'.

CATHERINE SINCLAIR (1800–64) was the daughter of a pompous and conceited Scottish politician, Sir John Sinclair. For twenty years she served as his secretary; his death when she was 34 freed her to become a successful novelist. Today her adult works are forgotten, but *Holiday House* (1839) remains a classic. It was the first children's book to present a real boy and girl who are, as Catherine Sinclair herself put it in the preface, 'noisy, frolicsome, [and] mischievous' but are neither disapproved of by the author, nor consigned to worldly failure and/or eternal damnation. Harry and Laura, however, do have to listen to a great many improving lectures from their relatives, including their Uncle David. His 'Nonsensical Story about Giants and Fairies', though amusingly told, is also heavily educational. It set a fashion for the moral fairy tale that continued into the next century.

ISAAC BASHEVIS SINGER (1904–91) was born in Poland, the son and grandson of rabbis. In 1935 he followed his older brother to America, where he became a journalist for the Yiddish-language paper, the *Daily Forward.* He soon began to publish stories and novels set in Poland, many of them drawing on Jewish mystical and supernatural tradition. Among his best-known works are *Satan in Goray* (1955), *The Magician of Lublin* (1960), and *Gimpel the Fool* (1957), a collection of tales. Many of his later stories and his novel *Enemies* (1970) were set in New York. Singer also published magical and humorous tales for children, collected in *Zlateh the Goat* (1966), *The Fools of Chelm* (1973), *When Shlemiel Went to Warsaw* (1968), which includes 'Menaseh's Dream', and other volumes. In 1978 he received the Nobel Prize for Literature.

ROBERT LOUIS STEVENSON (1850–94), one of the greatest nineteenth-century authors of children's books, was born and raised in Edinburgh. Though he studied law, his real interest was in literature, and he began to publish essays and travel pieces in his twenties. In 1880 he married the American Fanny Osbourne. The following year her 12-year-old son Lloyd drew the map of a desert island which led to Stevenson's writing the story that became *Treasure Island* (1883). It was followed by *A Child's Garden of Verses* (1885), *Kidnapped* (1886), *Dr. Jekyll and Mr. Hyde* (1886), and many other novels and tales for both children and adults. His *Fables*, which included 'The Song of the Morrow', seemed perverse to his wife, and they were not published until after Stevenson's death.

FRANK STOCKTON (1834–1902) was born and raised in Philadelphia, the seventh of thirteen children. Lame from birth, he entertained his brothers and sisters by telling fairy tales. As he wrote later, 'I did not dispense with monsters and enchanters or talking birds and beasts, but I obliged these creatures to infuse into their extraordinary actions a certain leaven of common sense.' He first became successful as the author of stories for children, collected in *Ting-a-Ling* (1870), *The Floating Prince* (1881), and *The Bee-Man of Orn* (1887), which contains 'The Griffin and the Minor Canon'. Stockton later wrote several popular humorous novels for adults, including *Rudder Grange* (1879) and *The Casting-Away of Mrs. Lecks and Mrs. Aleshine* (1886).

JAMES THURBER (1894–1961), American humorist and artist, was born in Columbus, Ohio. In the 1920s he began to publish satirical sketches, stories, and cartoons, most of them in the *New Yorker*. His work was collected in many volumes, including *The Owl in the Attic* (1931), *The Seal in the Bedroom* (1932), and *Fables for Our Time* (1940), an ironic modern-day version of Aesop which includes 'The Unicorn in the Garden'. Thurber also wrote five books for children, four of them based on fairy-tale motifs: *Many Moons* (1943), *The Great Quillow* (1944), *The White Deer* (1945), *The Wonderful O* (1957), and *The Thirteen Clocks* (1950).

SYLVIA TOWNSEND WARNER (1893–1978) was born in Harrow and trained as a musician, but was prevented from studying in Europe by the First World War. During the war, while working in a munitions factory, she began to write poems and stories. She became well known with the publication of her novel *Lolly Willowes* (1926), in which a middle-aged spinster moves to the depths of the country and becomes a witch. Warner, who also lived in the remote countryside, was one of the few close friends of T. H. White, whose biography she later wrote. She also published *The Cat's-Cradle Book* (1940), a collection of very unusual fairy tales told by a mother cat to her kittens, which includes 'Bluebeard's Daughter', and *Kingdoms of Elfin* (1977), a remarkable account of life among the elves.

H. G. WELLS (1866–1946), novelist, social critic, and science-fiction writer, was born in Kent, the son of an unsuccessful small tradesman. Working by day and studying at night, he won a scholarship to the Normal School of Science in London, and after many difficulties established himself as one of the most popular and influential authors of his time. He wrote such classic science-fiction fantasies as *The Time Machine* (1895), *The Invisible Man* (1897), and *The War of the Worlds* (1898), as well as realistic novels and works of scientific and political theory, including *The Shape of Things to Come* (1933). Wells published five volumes of stories, both realistic and fantastic; 'The Magic Shop' comes from *Twelve Stories and a Dream* (1903).

T. H. WHITE (1906–64) was born in India of British parents and educated at Cambridge University, where he gained a first-class honours degree in English Literature. His childhood was very unhappy, and he lived alone for the rest of his life with a menagerie of animals and birds to whom he was deeply devoted. He is best known today for his series of novels based on the legends of King Arthur: *The Sword in the Stone* (1938), *The Witch in the Wood* (1939), and *The Ill-Made Knight* (1940). They were later collected, together with an epilogue, in *The Once and Future King* (1958), and became the basis for the musical *Camelot*. White also wrote a book for children, *Mistress Masham's Repose* (1947), about the discovery of a colony of Lilliputians in England. 'The Troll' first appeared in *Gone to Ground* (1935).

OSCAR WILDE (1854–1900) was born in Dublin and educated at Trinity College and Oxford University, where he became famous as a classical scholar and self-proclaimed aesthete. After graduation he established himself in London as a dramatist, poet, critic, and wit. His best-known play, *The Importance of Being Earnest* (1895), is still widely performed, and his most famous aphorisms have passed into the public domain. Wilde published two books of poetic fairy tales, *The Happy Prince* (1888), which includes 'The Selfish Giant', and *A House of Pomegranates* (1891). These collections were strongly influenced by the tales of Hans Christian Andersen; and though they were based on stories he had told his two sons, they are also intended for adults.

JAY WILLIAMS (1914–78), at the age of 12, won a prize (a book) for telling the best ghost story at a campfire. According to him, he told stories to children ever after. Williams was born in Buffalo, New York, and educated at the University of Pennsylvania and Columbia. He worked as a vaudeville and night club comic, and then as a Hollywood press agent, before joining the US Army in 1941. After his discharge he began writing full-time. He is best known for his series of books about a young scientific genius called Danny Dunn, but he also published historical fiction for children and the collection of stories from which 'Petronella' is taken: *The Practical Princess and Other Liberating Fairy Tales* (1973).

JANE YOLEN (1939–) was born in New York City and educated at Smith College and the University of Massachusetts. She has worked as an editor and taught at Smith College; she now lives in Hatfield, Massachusetts, with her husband and three children. Although Yolen has written stories for adults, she is best known for her original fairy tales in the tradition of Hans Christian Andersen. Yolen is also a folk singer, and her style maintains the lyrical cadences of an oral tradition. Her tales are collected in *The Girl Who Cried Flowers* (1974), *The Hundredth Dove* (1978), *Neptune Rising* (1982), which contains 'The River Maid', and other volumes. She has also published *Touch Magic* (1983) a study of the fairy story.

PUBLISHER'S ACKNOWLEDGEMENTS

The editor and publisher are grateful for permission to include the following copyright stories in this collection:

Joan Aiken, 'The Man Who Had Seen the Rope Trick', from *A Bundle of Nerves* (Gollancz, 1976). Reprinted by permission of A. M. Heath for the author.

Donald Barthelme, 'The Glass Mountain', from *City Life*, © 1970 by Donald Barthelme, reprinted with the permission of Wylie, Aitken & Stone, Inc.

Angela Carter, 'The Courtship of Mr Lyon', from *The Bloody Chamber* (HarperCollins, 1979). Reprinted by permission of Victor Gollancz Ltd.

John Collier, 'The Chaser', from *Presenting Moonshine*, copyright 1940, © 1967 by John Collier. Reprinted by permission of Harold Matson Company Inc. First published in *The New Yorker*.

Walter de la Mare, 'The Lovely Myfanwy', from *Broomsticks and Other Tales* (1925), reprinted in *Collected Stories for Children* (1948). Reprinted by permission of The Literary Trustees of Walter de la Mare and The Society of Authors as their representative.

Jeanne Desy, 'The Princess Who Stood On Her Own Two Feet', from *Stories for Free Children* (1982). Copyright Jeanne Desy. Reprinted by permission of the author.

Philip K. Dick, 'The King of the Elves', from *The Collected Stories of Philip K. Dick* (Underwood-Miller, 1987). © Philip K. Dick 1987. Reprinted by permission of Scott Meredith Literary Agency.

Lord Dunsany, 'The Kith of the Elf-Folk', from *A Dreamer's Tales* (Boni & Liveright Inc., 1919). Copyright the Trustees of the Estate of Lord Dunsany and reprinted by permission of Curtis Brown Ltd., on their behalf.

Louise Erdrich, 'Old Man Potchikoo', first published in *Granta*, No. 27, Summer 1989. © Louise Erdrich 1989. Reprinted by permission of the author.

Richard Hughes, 'Gertrude's Child', from *Don't Blame Me and Other Stories* (Harper, 1940), reprinted in *The Wonder-Dog* (Greenwillow, 1977). © 1977 The Estate of the late Richard Hughes.

Richard Kennedy, 'The Porcelain Man', from *Collected Stories* (1987), © Richard Kennedy 1987. Reprinted by permission of the author.

Tanith Lee, 'Prince Amilec', from *Princess Hynchatti and Some Other Surprises.* Text © 1972 by Tanith Lee. Reprinted by permission of Farrar, Straus & Giroux, Inc., and Pan Macmillan Children's Books.

Ursula Le Guin, 'The Wife's Story', © 1982 by Ursula K. Le Guin, was written for *Changes*, edited by Michael Bishop and Ian Watson, and first appeared in *The Compass Rose.* Reprinted by permission of the author and the author's agent, Virginia Kidd and A. P. Watt Limited.

Bernard Malamud, 'The Jewbird' from *Idiots First* (Farrar, Straus & Giroux, Inc., 1963, Eyre & Spottiswood Ltd., 1964). Reprinted by permission of A. M. Heath on behalf of the estate of Bernard Malamud.

Naomi Mitchison, 'In the Family', from *Five Men and a Swan* (Allen & Unwin, 1957). Reprinted by permission of David Higham Associates Ltd.

Carl Sandburg, 'The Story of Blixie Bimber and the Power of the Gold Buckskin Whincher', from *Rootabaga Stories*, copyright 1922 and renewed 1950 by Carl Sandburg, reprinted by permission of Harcourt Brace Jovanovich, Inc.

Isaac Bashevis Singer, 'Menaseh's Dream', from *When Shlemiel Went to Warsaw & Other Stories.* Text © 1968 by Isaac Bashevis Singer. Reprinted by permission of Farrar, Straus & Giroux, Inc.

James Thurber, 'The Unicorn in the Garden', from *Fables For Our Time* (New York: Harper & Row; London: Hamish Hamilton, 1951), © 1940, 1951 James Thurber, © 1968 Helen Thurber. Reprinted by permission of Hamish Hamilton Ltd., and Rosemary A. Thurber.

Sylvia Townsend Warner, 'Bluebeard's Daughter', from *The Cat's Cradle.* Reprinted by permission of the Random Century Group Ltd. on behalf of Chatto & Windus as publisher and the Estate of Sylvia Townsend Warner.

H. G. Wells, 'The Magic Shop', from *Twelve Stories and a Dream* (1903), reprinted in *The Short Stories of H. G. Wells* (1927). Reprinted by permission of A. P. Watt Ltd. on behalf of The Literary Executors of the Estate of H. G. Wells.

T. H. White, 'The Troll', from *The Maharaja and Other Stories* (Macdonald). Reprinted by permission of David Higham Associates.

Jay Williams, 'Petronella', from *The Practical Princess and Other Liberating Fairy Tales* (1973). © Jay Williams 1973.

Jane Yolen, 'The River Maid', from *Neptune Rising* (Philomel, 1982). © Jane Yolen 1982. Reprinted by permission of Curtis Brown, Ltd.